Optimizing Air Pollution Control Equipment Performance

Optimizing Air Pollution Control Equipment Performance

Operation & Maintenance

Jay Richardson
Senior Combustion Engineer, Glastonbury, CT, USA

Louis Theodore
Consultant, Theodore Tutorials, East Williston, NY, USA

WILEY

Published by John Wiley & Sons, Inc., Hoboken, New Jersey.
Published simultaneously in Canada.

For general information on our other products and services or for technical support, please contact our Customer Care Department within the United States at (800) 762-2974, outside the United States at (317) 572-3993 or fax (317) 572-4002.

Wiley also publishes its books in a variety of electronic formats. Some content that appears in print may not be available in electronic formats. For more information about Wiley products, visit our web site at www.wiley.com.

Library of Congress Cataloging-in-Publication Data
Names: Richardson, Jay (Engineer), author. | Theodore, Louis, author.
Title: Optimizing air pollution equipment performance : operation & maintenance / Jay Richardson, Louis Theodore.
Description: Hoboken, New Jersey : Wiley, [2025] | Includes bibliographical references and index.
Identifiers: LCCN 2024035044 (print) | LCCN 2024035045 (ebook) | ISBN 9781394288656 (hardback) | ISBN 9781394288670 (adobe pdf) | ISBN 9781394288663 (epub)
Subjects: LCSH: Air–Purification–Equipment and supplies.
Classification: LCC TD889 .O68 2025 (print) | LCC TD889 (ebook) | DDC 628.5/3–dc23/eng/20240822
LC record available at https://lccn.loc.gov/2024035044
LC ebook record available at https://lccn.loc.gov/2024035045

Cover Design: Wiley
Cover Image: Courtesy of Jay Richardson

Set in 9.5/12.5pt STIXTwoText by Straive, Pondicherry, India

Dedication

Amelia Richardson
A loving wife who gave me the support to make this happen (JR)

Dr. John D. McKenna
The dearest of friends; a legend; and one of the premier authorities in our profession (LT)

Contents

About the Authors

Jay Richardson received his bachelor of science degree from the University of Connecticut in Chemical Engineering with a minor in chemistry. He has over 10 years' experience in the engineering, design, construction, commissioning, and testing of combustion equipment for industrial and utility boilers, process heaters, flares, thermal oxidizers, and catalyst activators. He has designed low and ultralow NO_x burners, over-fired air, flue gas recirculation, SCR/SNCR, and ignition systems. He has authored scientific journal articles on organic chemistry synthesis, magazine articles on physical combustion airflow modeling, and presented papers at international utility conferences on topics such as fuel conversions, NO_x reduction, biomass fuels, and hydrogen combustion. He has led the R&D testing of new ultralow NO_x burners and holds a patent (2022/0381433) on an AI supported ultralow NO_x burner system.

In addition to industrial combustion, Mr. Richardson contributed with the wildlife photographs in his son Joseph's nonfiction book about peregrine falcons (The Peregrine Falcon: A Wonderful Species, 2023).

Dr. Louis Theodore, son of poor, Greek immigrant parents, was born and raised in Hell's Kitchen. A graduate of Stuyvesant High School, he received degrees of MChE and EngScD from New York University and a BChE from The Cooper Union. For over 50 years, Dr. Theodore was a chemical engineering professor, as well as a graduate program director, researcher, professional innovator, and communicator in the engineering field. He has authored nearly 150 texts and reference books, nearly 200 technical papers, is an internationally recognized lecturer, and presented nearly 200 courses and seminars to industry, government, and academia. He also served as a consultant to the US EPA, DOE and DOJ, and Theodore Tutorials. Dr. Theodore is a member of Phi Lambda Upsilon, Sigma Xi, Tau Beta Pi, American Chemical Society, American Society of Engineering Education, Royal Hellenic Society, and a fellow of the Internation Air & Waste Management Association (AWMA). He is also the recipient of the AWMA's prestigious Ripperton award that is "presented to an outstanding educator who through example, dedication, and innovation has so inspired students to achieve excellence in their professional endeavors." He was also the recipient of the American Society of Engineering Education (ASEE) AT&T Foundation award for "excellence in the instruction of engineering students."

Dr. Theodore is a member of IAABO (International Association of Approved Basketball Officials) and certified to referee scholastic basketball games. He was honored in 2008 at Madison Square Garden for his contributions to basketball and the youth of America. He previously served on a Presidential Crime Commission under Gerald Ford and provided testimony as a representative of the parimutuel wagerer (horseplayer).

In addition to playing horses and gambling at casinos, Dr. Theodore's current technical interests include air pollution control, environmental management, risk management, virology, Monte Carlo applications, and developing potable water processes.

Foreword

Two fundamental reasons for the cleaning of gases in industry, particularly waste gases, are profit and protection. For example, profits may result from the utilization of blast furnace gases for heating and power generation, but impurities may have to be removed from the gases before they can be burned satisfactorily. Some impurities can be economically converted into sulfur, or solvent recovery systems can be installed to recover valuable hydrocarbon emissions. Protection of the health and welfare of the public in general, of the individual working in industry, and of property is another reason for cleaning gases.

The enactment of air pollution control regulations reflects the concern of the government for the protection of its people. For example, waste gases containing toxic constituents such as arsenic or lead fumes constitute a serious danger to the health of both plant operators and the surrounding population. Other waste gases, although not normally endangering health in the concentrations encountered, may kill plants, damage paintwork and buildings, or discolor wallpaper and curtains, thus making an industrial location a less pleasant area in which to live.

The extent to which industry cleans polluted gas streams depends largely on the limits imposed by four main considerations:

1) Concentration levels harmful to humans, physical structures, and plant and animal life
2) Legal limitations imposed by country, state, county, or city for the protection of the public health and welfare
3) Reduction of air pollution to establish civic good will
4) The reduction and/or elimination of potential liability concerns

These considerations are not necessarily independent. For example, the legal limits on emissions are also closely related to the degree of cost needed to prevent concentrations that can damage the ecosystem.

Once the optimum or near optimum level of an air pollution control device is achieved, it must be maintained day in and day out. However, such achievements rarely happen on their own in actual practice. High levels of control efficiency and performance result from the application of not only sound engineering practices but also operation/maintenance procedures, both of which are provided in this book. Control technology is self-defeating or if there are operation and/or maintenance problems, it creates undesirable side effects in meeting (limited) air pollution control objectives. Air pollution control should also be considered in terms of both the total technological system and ecological consequences. The former considers the technology that can be brought to bear on not only individual pieces of equipment but also the entire technological system. Consideration of ecological side effects must also take into account, e.g., the problem of disposal of possibly unmanageable accumulations of contaminants by other means. These may be concentrated in the collection process, such as groundwater pollution resulting from landfill practices or pollution of streams from the discharges of air pollution control systems.

Dr. John McKenna (1940–2022)

Preface

The atmosphere of the earth has undergone significant changes in the last 2 billion years. The original living cells could not have occurred or existed in an atmosphere containing free oxygen. Fossil records indicate that the carbon dioxide content has varied tremendously during this time. The life forms which are present today are the ones which have adapted to the current environment; the life forms which are now extinct are those that failed to adapt. The change in the environment is not the problem today; it is the abruptness of the change which causes the difficulties. Man-made air pollution is currently causing problems because it is producing rapid changes in the environment of the earth. Humans did not significantly affect the environment until relatively recent times. This is due to two reasons. First, the human population has been large for only a small part of Earth's history. Second, the bulk of man-made air pollution is intimately related to industrialization. In fact, humans did not begin to alter the environment until they began to live in communities. Fast forward to recent times.

In the last five decades, the technical community has expanded its responsibilities to society to include the environment, with particular emphasis on air pollution from industrial sources. Increasing numbers of engineers, technicians, and maintenance personnel are being confronted with problems in this most important area. The environmental engineer and scientist of today/tomorrow must develop proficiency and an improved understanding of air pollution control equipment in order to cope with these challenges.

This book serves two purposes. It may be used as a textbook for engineering students in an air pollution course. It may also be used as a reference book for practicing engineers, scientists, and technicians involved with air pollution control equipment. For this audience, it is assumed that the reader has already taken basic courses in physics, chemistry, and should have a minimum background in mathematics through calculus. The authors' aim is to offer the reader information on air pollution control equipment with appropriate practical applications and to provide an introduction to operation, maintenance, and inspection considerations. The reader is encouraged through references to continue his or her own development beyond the scope of the presented material.

The book is divided into three major parts: Prologue (Chapters 1–8), Air Pollution Control Equipment (Chapters 9–23), and Epilogue (Chapters 24–29). Following the introductory chapters, each chapter in Part II contains a short introduction to the control device, which is followed by operation, maintenance, inspection, and optimization details. Part III addresses engineering concerns. The Appendix contains writeups on the SI system and conversion constants.

Although design and selection information are presented, it is the primary intention of this book to discuss operation and maintenance topics and explore many of the repetitive problems that have plagued users of air pollution control equipment. The existence of these problems may be related to the complexity of the process or to a lack of well-defined operation techniques, among other reasons. In any event, this book also intends to emphasize where and how these factors can have a major impact on the maintenance problems of control devices.

Operation and maintenance problems have plagued users for nearly 100 years. The number and complexity of these problems have increased at a nearly exponential rate since the early 1970s. The 1950s and 1960s produced installations designed for the low to medium collection efficiency range. During the 1970s, designs of high efficiency were provided that often contained more component parts in more complex arrangements. The 1990s and 2010s, similarly, experienced advancements in pollution control equipment complexity and advancement. The time normally required for the training of field technicians and engineers on the problems (and solutions) with these newer installations for both gas and particulate control was not available at that time. Plants all over the world are struggling to find qualified maintenance and operations staff and turnover is currently at an all-time high. Although there has been a concentrated effort in recent years to better understand air pollution control equipment, the result – unfortunately – is akin to placing the cart before the horse.

The air pollution control device can represent a major portion of the investment in an industrial plant. Control equipment life and performance are essential to the economic operation of the plant; the need for a sound, well-planned operation, and maintenance program is therefore obvious. The success or failure of a particular installation will often rest on the implementation of this program, or its equivalent. Normal control equipment operation should include proper procedures for startup and shutdown, as well as guidelines for troubleshooting and improving operation. Once onstream, equipment failures need to be analyzed to examine major causes. Obtaining backup information for preventive maintenance programs and spare-parts requirements is also necessary. Although specific maintenance program requirements may vary from equipment to equipment, process to process, plant to plant, and from industry to industry, some basic steps and procedures are common to all. These include safety, inspection programs, and definition of inspection and maintenance responsibilities.

Knowledge of the information developed and presented in the various chapters is essential not only to the design and selection of industrial control equipment for atmospheric pollutants but more so to their proper operation and maintenance. It will enable the reader to obtain a better understanding of both the equipment itself and those factors affecting equipment performance.

Hopefully, the text is simple, clear, to the point, and imparts a basic understanding of the theory and mechanics of the calculations and applications in the air pollution control equipment field. It is also hoped that a meticulously accurate, articulate, and practical writing style has helped master the difficult task of explaining what was once a very complicated subject matter in a way that was easily understood. The authors feel that this delineates this text from others in the field.

As is usually the case in preparing a textbook, the problem of what to include and what to omit was particularly difficult. However, every attempt has been made to offer engineering course material to the reader at a level that they might better cope with some of the complex problems encountered in environmental service today.

Please note that reasonable care has been taken to assure the accuracy of the information contained in this book. However, the authors and the publisher cannot be responsible for erroneous omissions in the information presented or for any consequences arising from the use of the information published in the book. Accordingly, reference to original sources is encouraged. Reporting any errors or omissions is solicited in order to assure that appropriate changes may be made in future additions.

The authors are indebted to Dr. Ryan Dupont, Paul Farber, Dr. Walter Matystik, Emma Parente, Tony Buonicore, Mary K. Theodore, and Ronnie Zaglin for their technical support that provided invaluable assistance in the preparation of this edition. We are particularly grateful for the contributing authors Paul Farber, Sean Dooley, Vincenza Imperiale, Emma Parente, and Sarah Forster. Finally, the authors' sincere gratitude is due to all those who patiently assisted with the typing and proofreading of this manuscript.

Comments from Afar

As one of the original authors of an earlier book on this topic of air pollution control equipment, I feel a need to provide the reader with some general and background information regarding this book.

Tony Buonicore and I wrote an earlier version of this book. It was titled "Air Pollution Equipment, Operation, Maintenance, Inspection, and Design" (ISBN # 0-13-021154-0) and was published by Prentice-Hall (Upper Saddle River, NJ) in 1981. PH printed 1,250 copies. By 1985, it had sold less than 500 copies. Apparently, nobody was interested in an Operation and Maintenance (O&M) book and PH felt that they couldn't give it away. A decision was made to discontinue selling the book and PH graciously turned the copyright over to me while also selling the remaining stock to me at $2.50 per book. I in turn, gave Tony 100 copies who was then engaged with his first company, Buonicore, Cashman, & Associates. BCA was an air pollution control equipment consulting company that was soon sold and Tony became involved with his second company, Environmental Data Resources.

The earlier book shocked us by exhibiting a resiliency that is difficult to comprehend. I was soon selling over 100 copies per year at $80 per book. My former student, Dr. McKenna, later labeled it a "word of mouth" book that became not only the bible for those involved with the operation and maintenance of air pollution control equipment but also an excellent resource for those involved with air pollution.

I later gave Dr. John McKenna the right to also publish the book under his company's name (ISBN #882677-00-4) and added the book to his training institutes library. I continue to sell this early edition even to this day. Unbeknownst to me, John also granted Springer Verlag permission to sell the exact same book (ISBN # 878342851469). So, three different entities have been selling the same earlier book since the turn of the century.

Over 40 years have elapsed since the publication of the earlier book. Despite its continued success, I came to the conclusion that the book was in need of update and rewrite. With Tony gone, I needed help from an air pollution control equipment authority currently working in the field. Enter Jay Richardson. I was fortunate to meet and make his acquaintance at an AWMA (Air & Waste Management Association) annual meeting. My problems were solved as Jay proved to be a godsend when he agreed to co-author this book.

And what about Tony you ask? Tony and I continue to maintain close contact. Dinner – on Tony of course – at an Italian restaurant has become the order of the day at least twice a year. He has moved on to bigger and better and more financially rewarding activities in the energy field. He is one of my favorite students *and* unquestionably the best engineer and I graduated during a 50-year tenure as a professor of chemical engineering.

Finally, this earlier work remains one of my favorite books, and is second only is the three editions of "Introduction to Environmental Management"; the three books (1990, 2010, 2022) were co-authored by my favorite squeeze, Mary K. Theodore, and yours truly. I hope this revised work of the air pollution control equipment book becomes one of your favorites as well.

May, 2024

Louis Theodore
East Williston, NY

Part I

Prologue

Webster defines prologue as "the preface or introduction to a discourse... an introduction or prefatory act, event, etc." This part of the book provides a general introduction to our pollution equipment subject matter.

Optimizing Air Pollution Control Equipment Performance: Operation & Maintenance, First Edition. Jay Richardson and Louis Theodore.

1

Definitions/Glossary of Terms

As noted in the Preface, this book is concerned with air pollution control equipment and as such, this chapter primarily addresses air pollution control equipment-related terms. It has been written not only for academic use in colleges, and universities but also for both engineers and scientists who work in the air pollution control equipment field. This glossary may be used whenever and wherever information is needed about words and/or terms in air pollution.

Some additional points deserve mentioning:

1) Each definition avoids technical jargon.
2) No mathematical equations – in any form – are employed in the definition. In some instances, where necessary, common scientific and engineering units have been included.
3) Only keywords or terms used in practice are provided, as is the case in preparing a text, particularly a dictionary, the problems of what to include and what to omit have been particularly difficult.
4) Only one spelling is used for words with multiple accepted spellings, e.g., modeling versus modeling.
5) Some important acronyms are also included with a one-sentence definition.
6) As with nearly every glossary, the terms have been alphabetized.

This chapter defines many – but not all – of the terms that the reader will encounter in this book. The following list is therefore not a complete glossary of all terms that appear in this field. It should also be noted that many of the terms have come to mean slightly different things to different people; this will become evident as one delves deeper into the literature.[1–3]

1.1 Glossary of Terms

Absolute humidity. The amount of water vapor present in a unit mass of air, which is usually expressed in kg water vapor/kg dry air or lb water vapor/lb dry air.

Absolute pressure. The actual pressure exerted on a surface that is measured relative to zero pressure; it equals the gauge pressure plus the atmospheric pressure.

Absolute temperature. The temperature expressed in degrees of Kelvin or Rankine.

Absorbate. A substance that is taken up and retained by an absorbent.

Absorbent. Any substance that takes in or absorbs other substances.

Absorber. A device in which a gas is absorbed by contact with a liquid.

Absorption. The process in which one material (the absorbent) takes up and retains another (the absorbate) to form a homogeneous solution; it often involves the use of a liquid to remove certain gas components from a gaseous mixture.

Actual cubic feet per minute (acfm). A unit of flow rate measured under actual pressure and temperature conditions.

Acute (risk). Risks associated with short periods of time. For health risk, it usually represents short exposures to high concentrations of an agent.

Adiabatic. A term used to describe a system in which no gain or loss of heat is allowed to occur.

Adiabatic flame temperature. The maximum temperature that a combustion system can reach.

Optimizing Air Pollution Control Equipment Performance: Operation & Maintenance, First Edition. Jay Richardson and Louis Theodore.

Adiabatic lapse rate.	The rate at which the temperature of a moving air parcel decreases in the atmosphere as height above the surface increases when no heat is added or subtracted from the moving air parcel; the adiabatic lapse rate is 10 °C/km.
Adsorbent.	A substance (e.g., activated carbon, activated alumina, silica gel) that has the ability to condense or hold molecules of other substances on its surface.
Adsorber.	An apparatus in which molecules of gas or liquid are retained on the surface of an adsorbent.
Adsorption.	The physical or chemical bonding of molecules of gas, liquid, or dissolved solid to the external or internal (if porous) surface of a solid (adsorbent); it is an advanced method of treating waste that is employed to remove odor, color, or organic matter from a system.
Afterburner.	A secondary burner is located so that combustion gases from the primary incinerator are further burned to remove smoke, odors, and other pollutants.
Agent.	A biological, physical, or chemical entity capable of causing disease or adverse health effects.
Alpha (α) particle.	A positively charged helium nucleus (i.e., two protons and two neutrons) that is emitted spontaneously from the decay of radioactive elements.
Amines.	Organic compound related to ammonia.
Ammonia.	An alkaline gas composed of hydrogen and nitrogen; it has a strong odor when present in high concentrations.
Asphyxiant.	A vapor or gas that has little or no positive toxic effect but that can bring about unconsciousness and death by replacing air and thus depriving an organism of oxygen.
Aspirator.	A hydraulic device that creates a negative pressure by forcing liquid through a restriction, thus increasing the velocity head.
Atmosphere.	The general volume of air above the earth; it is the lower portion of the atmosphere in which pollution must be controlled.
Atmospheric dispersion.	The mixing of a gas or vapor (usually from a discharge point) with air in the lower atmosphere. The mixing is the result of convective motion and turbulent eddies.
Atmospheric stability.	A measure of the degree of atmospheric turbulence, often defined in terms of the vertical temperature gradient in the lower atmosphere.
Atomic fission.	The breaking down of a large atom into smaller atoms or elements, involving the liberation of heat, gamma rays, alpha particles, and beta particles.
Audit.	The examination of something with the intent to check, verify, or inspect a particular subject matter.
Auto-ignition temperature.	The lowest temperature at which a flammable gas in the air will ignite without an ignition source.
Auto-ignition.	The starting of a fire without the addition of an external source such as a flame, spark, or heat.
Average rate of death (ROD).	The average number of fatalities that can be expected per unit time (usually on an annual basis) from all possible risks and/or incidents.
Baffle.	A flow-regulating device usually consisting of a plate placed horizontally across a pipe or channel to restrict or divert the passage of a fluid, usually used for the purpose of providing a uniformly dispersed flow.
Ball joint.	A flexible pipe joint formed in the shape of a ball or a sphere.
Ball valve.	A nonreturn valve consisting of a ball resting on a cylindrical seat within a fluid passageway or pipe.
Barometric pressure.	The pressure of the ambient air in the atmosphere at a particular point on or above the surface of the earth.
Basic event.	A fault tree event (FTE) that is sufficiently basic that no further explanation or development of additional events is necessary.
Batch process.	A process that is not continuous; its operations are carried out with discrete quantities of material added and removed from it at appropriate time intervals.
Benzene.	Aromatic hydrocarbon formerly used in paints and varnishes, but now considered toxic for this purpose.
Beta (β) particle.	A charged particle emitted from a radioactive atomic nucleus; it has moderate penetrative power and is able to damage living tissue.

Bias. The systematic distortion of data; it is the tendency of a sample to be unrepresentative of all the cases involved in a study.

Bleeding. The gradual release of material and/or reduction of pressure from a system or process (e.g., by a valve or leak).

Blowdown. The cyclic or constant removal of a portion of any process flow to maintain the constituents of the flow at a desired level.

Blower. A fan employed to force or move air or gas under pressure.

Brownian movement. The constant, random movement of small, suspended particles due to collisions with other molecules.

Buffer. A solution containing both a weak acid and its conjugate weak base, which is employed to stabilize the pH value in a solution, and anything that acts to diminish and/or regulate changes in a system or process.

Bulk sample. A small portion of material that is collected and sent to a laboratory for analysis.

Bulk density. The mass per unit volume of a solid in a mixture such as a packed bed or soil mass; unlike the real solid particle density, the pore space is included in the volume for this calculation.

Burning. The chemical combination of oxygen and a fuel such as carbon, hydrogen, hydrocarbon, or other substances that combine with oxygen.

Butterfly damper. A plate or blade installed in a duct, flue, breeching, or stack that rotates on an axis to regulate the flow of gases.

Butterfly valve. A flow control valve containing a disk supported by a shaft on which it rotates.

Bypass. The avoiding of a particular portion of a process or pollution control system.

Bypass valve. A valve arranged to cause the fluid, which it controls to flow past some part of its normal path (e.g., to allow a liquid to avoid a filter through which it usually passes).

By-product. A material that is not one of the primary products of a production process and is not solely or separately produced by the production process.

C (ceiling). The term used to describe the maximum allowable exposure concentration of a hazardous agent related to industrial exposures to hazardous vapors.

CFM. Cubic feet per minute.

Calibration. The determination, checking, or adjustment of the accuracy of any instrument that gives quantitative measurements.

Cancer. A tumor formed by mutated cells.

Carbon dioxide. A compound formed by the complete combustion of carbon and oxygen, poisonous only at very high concentrations.

Carbonization. A pyrolysis or thermal breakdown of an organic compound or organic substance (called carbonaceous material because it contains carbon) to give a charcoal or other form of carbon.

Carcinogen. A cancer-causing chemical.

Carrier gas. A gas that acts as the mobile phase in gas chromatography.

Carryover. The entrainment of liquid or solid particles in the vapor evolved by a boiling liquid or from a process unit.

Catalysis. The process of catalyzing or promoting an action by a chemical agent that is not consumed in the reaction.

Chemical Abstract Service (CAS) numbers. CAS numbers are used to identify chemicals and mixtures of chemicals.

Catalyst. A substance whose presence changes (normally increasing) the rate of a chemical reaction without itself undergoing permanent change in its composition.

Catalytic combustion. The oxidation of organic compounds catalytically to cause the burning at a lower temperature than would otherwise be possible.

Catalytic cracking.	The breaking of a carbon–carbon bond with the acid of a catalyst; it is an essential process in the refining of petroleum.
Catastrophe.	A major loss in terms of death, injuries, and/or property damage.
Cause-consequence.	A method for determining the possible consequences or outcomes arising from a logical combination of input events or conditions that determine a cause.
Cavitation.	The formation of vapor bubbles in a liquid when subjected to localized low-pressure regions causing severe mechanical damage to the surface of metals exposed to it.
Centrifugal pump.	A device that increases the pressure of a liquid by using centrifugal force.
Chronic (risk).	Risks associated with long-term chemical exposure duration usually, at low concentrations.
Check valve.	A one-way valve that prevents flow through a pipe in an undesired direction; it opens in the direction of the normal flow and closes with a reversal of flow.
Chemical agent.	An element, compound, or mixture that coagulates, dissolves, neutralizes, solubilizes, oxidizes, concentrates, makes a pollutant mass more rigid, or otherwise facilitates the lessening of harmful effects or the removal of a pollutant from a fluid.
Chemical Process Quantitative Risk Analysis (CPQRA).	The process of hazard identification, followed by numerical evaluation of incident consequences and frequencies, and their combination into an overall measure of risk when applied to the chemical process industry. Ordinarily applied to episodic events. Related to Probabilistic Risk Assessment (PRA) used in the nuclear industry.
Chemical equation.	A representation of a chemical reaction using symbols to show the molar relationship between the reacting substances and the products.
Chlorine.	A chemical element generally in the form of a gas or dissolved in water used to sterilize it.
Chromatograph.	A method of analysis of separation based on adsorbing the gases, vapors, or substances of a mixture.
Clean Air Act.	A federal act empowering government agencies to control the cleanliness of the air by eliminating pollution.
Clean room.	Refers to a room in which the air is highly purified, particularly with regard to dust and other particulate matter. Clean room techniques are used in the processing of delicate materials and products.
Closed-loop recycling.	The reclaiming or reusing of wastewater for non-potable purposes in an enclosed process; it may also be applied to other streams including gaseous ones.
Closed-loop.	A term used to describe an enclosed, recirculating process.
Cocurrent.	A term used to describe a flow in which materials travel in the same direction.
Combustion.	A reaction at a high temperature with oxygen that produces carbon dioxide, water, and energy in the form of light and heat; it is a basic cause of air pollution.
Combustion products.	Chemical compounds in the form of gases, vapors, and particulates produced by the combustion action or a burning process, typical combustion gases contain products such as nitrogen, oxygen (less than in normal air), water vapor, carbon dioxide, carbon monoxide, nitrogen oxides, sulfur oxides, unburned fuel, carbon particles.
Concurrent.	A term used to describe a situation in which two or more controls or systems exist in an operated condition at the same time.
Condensate.	Any liquid resulting from the cooling of a gas or vapor.
Conditional probability.	The probability of occurrence of an event given that a precursor event has occurred.
Confidence interval.	A range of values of a variable with a specific probability that the true value of the variable lies within this range. The conventional confidence interval probability is the 95% confidence interval, defining the range of a variable in which its true value falls with 95% confidence.
Confidence limits.	The upper and lower range of values of a variable defining its specific confidence interval.
Consequences.	A measure of the expected effects of an incident outcome or cause.
Containment.	Something that is not desired and should be removed.
Continuous release.	Emissions that are of a continuous duration.
Control agency.	A governmental agency (city, country, state, or national) that deals with pollution and its control.
Control method.	Method of controlling or eliminating air or water pollutants, it may be a change in the process that produced the pollutant, but generally refers to a device specifically designed to eliminate the substance from the exhaust.

Convection. The transfer of heat through a fluid by the movement of the fluid; and the vertical movement of air leading to cooling.

Cooling tower. A hollow, vertical structure with internal baffles to disperse water so it is cooled by flowing air and by evaporation at ambient temperature.

Corona. An electrical discharge effect that causes ionization of oxygen and the formation of ozone; it is found in electrostatic precipitators.

Corrosion. The deterioration or destruction of a material by chemical action.

Countercurrent. A term used to describe the flow pattern within equipment in which two streams travel in opposite directions.

CPQRA. The acronym for chemical process quantitative risk analysis; it is analogous to a hazard risk assessment (HZRA).

Cracking. A refining process involving decomposition and molecular recombination of organic compounds to form molecules of smaller sizes that are suitable for fuels.

Cradle space. An area beneath the floor of a house or a building that allows access to utilities and other services.

Critical temperature. The temperature above which a gas or vapor cannot be liquefied by an increase in pressure alone.

Cryogenics. The production and utilization of extremely low temperatures.

Crystallization. The change of state of a substance from a liquid to a solid by the phenomenon of crystal formation by nucleation and accretion (e.g., the freezing of water onto ice).

Cutback. A coating substance or varnish that has been diluted or thinned.

Damper. A manually or automatically controlled valve or plate in a breeching, duct, or stack that is employed to regulate a draft or the rate of flow of air or other gases.

Dead time. The time interval, after a response to one signal or event, during which a system is unable to respond to another signal or event.

Dehumidifier. A device incorporated into many air conditioning systems to dry incoming air by passing it across a bed of a hygroscopic substance or through a spray of very cold water.

Delphi method. A polling of experts that involves the following: (i) select a group of experts (usually three or more); (ii) solicit, in isolation, their independent estimates of the value of a particular parameter and their reason for choice; (iii) provide initial analysis results to all experts and allow them to then revise their initial values; and (iv) use the average of the final estimates as the best estimate of the parameter and use the standard deviation of the estimates as a measure of uncertainty. The procedure is iterative with feedback between iterations.

Demister. A device composed of plastic threads, wire mesh, or glass fibers employed to remove liquid droplets entrained in a gas stream.

Dermal. Applied to the skin.

Desorption. A process of removing an absorbed material from a solid on which it is adsorbed by increasing the temperature or reducing the pressure, or both.

Detention time (detention period). The average time that a unit volume of a fluid is retained in a unit during a flow process.

Detoxification. The destruction of the toxic quality of a substance.

Dew point. The temperature at which the first droplet of water forms on the progressive cooling of a mixture of air and water vapor; at the dew point, the air becomes saturated with water.

Diffusion. The movement of individual molecules through narrow spaces.

Dike. An embankment that restricts the movement and provides the containment of a liquid.

Dilution ventilation. The mixing of contaminated air with uncontaminated air for the purpose of controlling potential airborne health hazards, fire and explosion conditions, odors, and nuisance-type contaminants.

Dispersion coefficient. The standard deviation, σ, in a specified direction used in a Gaussian plume atmospheric dispersion model.

Distillation.	A process of separating the constitutes of a liquid mixture by means of partial vaporization of the mixture and separate recovery of vapor and residue.
Domino effect.	The triggering of secondary events; usually considered when a significant escalation of the original incident could result.
Dose.	A weight (or volume) of a chemical agent; usually normalized to a unit of body weight.
Downcomer.	A pipe or flue that conveys gases, vapors, or condensate downward in blast furnaces, distillation towers, and refineries.
Downdraft.	A current of air in a stack with a bulk downward motion.
Downstream.	The direction in which a fluid stream is flowing.
Downtime.	The lost production time during which a piece of equipment is not operating correctly due to a breakdown, maintenance, or power failure.
Dry bulb.	A thermometer bulb maintained dry; it is used with a wet bulb to measure humidity.
Dry-bulb temperature.	The temperature of air; usually used in conjunction with the wet-bulb temperature to measure humidity.
Duct.	A round or rectangular conduit, usually metal or fiberglass, employed to transport fluids.
Economizer.	A device that uses the heat of combustion gas to raise the temperature of boiler feedwater prior to entry into the boiler.
Efficiency.	The degree of performance of a device or process; generally refers to the degree of purification or removal of contaminants when referring to pollution and its control.
Effluent.	A fluid, either gas or liquid, leaving equipment, process, building, or factory.
Ejector.	A device for moving a fluid or solid by entraining it in a high-velocity stream of air or water jet.
Elbow.	A pipe fitting that connects two pipes at a 90° angle.
Electrostatic field.	A region in which a stationary electrically charged particle is subjected to a force of attraction or repulsion as a result of another stationary electric charge.
Electrostatic precipitator (ESP).	A device that separates particles from a gas stream by passing the carrier gas between two electrodes across which a high voltage is applied.
Elutriator.	An air-sampling device, widely used in cotton-dust area air sampling, that uses gravitational forces to remove dust that is unfit for breathing from the air sample, and then collects the dust on a filter for further analysis.
Endothermic.	A term used to describe a process or change that occurs with the input of heat.
Entrainment.	The carryover of drops of liquid during a process such as distillation, absorption, scrubbing, or evaporation.
Environmental audit.	An independent assessment of the current status of a company's compliance with applicable environmental requirements.
Episode.	An air pollution incident in a given area caused by an increase in atmospheric pollutant concentration in response to meteorological conditions (inversion) that may result in a significant increase in illness or in death.
Episodic release.	A massive release of limited or short duration, usually associated with an accident.
Event.	An occurrence associated with an incident either as the cause or a contributing cause of the incident, or as a response to an initiating event.
Event sequence.	A specific sequence of events composed of initiating events and intermediate events that may lead to a hazard or an incident.
Event tree analysis (ETA).	A graphical logic model that identifies and attempts to quantify possible outcomes following an initiating and subsequent events.
Excursion.	An unintentional occurrence, such as a discharge of pollutants above the permitted amount, due to reasons beyond human control.
Exhaust.	A duct for the escape of gases, fumes, and odors from an enclosure.
Exfiltration.	The opposite of infiltration. The exhaust of gases from a building or structure due to wind velocity and thermal effects through defect in the structure, and to normal leakage around openings.

Exothermic. A term used to describe a process or change that occurs with the production of heat.

Expansion joint. A joint installed in a structure to provide for changes (often in length) due to expansion or contraction resulting from changes in temperature, without distortion of the structure.

Exposure period. The duration of an exposure.

External event. A natural or man-made event; often an accident that may serve as an initiating event in a sequence of subsequent events in a process leading to a release of hazardous materials.

Extraction. The process of dissolving and separating out particular constituents of a solid or liquid using an immiscible solvent.

Extrapolation. An estimation of unknown values by extending or projecting from known values.

Failure frequency. The frequency (relative to time) of failure.

Failure mode. A symptom, condition, or manner in which a failure occurs.

Failure probability. The probability that failure will occur, usually in a given time interval.

Failure rate. The number of failures divided by the total elapsed time during which these failures occur.

Fallout. The radioactive debris or material that settles to the earth after a nuclear explosion.

Fatal accident rate (FAR). The estimated number of fatalities per 10^8 exposure hours (roughly 1,000 employee working lifetimes).

Fatigue. The incremental weakening of a material as a result of repeated cycles of stresses that are far lower than its breaking load; the end result is failure.

Fault. A fracture in the earth along which there has been displacement parallel to the fault plane; an error or failure in a process that can lead to a hazardous event.

Fault tree. A method for representing the logical combinations of adverse hazard events that lead to a particular outcome (top event).

Fault tree analysis (FTA). A logic model that identifies and attempts to quantify possible causes of a hazard event.

Federal Register. A daily publication of laws and regulations promulgated by the US Federal Government.

Feedforward control system. A system in which changes are detected at the process input and a corrective signal is applied before process output is affected.

Fire point. The lowest temperature at which a liquid evolves vapor fast enough to support continuous combustion; it is usually close to the flash point.

Fission. The splitting of atomic nuclei into smaller nuclei, accompanied by the release of significant quantities of energy; it is induced by bombardment with neutrons from an external source and continued by the neutrons that are released.

Fixed-bed operation. An operation in which the additive material (e.g., catalyst, absorbent, filter media) remains stationary in the chemical reactor.

Fixed carbon. The ash-free, carbonaceous material that remains after volatile matter is driven off a dry solid sample.

Flammability. The ease with which a material (gas, liquid, or solid) will ignite spontaneously (auto-ignition) from exposure to a high-temperature environment, or from a spark or open flame.

Flange. A projecting rim, edge, lip, or rib used to connect piping or conduit.

Flap valve. A valve that is hinged at one edge and that opens and shuts by rotating about the hinge.

Flare. A tall stack that is employed to burn small, discrete quantities of undesirable gases.

Flash distillation. A separation process in which an appreciable portion of a liquid is quickly converted to vapor by changing the temperature and/or pressure; this method is widely employed for the desalination of seawater.

Flash point. The minimum temperature at which a liquid or solid gives off enough vapor to form a flammable mixture with the air near the surface of the liquid or solid.

Flotation. A treatment process using air, injected into a waste stream, that adheres to the solids and causes them to rise, allowing for their removal at a tank surface; it takes advantage of the differences in specific gravities of the air-associated solids and waste liquid.

Flow diagram. A chart or line drawing employed by engineers to indicate successive steps in the production of a chemical or treatment of a waste stream; it includes materials input and output, by-products, wastes, and other relevant data.

Flue gas. So called because it is discharged through a flue.

Fluidization. A technique in which a finely divided solid is caused to behave like a fluid by suspending it in a moving gas or liquid.

Fluoride. A compound of fluorine with another chemical element.

Fluorine. A chemical element with the symbol F; this is very reactive.

Fog. Impedes visibility but has no harmful effects with regard to pollution.

Forced draft. The positive pressure created by the action of a fan or blower.

Free convection. The motion and mixing of a fluid caused solely by density differences within the fluid.

Frequency. Number of occurrences of an event per unit time.

Friction head. The pressure differential required to overcome the frictional resistance to flow.

Fugitive source. Any source of emissions that is not confined to an identifiable point of discharge.

Fume. An aerosol or suspension of fine solid or liquid particles that may be visible and that may combine with unwanted gases, vapors, or odors.

Fusion. The process in which the nuclei of two light elements (chiefly the hydrogen isotopes) combine to form the nucleus of a heavier element with the release of substantial amounts of energy.

Gamma decay. A nuclear reaction in which the nucleus of an atom emits a pulse of energy in the form of gamma radiation.

Gas. In gaseous or vapor form, such as air, flue gas, or natural gas, this is a term that refers to any material in gaseous form.

Gate valve. A valve in which the closing element consists of a disk that slides vertically over the opening or cross-sectional area through which liquid passes; it is suitable for on–off control but not for throttling.

Gauge. An instrument for measuring and indicating such process variables as pressure, liquid level, thickness, volumetric flow, etc.

Gaussian model. A plume dispersion model of mixing and turbulence in the lower atmosphere based on the assumption that a pollutant concentration within the plume is normally distributed vertically when sampled perpendicular to the direction of flow.

Grab sample. A single sample that is collected at such a time and place so that it is ideally representative of a total discharge.

Gradient. The rate of change of a quantity or variable.

Guide vane. A fixed or adjustable device intended to direct fluid flow in a conduit channel.

Half-life. The time required for a chemical concentration or quantity to decrease by half of its current value.

Hazard (problem). An event associated with an accident that has the potential for causing damage to people, property, or the environment.

Hazard and operability study (HAZOP). A technique to identify process hazards and potential operating problems using a series of guide words that are key to process deviations.

Hazard risk assessment (HZRA). A technique associated with quantifying the risk of a hazard employing probability and consequence information.

Head. The height of the free surface of fluid above any point in a hydraulic system; it is a measure of the pressure or force exerted by a fluid.

Health (problem). A problem normally associated with and arising from the continuous emission of a chemical into the environment causing adverse health effects in humans.

Health risk assessment (HRA). A technique associated with quantifying the risk of a health problem employing toxicology (dose-response) and exposure information.

Hearth. The bottom of a furnace on which waste materials or fuels are exposed to a flame.

Heat of combustion. The heat liberated in a combustion reaction.

Heat of vaporization. The heat required to convert a liquid or solid to a vapor.

Heterogeneous. A term used to describe a mixture of different phases (e.g., liquid–vapor, liquid–vapor–solid).

Holdup. A volume of material held or contained in a process vessel or line.

Hood. A canopy or collecting structure over a process area or piece of equipment; it is employed to collect fumes, gases, vapors, or fine particulates that are then exhausted or treated in some manner.

Human error. Actions by engineers, operators, managers, and others that may contribute to or result in accidents.

Human error probability The ratio between the number of human errors and the number of opportunities for human error.

Human factors. Factors attempting to match human capacities and limitations.

Human reliability. A measure of human errors.

Humidification. A process for increasing the water content of air; it is usually incorporated into an air conditioning system.

Humidity. The measure of the amount of water vapor in the air at any given time.

Hybrid. An offspring of two organisms that differ in at least one gene, and any equipment that differs from another in at least one respect, or accomplishes two or more objectives in a process.

Hydrocarbons. Organic compounds of carbon and hydrogen.

Hydroelectric power. The electricity produced by turbines capturing the energy of moving water.

Hydrogen cracking. The decomposition of petroleum or other hydrocarbons by heat to give lower-boiling materials employed as gasoline, kerosene, other motor fuels, domestic fuel oil, and other products.

Hydrogen sulfide. Common air pollutant with characteristic rotten egg odor that causes tarnishing of silver, copper, and other metals.

In situ. A term used to describe any reaction or analysis occurring in place; and a term used to describe a fossil, mineral, or rock found in its original place of deposition, growth, or formation.

Incident. An event.

Incineration. The oxidation by burner of gases, vapors, liquids, or solids, to eliminate them and control pollution.

Incubation. Controlled environmental conditions (e.g., temperature and moisture) for the growth and development of microorganisms.

Individual risk. The risk to an individual from a hazardous chemical or event.

Induced draft. The negative pressure created by the action of a fan, blower, or other gas-moving device.

Infiltration. Air that enters a building through cracks and windows due to wind velocity or stack action of the building.

Inert. A term used to describe a material that is chemically inactive; they can be ingredients added to mixtures chiefly for bulk and weight purposes.

Ingestion. The intake of a chemical through the mouth.

Initiating event. The first event in an event sequence.

Instantaneous release. Emissions that occur over a very short duration.

Intermediate event. An event that propagates or mitigates the initiating event during an event sequence.

Ion. A charged molecule or atom that has gained or lost one or more electrons; the migration of an ion affects the transport of electricity through an electrolyte or, to a certain extent, through a gas.

Isobaric. A term used to describe a process that occurs at constant pressure.

Isokinetic sampling. A technique for collecting air pollutants; it is constructed so that the gas entering the sampling nozzle is at the same speed as the surrounding air.

Isopleth. A plot of uniform pollutant concentrations, usually downwind from a release source.

Isotherm. A line connecting point of equal temperature that is employed on climactic maps or in groups of thermodynamic relations.

Isotropic. A term used to describe anything that exhibits uniform physical properties throughout and in all directions.

Laminar flow. One of the two types of flow that occur in fluids (the other being turbulent) in which the fluid follows a smooth, well-defined path; this type of flow occurs at low Reynolds numbers.

Landfill.	A method of dumping solid wastes and using them to fill depressions in land.
Lapse rate.	The rate at which temperature decreases in the atmosphere as height above the surface increases; the adiabatic lapse rate is 10 °C/km.
Latent heat.	The heat released or absorbed by a change of state.
Leaching.	The process by which soluble material dissolves and is removed from a solid in a solution.
Leakage.	An undesired and gradual escape or entry of a quantity.
Lethal concentration (LC).	The concentration of a chemical that will kill test animals, usually based on 1–4 hr exposure duration.
Lethal concentration 50 (LC_{50}).	The concentration of a chemical that will kill 50% of test animals; usually based on 1–4 hr exposure duration.
Lethal dose (LD).	The quantity of a chemical that will kill a test animal; usually normalized to a unit of body weight.
Lethal dose 50 (LD_{50}).	The quantity of a chemical that will kill 50% of test animals; usually normalized to a unit of body weight.
Level of concern (LOC).	The concentration of a chemical above which there may be adverse human health effects.
Life cycle.	The phases, changes, or stages that an organism passes through during its lifetime; the duration used to evaluate the costs of a process based on its operational life prior to replacement.
Likelihood.	A measure of the expected probability or frequency of the occurrence of an event.
Liner.	A relatively impermeable barrier designed to prevent leachate from leaking from a landfill; and an insert or sleeve for pipes or conduits to prevent leakage or infiltration.
Liquid trap.	The sumps, well cellars, and other traps employed in association with oil and gas production, gathering, and extraction operations for the purpose of collecting oil, water, and other liquids.
Liquid-level control.	A control device used to monitor liquid level in a reactor or container designed to actuate to limit fluid entering the reactor or container to maintain fluid level at a set value.
Maximum allowable concentration (MAC).	A maximum concentration that should be used for a particular contaminate.
Mach number.	The ratio of the speed of an object to the speed of sound in the undisturbed medium through which the object is traveling; anything traveling supersonically has a Mach number greater than one.
Make-up solvent.	The solvent introduced into a process that compensates for solvent lost from the process during operation.
Malfunction.	A failure in the normal operation of a system or process.
Malignant.	A cancerous tumor.
Manifest.	A document employed for identifying the quantity, composition, origin, routing, and destination of material during disposal, treatment, or storage.
Manifold.	A pipe fitting with numerous branches to convey fluids among a large pipe and several smaller pipes or to permit the choice of diverting flow from one of the several sources or to one of the many discharge points.
Manometer.	An instrument for measuring pressure that usually consists of a U-shaped tube containing a liquid.
Masking.	Hiding one odor with another that is more powerful or more desirable.
Mathematical model.	A mathematical representation of a real process, expressed as a set (1 or more) of equations or algorithms.
Maximum individual risk.	The highest individual risk in an exposed population subjected to a hazard risk.

Mercaptan. An organic sulfur compound having an extremely low odor threshold and requiring very little to produce a strong odor in a large volume of air, often used to odorize natural gas.

Metastable. A term used to describe an intermediate state in between stability and instability.

Methane. Lowest molecular weight saturated hydrocarbon, CH_4, the main constituent of natural gas.

Miscibility. The ability of a liquid or gas to dissolve uniformly in another liquid or gas.

Mist eliminator. A control device employed to remove water/liquid droplets entrained in a gas stream.

Mole fraction. A ratio employed in expressing concentrations of solutions and mixtures; the mole fraction of any component of a mixture is defined as the number of moles of that component divided by the total number of moles of the mixture.

Molecular sieve. A microporous structure composed of crystalline aluminosilicates that are chemically similar to clays and feldspars and belong to a class of materials known as zeolites.

Monomer. A molecule that is capable of conversion to polymers, synthetic resins, or elastomers by combination with itself or other similar molecules; it usually contains carbon and is of relatively low molecular weight.

Monte Carlo method. A method that constructs an artificial stochastic model of a process on which sampling experiments are performed; it usually involves the use of random numbers in its calculations.

Mutagen. A chemical capable of changing a living cell.

Natural draft. The negative pressure created by the height of a stack or chimney and/or the difference in temperature between flue gases and the atmosphere.

Neutral. A term used to describe a particle without an electric charge; and a term used to describe a solution that is neither acidic nor basic.

Nitrogen oxides. The various chemical compounds of nitrogen and oxygen; the most highest form is NO_2, which has a brown color.

Noble. A term used to describe an element that is completely unreactive or reacts only to a limited extent with other elements.

Nonconservative. A term used to describe a substance that can react.

Normal boiling point. The boiling point when the ambient pressure is 1 atmosphere.

Nuclear reaction. A reaction that involves a change in the nucleus of an atom, as opposed to a chemical reaction in which only the electrons take part.

Nucleus. The small, positively charged central mass of an atom that contains essentially the entire mass of the atom in the form of protons and neutrons; the central portion of a living cell that controls its functions; and any small particle that can serve as the basis for crystal growth.

Off-gas. A gaseous product from the chemical, biological or thermal decomposition of a material.

Optimization. The adjustment of a system or process to make it as effective or functional as possible.

Organic. A term used to describe anything that contains carbon and is thus derived from living organisms.

Orifice. An opening with a closed perimeter in a plate, wall, or partition, through which fluid may flow; it is generally employed for the purpose of measuring or controlling a fluid.

Osmosis. The passage of a pure liquid (usually water) through a semipermeable membrane from a solution of low concentration into a solution of a higher concentration (e.g., the flow of pure water into a solution of salt and water).

Oxidizing agent. Any material that removes hydrogen or electrons from or adds oxygen to an element or compound.

Ozone. Oxygen in triatomic form. Oxygen can be converted to ozone by an electrical discharge and serves as a powerful oxidizing agent.

Packed tower. A device that forces gas through a tower packed with metal, plastic, ceramic, or crushed rock, while liquid is sprayed over the packing material.

Packing. A collar or gasket employed to seal mechanical devices to prevent leakage of liquid, and the inert material employed in absorption towers and distillation columns.

Parallel. A term used to describe two straight lines that are everywhere equidistant from each other, and the arrangement of two or more units that perform the same function and share a common input and a common output, e.g., two pumps operating in parallel.

Parameter.	A quantitative or characteristic property that describes the physical, chemical, or biological conditions of a system.
Partial pressure.	The pressure exerted by one of the several components of a gaseous mixture.
Partial vacuum.	The description of a space condition in which the pressure is less than atmospheric.
Partial volume (pure component volume).	The volume that would be occupied by one component of a gaseous mixture if it were alone and under the total pressure and temperature of the mixture.
Parts per billion (ppb).	The fraction (ppb_m for mass fraction and ppb_v for volume fraction) multiplied by 10^9; it is a unit used to measure extremely small quantities of a substance.
Parts per million (ppm).	The fraction (ppm_m for mass fraction and ppm_v for volume fraction) multiplied by 10^6; it is a unit used to measure extremely small quantities of a substance.
PEL.	The acronym for permissible exposure limit; the permissible exposure limit of a chemical in air, established by the Occupational Safety and Health Administration (OSHA).
Periodic.	Phenomena that repeat after regular intervals of time.
Peroxide.	A highly oxidized chemical compound.
Personal protection equipment (PPE).	Material/equipment worn to protect a worker from exposure to hazardous agents.
Phenol.	An organic compound used as a disinfectant that is also known as carbolic acid.
Photochemical.	Catalyzed by or promoted by the action of light.
Photochemical smog.	A type of smog that is caused by a combination of sunlight, ozone, nitrogen oxides, and various organic vapors.
Photon.	A quantum of electromagnetic radiation with zero mass; it is the unit particle of light.
Physical model.	The simulation of real process by a physical experiment that models the important features of the original process that are the object of study.
Pilot plant.	A trial assembly of small-scale reactions and processing equipment.
Pitot tube.	A device that is inserted into a pipe to measure the velocity of a flowing fluid.
Plume.	A visible discharge of steam, smoke, or fumes that may or may not be objectionable because of contaminant content but that may be objectionable because of its visibility.
Pneumatic.	A term used to describe anything of or containing wind, air, or gases.
Polymerization.	A chemical reaction usually carried out with a catalyst, heat, and often under high pressure, in which a large number of relatively simple molecules combine to form chain-like macromolecules.
Pore.	Any interstitial or void space in a solid material.
Porosity.	The ratio of the volume of open space in a bed to the total bed volume.
Potable water.	Any water that is safe for drinking.
Power plant.	A unit constructed for the conversion of stored energy (usually in fossil fuels) into another desired form of energy.
PPM.	Parts per million is a measure of concentration that generally refers to parts per million by volume, although sometimes (ordinarily specified) by weight.
Preheater.	A unit employed to heat the air needed for combustion by absorbing heat from the hot flue gases.
Precision.	The degree of "exactness" of repeated measurements.
Pressure drop.	A measure of the difference in pressures measured immediately upstream and downstream of a unit or process.
Pressure relief valve.	A valve that opens automatically to a sufficient area when the pressure reaches an assigned limit in order to relieve the stress on a pipeline or vessel.
Primary pollutant.	A pollutant emitted directly from a process stack.

Probability. An expression for the likelihood of occurrence of an event or an event sequence, usually over an interval of time.

Probe. A tube employed to measure pressures at a distance from the actual measuring equipment, and a general term used to describe a sampling port.

Process. A manufacturing procedure or the equipment employed in the chemical, pharmaceutical, food, metalworking, or other industry.

Process control. The manipulation of the conditions of a process to effect a desired change in the output characteristics of the process.

Process safety audits. An inspection of a plant or process unit, drawings, procedures, emergency plans, and/or management systems, etc., usually by an independent, impartial team.

Process Safety Management (PSM). A management system that is focused on prevention of, preparedness for, mitigation of, response to, and restoration from catastrophic releases of chemicals or energy from a process associated with a facility.

Process vent. Any open-ended pipe in a process or stack that is vented to the atmosphere.

Protective system. Systems, such as pressure vessel relief valves, that function to prevent or mitigate the occurrence of an accident or incident.

Protocol. The plan and procedures that are to be followed in conducting a test or project.

Proton. The positively charged fundamental unit of matter with about the same mass as a neutron; it is present in all atomic nuclei.

ppm. The parts per million of a chemical in air, water, or solid – almost always on a volume basis in air designated as ppm_v as opposed to on a mass basis in water and solid designated as ppm_m.

ppb. The parts per billion of a chemical in air, water, or solid – almost always on a volume basis in air, designated as ppb_v as opposed to on a mass basis in water and solid designated as ppm_m.

Psychrometric chart. A chart employed to determine the properties of moist air.

Purging. A cleansing or removal of impurities, foreign matter, or undesirable elements from a vessel or reactor.

Pyrolysis. The chemical decomposition of organic matter through the application of heat in the absence of oxygen.

Quality assurance/quality control (QA/QC). A system of procedures, checks, audits, and corrective actions to ensure that all research, design and performance, environmental monitoring and sampling, and other technical and reporting activities are of the highest achievable quality.

Quenching. A rapid cooling of metals or alloys by immersion in cold water or oil, or air by exposure to cold water.

Quiescent. A term used to describe a body at rest.

Radiation. The emission and propagation of energy through space or through a material medium, usually in the form of electromagnetic waves.

Radioactive. A term used to describe and substance that emits radiation either naturally or as a result of chemical manipulation.

Reboiler. An auxiliary heating unit for a distillation column, designed to supply additional heat to the lower portion of the column.

Recirculation. The repeated flow of a fluid around a closed system.

Rectification. The enrichment or purification of a vapor during a distillation process by contact or interaction with a countercurrent stream of liquid condensed from the vapor.

Reducing agent. Any material that adds hydrogen or electrons to an element or compound.

Refining. A separation process in which undesirable components are removed from various types of mixtures to yield a more purified product.

Reflux. The liquid that has condensed from a rising vapor and is allowed to flow back down a column in a distillation process.

Refractory. An inert, ceramic material with the ability to retain its physical shape and chemical identity when subjected to extremely high temperatures.

Refrigerant. A substance that is suitable as the working medium of a cycle of operations producing refrigeration (e.g., liquid ammonia, Freon).

Regression. A data analysis method used to measure the extent to which two variables increase or decrease together or of the extent to which one increases as the other decreases.

Relief valve. A pressure relief device such as pressure valves, rupture disks, and other pressure relief systems employed to protect process components from possible explosions or other damage.

Retrofit. The addition or removal of a piece of equipment, or a required adjustment, connection, or disconnection of an existing piece of equipment, often for the purpose of reducing emissions or optimizing a process.

Reverse flow. A flow in a direction opposite to normal flow.

Reverse osmosis. A water treatment process employed to separate water from pollutants by the application of pressure to force the water through a semipermeable membrane against the osmotic pressure of the solution.

Reynolds number. A dimensionless number used in fluid flow calculations to describe the level of turbulence within a system.

Risk. A measure of economic loss or human injury in terms of both the incident likelihood and the magnitude of the loss or injury.

Risk analysis. The structured evaluation of incident consequences, frequencies, and risk assessment results.

Risk assessment. The process by which risk estimates are made.

Risk contour. Lines on a risk graph that connect points of equal risk.

Risk estimation. Combining the estimated consequences and likelihood of a risk.

Risk management. The application of management policies, procedures, and practices in analyzing, assessing, and controlling risk.

Risk perception. The perception of risk that is a function of age, race, sex, personal history and background, familiarity with the potential risk, dread factors, perceived benefits of the risk-causing action, marital status, residence, etc.

Rotameter. A device composed of a float inside a tapered glass tube employed for measuring fluid flow.

Rotary dryer. A long, steel cylinder, slowly rotating, with a slightly inclined axis, through which a solid or semisolid material passes to be dried by hot air.

Rotary valve. A valve consisting of a casing, more or less spherical in shape, and a gate that turns through 90° when opening or closing; it has a cylindrical opening of the same diameter as that of the pipe it serves.

Rupture disk. A thin piece of metal between flanges that breaks at a certain pressure to prevent dangerous pressure buildup within a pressure vessel.

Safety data sheet (SDS). A compilation of information required under OSHA communication standard on the identity of hazardous chemicals, health and physical hazards, exposure limits, and precautions.

Safety valve. A valve that automatically opens when prescribed conditions, usually of pressure, are exceeded in a pipeline or other closed receptacle containing liquid and/or gases.

Salinity. The amount of salt in water.

Salting out. A reduction in the water solubility of a solid, liquid, or gas by adding a salt to an aqueous solution of the substance.

Saturated steam. Steam that is in equilibrium with liquid water at a given temperature or pressure.

Scale-up. The calculations and planning involved in increasing operations from the pilot plant stage to the large-scale production stage.

SCFM. Standard cubic feet per minute, flow rate corrected to standard pressure and temperature.

Scheduled maintenance. Any periodic procedure that is necessary to maintain the integrity or reliability of a system, which can be anticipated and scheduled in advance.

Scope of work. A document detailing program requirements.

Scrubber.	A device that uses a liquid spray to remove particulates and gaseous components from an air stream; the gases are removed by absorption or a chemical reaction and the particulates are removed through contact and capture by the liquid droplets.
Sensible heat.	The heat that, when added or removed, results in a change of temperature.
Serendipity.	An unexpected scientific discovery that turns out to be more important than the project being researched.
Sewer.	A collection and disposal pipe or conduit for liquid wastes.
Shutdown.	The cessation of operation of an affected facility for any purpose.
Side reaction.	A secondary reaction accompanying or following a primary reaction.
Sink.	A receptacle for the materials moved through a system.
Site inspection.	An on-site investigation to determine whether there is a release or potential release and the extent and severity of hazards posed by the release.
Slaking.	The process of mixing with water so that a chemical combination occurs, as in slaking of lime.
Sludge.	The thick, solid, or semisolid waste that accumulates as a result of the settling which occurs during water and wastewater treatment processes, especially sedimentation.
Slurry.	A mixture of solid matter and liquid.
Smelter.	A facility that melts or fuses ore.
Smog.	Originally meant to consider a combination of smoke and fog.
Societal risk	A measure of risk to a group of individuals.
Solute.	The substance that is dissolved in a solvent to form a solution.
Solution.	A homogeneous mixture of two or more substances constituting a single phase.
Source term.	The estimation of the mass or volumetric release of a (hazardous) agent from a specific source.
Space velocity.	The volume of gas or liquid, measured at a specified temperature and pressure, passing through a unit volume in a certain unit time.
Sparger.	A perforated pipe through which steam, air, water, or other fluid is injected into a liquid during a reaction (e.g., during fermentation).
Spent.	A term used to describe any material that has been used and, as a result of contamination, can no longer serve the purpose for which it was produced without further processing.
Stack.	A vertical duct or conduit for feeding exhausts into the atmosphere.
Stand-alone system.	An independent system or process.
State.	One of the three forms that matter can take: solid, liquid, and gas.
Static head.	The pressure in a fluid due to the height of fluid above a given point.
Steady flow.	A flow that does not vary with time; the mass flow rate is constant and all other quantities (e.g., temperature, pressure, velocity) are independent of time.
Steam drum.	A vessel in a boiler in which the saturated steam is separated from the steam-water mixture and into which the feedwater is introduced.
Stoichiometric.	A term used to describe the elements of a compound in exactly the proportion represented by the compound's chemical formula, and a term used to describe the minimum amount of a chemical necessary to completely react with another reactant.
Stoker.	A mechanical device employed to feed solid fuel or solid waste to a furnace.
Stop valve.	A valve installed in a pipeline to shut off flow for the purpose of inspection or repair.
Stripper.	A column in which one or more components of a liquid stream are removed by being transferred into a gas stream.
Sump.	A pit or tank that catches liquid runoff for drainage or disposal.
Supercooled liquid.	A liquid cooled below its normal freezing point without solidification.
Superficial mass velocity.	The quantity obtained when the mass rate of flow is divided by the total cross-sectional area, regardless of the presence of any obstruction to flow.
Superheated steam.	Steam at a temperature above its boiling point.
Synergism.	The cooperative interaction of two or more chemicals or other phenomena producing a greater total effect than the sum of their individual effects.
Synthesis.	The combination of parts to form a whole; and creation of a substance that either duplicates a natural product or is a unique material not found in nature.

Systemic. A term used to describe something that affects the entire system or body.

Tailings. The residues of raw materials or waste separated out during the processing of mineral ores.

Tee. A pipe fitting that has the outlets at right angles to the inlet.

Thermal pollution. The addition of large amounts of heat to air, water, or land to the degree that the added temperature may have harmful effects.

Theoretical air. The quantity of air, calculated from the chemical composition of a waste, that is required to burn/combust a waste/feedstock completely so that no oxygen remains.

Throat velocity. The gas or liquid velocity through the throat of a contraction.

Throttle valve. A valve designed to control the rate of fluid flow.

Time of failure. The time when a duty or intended function associated with a component or entire system is no longer able to be performed.

TLV. The acronym for the threshold limit value (established by the American Council of Government Industrial Hygienists, ACGIH). The concentration of a chemical in air that may be breathed in without harmful effects for five consecutive eight-hour working days.

TLV-TWA. The allowable time-weighted average concentration of a chemical in air for an 8-hr workday/40-hr work week that produces no adverse health effects in exposed individuals.

TLV-STEL. The acronym for short-term exposure limit, a 15-minute, time-weighted average concentration to which workers may be exposed up to four times per day with at least 60 minutes between successive exposures with no ill effect if the TLV-TWA is not exceeded.

TLV-C. The ceiling exposure limit representing the maximum concentration of a chemical in air that should never be exceeded in any part of the working exposure.

Top event. The accident, event, or incident at the "top" of a fault tree that is traced downward to more basic failures using logic gates to determine their causes.

Toxic dose. The combination of concentration and exposure duration for a toxic agent to produce a specific harmful effect.

Transient. A term used to describe anything that changes with time.

Triple point. The temperature and pressure at which all three states (i.e., solid, liquid, gas) of a substance exist together.

Tsunami. A sea wave caused by an underwater seismic disturbance such as sudden faulting, land sliding, or volcanic activity.

Turbine. A machine that converts the energy in a stream of fluid into mechanical energy by passing the stream through a system of fan-like blades, causing them to rotate.

Turbulence. The fluid property that is characterized by irregular variations in the speed and direction of the movement of individual particles or elements of flow.

UEL/UFL. The upper explosive/flammability limit, the highest concentration of a chemical in air that will produce an explosion or flame if ignited.

Uncertainty. A measure, often quantitative, of the degree of doubt or lack of certainty associated with an estimate.

Unit operation. One of many operations employed in chemical engineering in the industrial production of various chemicals in which a chemical change takes place.

Unit process. One of many operations employed in chemical engineering in the industrial production of various chemicals in which only physical changes and no chemical changes take place.

Unstable. A term used to describe a chemical that tends to move toward decomposition or other unwanted chemical change during normal handling or storage.

Unsteady flow. A flow that is characterized by a mass flow rate and/or other quantities that vary with time.

Upset. A disturbance in the functioning, fulfillment, or completion of a process or material.

Upstream. The direction from which a stream is flowing.

Uptake. The act of taking up, drawing up, or absorbing.

Useful life. The period during which a piece of equipment or a process operates as per its intended function.

Valve. A device for controlling the flow of a fluid through a pipe or tube.

Vane. A device that pivots on a rooftop or other elevated location for the purpose of determining the wind direction.

Vapor. A liquid which has been evaporated and exists in gaseous form.

Vent. An opening through which a fluid or gas is ejected from a system.

Venturi scrubber. A unit using a liquid, usually water, to remove particulate and gaseous pollutants from the air.

Viscosity. The internal resistance to flow exhibited by a fluid; it is the ratio of shearing stress to rate of shear.

Watershed. The area surrounding a stream that supplies it with water runoff.

Weir box. A box installed in a narrow open channel upstream from a weir to provide an enlargement of the flow area; the velocity of approach of the weir is consequently reduced.

Wet bulb. A thermometer bulb maintained wet with distilled water; it is employed with the dry bulb to measure humidity.

Zeolite. Any of the hydrated aluminum complex silicates, either natural or synthetic, that possesses cation exchange properties.

References

1 **L. Theodore, J. Reynolds, and K. Morris**, *Dictionary of Concise Environmental Terms*, Gordon and Breach Science Publishers, Amsterdam, The Netherlands, 1997.

2 **M.K. Theodore, and L. Theodore**, *Introduction to Environmental Management*, 3rd Ed., CRC Press/Taylor & Francis Group, Boca Raton, FL, 2021.

3 **L. Thoedore**, *Chemical Engineering: The Essential Reference*, McGraw-Hill, New York City, NY, 2014.

2

The Air Pollution Problem

The atmosphere of the earth has undergone significant changes in the last 3.5 billion years. Fossil records indicate that the carbon dioxide content has varied tremendously during this time. The life-forms that are present today are the ones which have adapted to the current environment; the life-forms that are now extinct are those that failed to adapt. The change in the environment is not the problem today; it is the abruptness of the change that causes the difficulties. Man-made air pollution may be currently causing problems because it is producing rapid changes in the environment of the earth.

Humans did not significantly affect the environment until relatively recent times. This is due to two reasons. First, the human population has been large for only a small part of its history. Second, the bulk of man-made air pollution is intimately related to industrialization. In fact, humans did not begin to alter the environment until they began to live in communities.

From the fourteenth century until recently, the primary air pollutants have been coal smoke and gases released in industrialized areas. Unfortunately, control of pollutants rarely takes place prior to public outcry, even though the technology for controlling pollutants may be available. Early recognition of pollutants as health hazards has not resulted in pollution reduction; only when personal survival is at stake has effective action been taken against pollutants.

The average person breathes 35 pounds of air each day – six times as much as the food and drink normally consumed in the same period of time. However, a half-century ago over 200 million tons of waste products were being released into the air annually. Slightly over half of the pollution came from the internal combustion engines of cars and other motor vehicles. Roughly 22% came from fuel burned at stationary sources such as power generating plants, and another 15% was emitted from industrial processes.

2.1 Early History

During the reign of the English King Edward I (1272–1307), there was a protest by the nobility against the use of "sea" coal. In the succeeding reign of Edward II (1307–1327), a man was put to torture ostensibly for filling the air with a "pestilential coal" resulting from the use of coal. Under Richard III (1377–1399), and later under Henry V (1413–1422), England took steps to regulate and restrict the use of coal. Both taxation and regulation of the movement of coal in London were employed.

Other legislation, parliamentary studies, and literary comments appeared sporadically during the next 250 years in England. In 1661, a pamphlet was published by the Royal Command of Charles II entitled "Fumifugium; or the Inconveniences of Air and Smoke of London Dissipated; Together with Some Remedies Humbly Proposed." The paper was written by John Evelyn, one of the founding fathers of the Royal Society. In 1819, a Select Committee of the British Parliament was formed to study smoke abatement. As in the case of most civic actions, by the time the committee submitted its report, the problem had subsided, and no action was taken. The smoke problem in London reached its peak in December 1952; during this "air pollution episode" 4,000 people died, primarily of respiratory problems.

Optimizing Air Pollution Control Equipment Performance: Operation & Maintenance, First Edition. Jay Richardson and Louis Theodore.

The term "smog" originated in Great Britain, where it was used to describe the 1,063 smoke-fog deaths that occurred in Glasgow, Scotland in 1909. In 1948, 20 people died and several hundred became ill in the industrial town of Donora, Pennsylvania. St. Louis, Cincinnati, and Pittsburgh have had similar smoke problems, all of which have been significantly reduced because of public outcry. Additional details of the often-referenced episodes are summarized below:[1]

1) On Friday, December 5, 1952, static weather conditions turned the air of London, England, into a deadly menace. A prolonged temperature inversion held the city's air close to the ground and an anticyclonic high-pressure system prevented the formation of winds that would have dispersed the pollutants that were accumulating heavily at ground level. For five days, the Greater London area was blanketed in airborne muck. Few realized it at the time, but there were 4,000 more deaths than normal for a 5-day period, hospital admissions were 48% higher, and sickness claims to the national health insurance system were 108% above average. Eighty-four percent of those who died had preexisting heart or lung diseases. Hospital admissions for respiratory illness increased threefold and deaths due to chronic respiratory disease increased tenfold.
2) The same static atmospheric condition in (1) had caused a similar incident in Donora, Pennsylvania in 1948. A town of only 14,000, it had 15–20 more deaths than normal during the episode. More than 6,000 of its residents were adversely affected, 10% of them were serious. Among those with preexisting illnesses, 88% of the asthmatics, 77% of those with heart diseases, and 79% of those with chronic bronchitis and emphysema were adversely affected. Allowing for the great difference in population, Donora paid a much higher price for air pollution than London.
3) New York City has experienced similar periods of atmospheric stagnation on numerous occasions during the last half-century. During one such episode, in 1953, the city reported more than 200 deaths above normal. Birmingham, Alabama is another high-exposure area whose residents have frequently exhibited a greater-than-average incidence of respiratory irritation symptoms such as coughing, burning throats or lungs, and shortness of breath. Environmental Protection Agency (EPA) monitoring studies indicated that nonsmokers in these two cities developed respiratory symptoms two or three times more frequently than nonsmokers in cleaner communities.
4) In the early 1900s, gases from short stacks at two copper smelters near the Georgia border of Tennessee caused widespread damage to vegetation in the surrounding countryside. When taller stacks were built, damage extended 30 miles into the forests of Georgia. An interstate suit resulted which was finally carried to the United States Supreme Court. The problem was eventually solved by means of a by-product sulfur dioxide recovery plant.
5) Two decades later, a similar case involved the lead and zinc smelter of the Consolidated Mining and Smelting Company of Canada at Trail, B.B. The smelter was located on the west bank of the Columbia River, 11 miles north of the international boundary between Canada and the United States. When extensive damage to vegetation occurred on the US side of the border, a damage suit was filed and finally settled by an international tribunal. In this case, after damages were assessed, the problem was solved partly by sulfur recovery and partly by operating the smelter according to a plan based on meteorological considerations.
6) Unfortunately, the climatic conditions and human activities that combine to form critical buildups of pollutants are by no means uncommon in the United States. They occur periodically in various parts of the country and will continue to threaten public health as long as air pollutants are emitted into the atmosphere in amounts sufficient to accumulate to dangerous levels.

Smoke (particulates) and acid gases produced by the burning of coal have been significant air pollutants for more than 400 years. The use of oil and gas has reduced this problem somewhat; however, local natural resources (and hence costs) currently often dictate the choice between coal, oil, and gas for stationary energy conversion and electricity production.

The internal combustion engine in the automobile is a more recent but no less significant air pollution. The use of oil and gas has reduced this problem somewhat; however, 30 years passed before action was taken to reduce automobile pollution. Typical smog problems are related to the automobile industry, and the pollution problems are all concentrated in urban areas.

2.2 Sources and Classifications of Air Pollution

Air pollutants may be divided into two broad categories, natural and man-made. Natural sources of air pollutants include the following:

- Wind-blown dust
- Volcanic ash and gases
- Ozone from lightning and the ozone layer
- Esters and terpenes from vegetation
- Smoke, gases, and fly ash from forest fires
- Pollens and other aeroallergens
- Gases and odors from natural decomposition
- Natural radioactivity

Such sources constitute background pollution and that portion of the pollution problem over which control activities can have little, if any, effect. Man-made sources cover a wide spectrum of chemical and physical activities, and are the major contributors to urban air pollution. Air pollutants in the United States pour out from over 200 million vehicles, from the refuse of over 350 million people, from the generation of billions of kilowatts of electricity, and from the production of innumerable products demanded by everyday living. Over 300 million tons of air pollutants are generated annually in the United States alone.

Air pollutants may also be classified by origin and state of matter. Under classification by origin, the following subdivisions pertain: (i) primary – emitted to the atmosphere from a process and (ii) secondary – formed in the atmosphere as a result of a chemical reaction. Under state of matter, there exist the classifications: particulate and gas (true gases, such as sulfur dioxide, nitrogen oxides, ozone, and carbon monoxide, and vapors, such as gasoline, paint solvents, and dry-cleaning agents). Particulates may be defined as solid or liquid matter whose effective diameter is larger than a molecule but smaller than approximately 1,000 μm (microns). Particulates dispersed in a gaseous medium are collectively termed an aerosol. The terms smoke, fog, haze, and dust are commonly used to describe particular types of aerosols, depending on the size, shape, and characteristic behavior of the dispersed particles. Aerosols are rather difficult to classify on a scientific basis in terms of their fundamental properties such as settling rate under the influence of external forces (such as gravity), optical activity, ability to absorb electric charge, particle size and structure, surface-to-volume ratio, reaction activity, and physiological action; in general, particle size and settling rate have been the most characteristic properties for many purposes. For example, particles larger than 100 μm (micrometers) may be excluded from the category of dispersions because they settle too rapidly. On the other hand, particles on the order of 1 μm or less settle so slowly that, for all practical purposes, they are regarded as permanent suspensions.

Despite possible advantages of scientific classification schemes, the use of popular descriptive terms such as smoke, dust, and mist, which are essentially based on the mode of formation, appears to be a satisfactory and convenient method of classification. In addition, this approach is so well established and understood that it undoubtedly would be difficult to change. Additional details on the two topics are available in the next chapter.

2.3 The Need for Control

Two fundamental reasons for the cleaning of gases in industry, particularly waste gases, are profit and protection. For example, profits may result from the utilization of blast furnace gases for heating and power generation, but impurities have to be removed from the gases before they can be burned satisfactorily. Some impurities can be economically converted into sulfur, or solvent recovery systems can be installed to recover valuable hydrocarbon emissions. Protection of the health and welfare of the public in general, or the individual working in industry, and of property is another reason for cleaning gases. The enactment of air pollution control regulations reflects the concern of the government for the protection of its people. Waste gases containing toxic constituents such as arsenic or lead fumes constitute a serious danger to the health of plant operators and the surrounding population. Other waste gases, although not normally endangering health in the concentrations encountered, may kill plants, damage paintwork and buildings, or discolor wallpaper and curtains, thus making an industrial town a less pleasant place in which to live and work.

The extent to which industry cleans polluted gas steams depends largely upon the limits imposed by four main considerations:

1) Concentration levels harmful to humans, physical structures, and plant and animal life.
2) Legal limitation imposed by the country, state, county, village, or city for the protection of the public health and welfare.
3) Reduction of air pollution to establish civic goodwill.
4) The reduction and/or elimination of potential liability patterns.

These considerations are not independent. For example, the legal limits on emissions are also closely related to the degree of cost needed to prevent concentrations that would damage the ecosystem.

There seems to be little question that, during many of the more serious pollution episodes described earlier, air pollution can be a deadly killer. As discussed in the previous section, hundreds of excess deaths have been attributed to incidents in London in 1952, 1956, 1957, and 1962, in Donora, Pennsylvania in 1948, and in New York City in 1953, 1963, and 1966. Many of these people were in failing health and were generally those suffering from lung conditions. Hundreds of thousands have suffered from serious discomfort and inconvenience, including eye irritation and chest pains, during these and other such incidents. These acute problems are actually the lesser of the health problems. There is considerable evidence of the chronic threat to human health from air pollution. This evidence ranges from the rapid rise of emphysema as a major health problem, through the identification of carcinogenic compounds in Los Angeles smog, to statistical evidence that people exposed to polluted atmosphere over extended periods of time suffer from a number of ailments and a reduction in their lifespan.

Property damage is presently the best-documented example of the pollution effects. It disintegrates nylon stocking in Chicago and Los Angeles; ruins historic statues and buildings in Venice, Athens, and Cologne; affects visibility on roadways and at airports; destroys fruit, citrus trees, and vegetable crops in California, New Jersey, Georgia, Florida, Washington, and many other states; and affects cattle health and growth. Cleopatra's Needle, standing in New York City's Central Park, has deteriorated more in 80 years in that park than in 3,000 years in Egypt. Point sources, such as chemical plants, can produce objectionable odors or chemical mists capable of blackening the paint on cars and houses.

The greatest long-term need for a deeper understanding of the atmospheric environment lies in a crucial area: the modes of action and effects of pollutants on humans, animals, plants, and inanimate objects. Existing knowledge, buttressed by the clear need for haste, provides the current basis for establishment of air quality criteria. However, the ability to refine and augment such criteria and standards, to predict the effects of pollutants, and to detect such effects at an early stage will require much more penetrating knowledge than now exists of the effects themselves and of the mechanisms of contaminant action.

2.4 Estimating Pollutant Emissions

It is important for the design engineer to be able to estimate the uncontrolled pollutant emissions from a source. However, adequate information on all emission sources is seldom available in a manner that it can be readily used. In these cases, one has to investigate other resources for generating emissions estimates. One such information source is EPA's Compilation of Air Pollutant Emissions Factors AP-42,[2] which provides emission factors for air pollutants for several industrial activities. This compilation provides emissions factors for use in some instances for both controlled and uncontrolled emissions. AP-42 also provides a methodology for calculating emissions from specific sources, e.g., losses from storage tanks and loading losses. Mass estimates of releases computed using AP-42 emissions factors may be overly conservative in some cases and should be validated for site-specific conditions, if possible.[3]

The aforementioned emissions factors are average values that relate the quantity of a pollutant released to the atmosphere with the activity associated with the release of that pollutant. It is usually expressed as the weight of a pollutant divided by a unit weight, volume, distance, or duration of the activity that emits the pollutant (e.g., kilograms of particulate emitted per megagram of coal combusted). Using such factors permits the estimation of emissions from various sources of air pollution. In most cases, these factors are simply averages of all available data of acceptable quality, generally without consideration of the influence of various process parameters such as temperature and reactant concentrations. For a few cases, however, such as in the estimation of volatile organic emissions from petroleum storage tanks, the AP-42 document contains empirical formulas, which can relate emissions to such variables as tank diameter, liquid temperature, and wind velocity. Emission factors correlated with such variables tend to yield more precise estimates than would factors derived from broader statistical averages.

As indicated above, emission factors are very useful tools for estimating emissions of air pollutants. However, because such factors are averages obtained from data of wide range and varying degrees of accuracy, emissions calculated this way for a given facility are likely to differ from that facility's actual emissions. Because they are averages, those factors will indicate higher emission estimates that are actual for some sources and lower for others. Only specific source measurements can determine the actual pollutant contribution from a source, under conditions existing at the time of the test (see more in Part III). For the most accurate emissions estimate, it is recommended that source-specific data be obtained whenever possible. Emission factors are more appropriately used to estimate the collective emissions of a number of sources, such as is done in the emissions inventory efforts for a particular geographic area.

If factors are used to predict emission from new or proposed sources, users should review the latest literature and technology to determine if such sources would likely exhibit emissions characteristics different from those of typical existing sources.

AP-42 lists emission factors for processes and operations too numerous to detail here. Listed below are only a few of the sources for which information is provided[2]:

1) Bituminous and subbituminous coal combustion
2) Anthracite coal combustion
3) Fuel oil combustion
4) Natural gas combustion
5) Primary aluminum production
6) Primary copper smelting
7) Iron and steel production
8) Zinc smelting
9) Gray iron foundries
10) Portland cement manufacturing
11) Taconite ore processing
12) Industrial paved roads
13) Residential wood stoves
14) Waste oil combustion
15) Refuse combustion
16) Sewage sludge incineration
17) Surface coating
18) Soap and detergents
19) Residential fireplaces
20) Waste water collection, treatment, and storage
21) Ammonium phosphates
22) Industrial flares
23) Particle size distribution data and sized emission factors for selected sources
24) Generalized particle size distributions

2.5 Measurement Methods

An accurate quantitative analysis of the discharge of pollutants from a process must be determined prior to the design and/or selection of control equipment. If the unit is properly engineered, by utilizing the emission data as input to the control device and the code requirements as maximum effluent limitations, most particulate pollutants can be successfully controlled by one or a combination of the methods to be discussed in later chapters.

The objective of source testing is to obtain data representative of the process being sampled. The steps followed in obtaining representative data from a large source are:

1) Obtaining a measurement that reflects the time and magnitude of the characteristic being measured at the location where the measurement is made.
2) Taking a number of measurements in such a manner that the data obtained from these measurements are representative of the source.

Sampling is the keystone of source analysis. Sampling methods and tools vary in their complexity according to the specific task; therefore, a degree of both technical knowledge and common sense is needed to design a sampling function. Sampling is performed to measure quantities or concentrations of pollutants in effluent gas streams, to measure the efficiency of a pollution abatement device, to guide the designer of pollution control equipment and facilities, and/or to appraise contamination from a process or source. A complete measurement requires the determination of the concentration and contaminant characteristics, as well as the associated gas flow. Most statutory limitations require mass rates of emission; both concentration and volumetric flow rate data are therefore required. However, the description of specific methods of analysis, i.e., EPA Method 5, Orsat, and gas chromatography, is beyond the scope of this chapter. The reader is referred to the voluminous literature available on the subject. Additional details on this topic are provided in Part III, Chapter 24.

References

1 **L. Theodore**, *Air Pollution Control and Waste Incineration for Hospitals and Other Medical Facilities*, Van Nostrand Reinhold, New York, NY, 1990.

2 USEPA, *Compilation of Air Pollutant Emission Factors, AP-42, US*, Environmental Protection Agency, Washington, DC, 1985.

3 **R. Sober, and D. Paul**, Less-subjective odor assessment, *Chem. Eng.*, 130, 1992.

3

Classifications, Sources, and Effects of Air Pollution

Not long ago, the nation's national resources were exploited indiscriminately. Stacks dispersed smoke from factories and power plants; waterways served as industrial pollution sinks; and the land proved to be a cheap and convenient place to dump industrial and urban wastes. However, society is now more aware of the environment and the need to protect it.

The American people have been involved in a great social movement known broadly as "environmentalism." Society has been concerned with the quality of the air one breathes, the water one drinks, and the land on which one lives and works. While economic growth and prosperity are still important goals, opinion polls show overwhelming public support for pollution controls and a pronounced willingness to pay for them. This chapter presents the reader with information on pollutants and categorizes their sources by the media they threaten.

Since the Clean Air Act was passed in 1970, the United States has made impressive strides in improving and protecting air quality. As directed by this Act, the Environmental Protection Agency (EPA) set National Ambient Air Quality Standards (NAAQS) for those pollutants commonly found throughout the country that posed the greatest overall threats to air quality. These pollutants, termed "criteria pollutants" under the Act, including ozone, carbon monoxide, airborne particulates, sulfur dioxide, lead, and nitrogen oxide are currently a major concern in a number of areas in the country. The following section focuses on a number of the most significant air quality challenges: ozone and carbon monoxide, airborne particulates, airborne toxics, sulfur dioxide, acid deposition, and indoor air pollutants.

3.1 Sources of Air Pollutants

As mentioned in the previous chapter, air pollutants may be divided into two broad categories, *natural* and *man-made*. Natural sources of air pollutants constitute background pollution and that portion of the pollution problem over which control activities can have little, if any, effect. Man-made sources cover a wide spectrum of chemical and physical activities and are the major contributors to urban air pollution. Air pollutants in the United States pour out from over 200 million vehicles, from the refuse of over 300 million people, from the generation of billions of kilowatts of electricity, and from the production of innumerable products demanded by everyday living.

Under classification by origin, the following subdivisions pertain: *primary* – emitted to the atmosphere from a process, and *secondary* – formed in the atmosphere as a result of a chemical reaction. Under the state of matter, there exist the classifications *particulate* and *gaseous* (true gases such as sulfur dioxide, nitrogen oxides, ozone, and carbon monoxide, and vapors such as gasoline, paint solvents, and dry-cleaning agents). Particulates may be defined as solid or liquid matter whose effective diameter is larger than a molecule but smaller than approximately 1,000 μm. Particulates dispersed in a gaseous medium are collectively termed an aerosol. The terms smoke, fog, haze, and dust are commonly used to describe particulate types of aerosols, depending on the size, shape, and characteristic behavior of the dispersed particulates. Aerosols are rather difficult to classify on a scientific basis, i.e., in terms of their fundamental properties, such as their settling rate, under the influence of external forces, optical activity, ability to absorb electrical charge, particle size and structure, surface to volume ratio, reaction activity, physiological, and action. In general, the particle size and settling rate have been the most characteristic properties for many purposes. For example, particles, larger than 100 μm may be excluded from the category of dispersions, because they settle too rapidly. On the other hand, particles on the order

Optimizing Air Pollution Control Equipment Performance: Operation & Maintenance, First Edition. Jay Richardson and Louis Theodore.

of 1 μm or less settle so slowly that, for all practical purposes, they are regarded as permanent suspensions. Despite the possible advantages of scientific classification schemes, the use of popular descriptive terms, such as smoke, dust, and mists, which are essentially based on the mode of formation, appears to be a satisfactory and convenient method of classification. In addition, this approach is so well-established and understood that it undoubtedly would be difficult to change.

Dust is typically formed by the pulverization or mechanical disintegration of solid matter in particles of smaller size by processes, such as grinding, crushing, and drilling. Particle sizes of dust range from a lower limit of 1 μm up to about 100 or 200 μm and larger. Dust particles are usually irregular in shape, and the particle size refers to some average dimension for any given particle. Common examples include fly ash, rock dust, and ordinary flour.

Smoke implies a certain degree of optical density, it is typically derived from the burning of organic materials such as wood, coal, and tobacco. Smoke particles are very fine, ranging in size from less than 0.01 μm up to 1 μm. They are usually spherical in shape if of liquid or tarry composition and irregular in shape, if of solid composition.

Fumes are typically formed by processes, such as sublimation, condensation, or combustion, generally at relatively high temperatures. The range in particle size from less than 0.1 μm to 1 μm. Similar to smoke, they settle very slowly and exhibit strong Brownian motion.[1]

Mists or fogs are typically formed, either by the condensation of water, or some other vapor on suitable nuclei, giving a suspension of natural fogs in mists, which usually range in size between 2 and 200 μm. Droplets larger than 200 μm are more properly classified as drizzle or rain.

When a liquid or solid substance is broken up into smaller and smaller particles, more of its surface area is exposed to air. Under certain circumstances, the substance, whatever its chemical composition, tends to physically or chemically combine with other particles or gases in the atmosphere. The resulting combinations are frequently unpredictable. Very small aerosol particles (from 0.001 μm to 0.1 μm) can act as condensation nuclei to facilitate the condensation of water vapor that can promote the formation of fog and ground mist. Particles, less than 2 or 3 μm in size, about half (by weight) of the particle suspension of urban air – can penetrate into mucous membranes and attract and convey harmful chemicals such as sulfur dioxide. By virtue of the increased surface area of the small aerosol particles, and as a result of absorption into gas molecules, or other such properties that are able to facilitate chemical reactions, aerosols tend to exhibit greatly enhanced surface activity.

Many substances that oxidize slowly in a given state can oxidize extremely quickly, or possibly even explode when dispersed as fine particles in the air. Dust explosions, for example, are often caused by the unstable burning or oxidation of combustible particles, brought about by relatively large specific areas. Absorption in catalytic phenomena can also be extremely important in analyzing and understanding particulate pollution problems. The conversion of sulfur dioxide to corrosive sulfuric acid, assisted by the catalytic action of iron oxide particles, for example, demonstrates the catalytic nature of certain types of particles in the atmosphere.

3.2 Atmospheric Air Pollutants

3.2.1 Ozone and Carbon Monoxide

Ozone is one of the most interactive and widespread environmental problems. Chemically, ozone is a form of oxygen with three oxygen atoms instead of two found in regular oxygen. This makes it very reactive so that it combines with practically every material with which it comes in contact. In the upper atmosphere, where ozone is needed to protect people from ultraviolet radiation, the ozone is being destroyed by man-made chemicals, but at ground level, ozone can be a harmful pollutant.

Ozone is produced in the atmosphere when sunlight triggers chemical reactions between naturally occurring atmospheric gases and pollutants such as volatile organic compounds (VOCs) and nitrogen oxides. The main sources of VOCs and nitrogen oxides are combustion sources such as motor vehicle traffic.

Carbon monoxide is an invisible, odorless product of incomplete fuel combustion. As with ozone, motor vehicles are the main contributor of carbon monoxide formation. Other sources include wood-burning stoves, incinerators, and other industrial processes. Since auto travel and the number of small sources of VOCs are expected to increase, even extensive efforts may not sufficiently reduce emissions of ozone and carbon monoxide.[2]

3.3 Airborne Particulates

Particulates in the air include dust, smoke, metals, and aerosols. Major sources include steel mills, power plants, cement plants, smelters, and diesel engines. Other sources are grain storage elevators, industrial roads, construction work, and demolition. Wood-burning stoves in fireplaces can also be significant sources of particulates. Urban areas are likely to have windblown dust from roads, parking lots, and construction work.[2]

3.4 Airborne Toxins

Toxic pollutants are one of today's most serious, emerging problems that are found in all media. Many sources emit toxic chemicals into the atmosphere: industrial and manufacturing processes, solvent use, sewage treatment plants, hazardous waste handling in disposal sites, municipal waste sites, incinerators, and motor vehicles. Smelters, metal refiners, manufacturing processes, and stationary fuel combustion sources emit such toxic metals as cadmium, lead, arsenic, chromium, mercury, and beryllium. Toxic organics such as vinyl chloride and benzene are released by a variety of sources such as plastics in chemical manufacturing plants and gas stations. Chlorinated dioxins are emitted by some chemical processes in the high-temperature burning of plastics and incinerators.[2]

3.5 Sulfur Dioxide and Acid Deposition

Sulfur dioxide can be transported long distances in the atmosphere due to its ability to bond to particulates. After traveling, sulfur dioxide usually combines with water vapor to form acid rain. Sulfur dioxide is released into the air primarily through the burning of coal and fuel oils. Today, approximately 2/3 of all natural sulfur dioxide emissions come from electric power plants. Other sources of sulfur dioxide include refineries, pulp and paper mills, smelters, steel and chemical plants, in energy facilities related to oil shale, synthetic fuels, and oil and gas production. Home furnaces and coal-burning stoves are sources that directly affect residential neighborhoods.

Acid deposition is a serious environmental concern in many parts of the country. The process of acid deposition begins with the emissions of sulfur dioxide (primarily from coal-burning power plants) and nitrogen oxides (primarily from motor vehicles and coal-burning power plants). These pollutants interact with sunlight and water vapor in the upper atmosphere to form acidic compounds. During a storm, these compounds fall to the earth as acid rain or snow: the compounds may join dust, or other dry airborne particles and fall as "dry deposition."[3]

3.6 Indoor Air Pollutants

Indoor air pollution is rapidly becoming a major health issue in the United States. Indoor pollution levels are quite often higher than outdoors, particularly where buildings are tightly constructed to save energy. Since most people spend 90% of their time indoors, exposure to unhealthy concentrations of indoor air pollutants is often inevitable. The degree of risk associated with exposure to indoor pollutants depends on how well buildings are ventilated and the type, mixture, and the amounts of pollutants in the building.[4] Indoor air pollutants of special concern are described below. More detailed information on indoor air quality is available in the literature.[4–6]

3.6.1 Radon

Radon is a unique environmental problem because it occurs naturally. Radon results from the radioactive decay of radium 226, found in many types of rocks and soil. Most indoor radon comes from the rock and soil around the building and enters structures through cracks that are openings in the foundation or basement. Secondary sources of indoor radon are well water and building materials.[4]

3.6.2 Environmental Tobacco Smoke

Environmental tobacco smoke is smoke that non-smokers are exposed to from smokers. The smoke has been judged by the Surgeon General, the National Research Council, and the International Agency for Research on Cancer to pose a risk of lung cancer to non-smokers. Tobacco smoke contains a number of pollutants, including inorganic gases, heavy metals, particulates, VOCs, and products of incomplete burning, such as polynuclear aromatic hydrocarbons.[3] While this is less of an issue in the United States due to increased regulations on smoking, it still poses a significant issue worldwide.

3.6.3 Asbestos

Asbestos has been used in the past in a variety of building materials, including many types of insulation, fireproofing, wallboard, ceiling tiles, and floor tiles. The remodeling or demolition of buildings with asbestos-containing materials frees tiny asbestos fibers in clumps or clouds of dust. Even with normal aging, materials may deteriorate and release asbestos fibers. Once released, these asbestos fibers can be inhaled into the lungs and accumulate.[3]

3.6.4 Formaldehyde and Other Volatile Organic Compounds

The EPA has found formaldehyde to be a probable human carcinogen. The use of formaldehyde in furniture, foam insulation, and pressed wood products such as plywood, particleboard, and fire board, makes formaldehyde a major indoor air pollutant. VOCs commonly found indoors include benzene from tobacco smoke and perchloroethylene emitted from dry-cleaned clothes. Paints and stored chemicals, including certain cleaning compounds, are also major sources of VOCs. VOCs can also be emitted from drinking water; 20% of water supply systems have detectable amounts of VOCs.[4]

3.6.5 Pesticides

Indoor and outdoor uses of pesticides, including termiticides, and wood preservatives are another cause of concern. Even when used as directed, pesticides may release VOCs. In addition, there are about 1,200 inert ingredients added to pesticide products for a variety of purposes. While not "active" in attacking the particular pest, some inert ingredients are chemically or biologically active and may cause health problems. EPA researchers are presently investigating whether indoor use of pesticides and subsurface soil injection of termiticides can lead to hazardous exposure. A notable example of harmful pesticides is the effects of DDT on the peregrine falcon. The pesticide was extremely effective and used worldwide in the 1940s. In 1962, it was discovered that the pesticide was moving up the food chain and damaging peregrine falcon eggs, preventing them from reproducing and brought them to the brink of extinction. Eliminating the use of this pesticide and other significant efforts have brought the species back to self-sustaining levels.

3.7 Water and Land Pollutants

3.7.1 Water Pollutants

The EPA, in partnership with state and local governments, is also responsible for improving and maintaining water quality.[1] These efforts are organized around three themes. The first is maintaining quality drinking water. This is addressed by monitoring and treating drinking water prior to consumption and by minimizing the contamination of the surface water and protecting against contamination of groundwater needed for human consumption. The second is preventing the degradation and destruction of critical aquatic habitats including wetlands, nearshore coastal waters, oceans, and lakes. The third is reducing the pollution of free-flowing surface waters and protecting their uses.[7]

3.7.2 Land Pollutants

Land has been used as dumping grounds for wastes. Improper handling, storage, and disposal of chemicals can cause serious problems. Several types of wastes that are placed in the land are industrial hazardous waste, municipal waste, mining wastes, radioactive wastes, and leakage from underground storage tanks.

3.8 Effects of Air Pollution

There seems to be a little question that during many of the more serious pollution episodes, air pollution was a deadly killer. As discussed in the previous chapters, hundreds of excess deaths have been attributed to incidents in London in 1952, 1956, 1957, and 1962, in Donora; Pennsylvania in 1948; and in New York City in 1953, 1963, and 1966. Many of these people were in failing health, generally those suffering from lung conditions. Hundreds of thousands have suffered from serious discomfort and inconvenience, including irritation and chest pains during these and other such incidents. These acute problems are actually the lesser of the health problems.

There is considerable evidence of the chronic threat to human health from air pollution. This evidence ranges from the rapid rise of emphysema as a major health problem through the identification of carcinogenic compounds in Los Angeles smog to the statistical evidence that people are exposed to polluted atmospheres over extended periods of time suffer from a number of ailments and a reduction in their life span.

Two definitions are now formulated. An air *pollutant* is a gas, liquid, or solid (excluding water vapor) which results in a deviation in the composition of the dry, clean, and sea-level air. It is therefore necessary to expand upon the definition of an air pollutant. A *harmful* air pollutant is an air pollutant that is present in sufficient quantities so as to detract from the quality of life. Thus, this harmful air can result in injuries to human, animals, plant, life, or property, or can unreasonably interfere with the comfortable enjoyment of life, or property, or the conduct of business. Two related factors are needed to distinguish between a "pollutant" and a "harmful pollutant;" these are the quantity of pollution present and the duration of exposure to the pollutant.

Vegetation requires carbon dioxide; the nitrogen in atmosphere acts as a diluent; the oxygen is required for animal life. In other words, the earth's atmosphere is suited for life as we know it; more properly, the life present on earth today is that which has adapted itself to the current atmosphere. Numerous prehistoric animals failed to adjust to the changing environment and are extinct; the process is continuing today.

Biomedical investigations are used to determine quantities and duration that result in harmful effects from pollutants. The studies involving vegetation are fairly advanced because of the annual cycle of most plants. Conclusive results can be obtained from relatively short time period laboratory studies.

3.8.1 Bodily Responses to Air Pollution

The effects of air pollution on humans are very difficult to establish; this is primarily due to the problems in simulating the entire lifetime of a person in the laboratory in a significantly shorter time period. The effects on humans generally involve the eyes and the respiratory tract.

The eye is sensitive to certain gases and particulate matter; tearing normally flushes these foreign materials from the eyes. No lasting damage to the eyes has been shown to result from air pollution.

The respiratory system removes impurities from the inhaled air, warms and humidifies the air, and then supplies oxygen to the circulatory system; the circulatory system releases carbon dioxide, and is exhaled. The three major portions of the respiratory system are the nasopharyngeal (which includes the nasal cavity and the pharynx), the respiratory ducts, and the lungs.

Particles, larger than 10 μm are removed in the nasal pharyngeal section. The nasal hairs block large particles. Some particles impinge upon the surface of the pharynx, which is covered with mucus, and collected there. Mucus is then swallowed, which may lead to other problems. Some gases, absorbed by the mucus, may cause irritation of the pharynx.

The respiratory ducts are also covered with mucus; the mucus moves in the direction of the pharynx. If a large amount of foreign material becomes embedded in the mucus, the mucus stops moving and air can no longer be cleaned effectively.

The oxygen and carbon dioxide exchange between the blood in the air takes place within the lungs. Particles less than 10 μm in diameter may enter the lung; particles smaller than one micron may remain in tiny sacs within the lung, thereby decreasing its efficiency. Smaller particles may pass through the lung membrane and enter the bloodstream. Phagocyte cells in the lung engulf the particles trapped in sacks and carry them off to the mucus or the bloodstream for removal from the body.

Gases in the air may irritate the respiratory tract, irritate the tissues in the lungs, or enter into the bloodstream. Once in the bloodstream, the dissolved gases may be deposited in various organs of the body where they may have an effect. Unusually severe pollution levels may impair the heart.

Table 3.1 Maximum breathing capacity (l/min).

	Low Pollution		High Pollution	
Age	**Male**	**Female**	**Male**	**Female**
All ages	125.9	85.1	119.3	83.1
30–44	151.4	97.2	144.9	93.8
45–64	112.9	84.9	110.5	78.9
65+	87.5	63.7	80.3	55.6

The effects of air pollution on man are difficult to assess. Some problems become apparent only after a lifetime, but answers are needed now. The interaction between man and pollution may be described using several classifications. Epidemiological, statistical, pathophysiological, and biological classifications have been used. Both mortality rates and morbidity can be related to air pollution levels. Morbidity studies include the incidence and prevalence of disease, especially respiratory systems.[8]

Breathing capacity is affected by air pollution. This is readily seen in Table 3.1. Note the ambiguity in typical results from the medical study.

In summary, air pollution affects human health. Greater than expected mortality rates have been correlated with pollution levels. The incidence and prevalence of disease, lung impairment, increased airways resistance, and numerous other respiratory problems have been related to air pollutant concentrations.

References

1 **L. Stander, and L. Theodore**, *Environmental Regulatory Calculations Handbook*, CRC Press Taylor & Francis Group, Boca Raton, FL, 2012.

2 **M.K. Theodore, and L. Theodore**, *Introduction to Environmental Management*, CRC Press Taylor & Francis Group, Boca Raton, FL, 2021.

3 **J. Mycock, J. McKenna, and L. Theodore**, *Handbook of Air Pollution Control Equipment Technology*, CRC Press Taylor & Francis Group, Boca Raton, FL, 2008.

4 **G. Burke, B. Singh, and L. Theodore**, *Handbook of Environmental Management and Technology*, 2nd Ed., John Wiley & Sons, Hoboken, NJ, 2000.

5 **L. Theodore**, *Air Pollution Control for Hospitals and Other Medical Facilities*, Garland Press, New York City, NY, 1981.

6 **D. Green, and M. Southard**, *Perry's Chemical Engineering Handbook*, 9th Ed., McGraw Hill, New York City, NY, 2019.

7 **L. Theodore, and R.R. Dupont**, *Water Resource Management Issues: Basic Principles and Application*, CRC Press Taylor & Francis Group, Boca Raton, FL, 2020.

8 **M. Reynolds, and L. Theodore**, *A Guide to Virology for Engineers and Applied Scientists: Epidemiology, Emergency Management, and Optimization*, John Wiley & Sons, Hoboken, NJ, 2023.

4

Multimedia Concerns

It is now increasingly clear that some treatment technologies (as will be discussed in later chapters), while solving one pollution problem, have created others. Most contaminants, particularly toxics, present problems in more than one medium. Since nature does not recognize neat jurisdictional compartments, these same contaminants are often transferred across media. Air pollution control devices or industrial wastewater treatment plants prevent wastes from going into the air or water, respectively, but the toxic ash and sludge that these systems produce can become hazardous waste problems themselves. For example, removing trace metals from a flue gas usually transfers the products into a liquid or solid phase. Does this exchange an air quality problem for a liquid or solid waste management problem? Wastes disposed of on land or in deep wells may contaminate groundwater and evaporate from ponds and lagoons can convert solid or liquid wastes into air pollution.[1]

Control of cross-media pollutants cycling in the environment is therefore an important step in the management of environmental quality. Pollutants that do not remain where they are released or where they are deposited move from a source to receptors by many routes, including air, water, and land. Unless information is available on how pollutants are transported, transformed, and accumulated after they enter the environment, they cannot effectively be controlled. A better understanding of the cross-media nature of pollutants in their major environmental processes – physical, chemical, and biological – is required.

A multimedia approach to air pollution control was long overdue. As described above, it integrates air, water, and land into a single concern and seeks a solution to pollution that does not endanger society or the environment. The challenges for the future environmental professional include:

1) Conservation of natural resources
2) Control of air–water–land pollution
3) Regulation of toxics and disposal of hazardous waste
4) Improvement of quality of life

Other examples include acid deposition, residue management, water reuse, and hazardous waste treatment and/or disposal.

EPA's own single-medium offices, often created sequentially as individual environmental problems were identified and responded to and legislation, have played a role in impeding the development of cost-effective multimedia prevention strategies. In the past, innovative cross-media agreements involving or promoting pollution prevention and voluntary arrangements for overall reductions in releases have not been encouraged. However, new initiatives are characterized by the use of a wide range of tools, including market incentives, public education and information, small business grants, technical assistance, research and technology applications, as well as more traditional regulations and enforcement.

The reader should also note that the federal government through its military arm is responsible for some major environmental problems. It has further compounded these problems by failing to apply multimedia or "multi agency" approaches. The following are excerpts from a front-page article by Keith Schneider in the August 5, 1991, *New York Times*.[2] "A new strategic goal for the military is aimed at restoring the environment and reducing pollution at thousands of military and other government military industrial installations in the United States and abroad ... the result of environmental contamination on a scale almost unimaginable. The environmental projects are spread through four federal agencies and

Optimizing Air Pollution Control Equipment Performance: Operation & Maintenance, First Edition. Jay Richardson and Louis Theodore.

three military services and are directed primarily by deputy assistant secretaries. Many of the military industrial officials interviewed for this article said the scattered environmental offices are not sharing information well, while suffering at times from duplicated efforts and might not be supervising research or contractors closely enough. Environmental groups state agencies and the Environmental Protection Agency began to raise concern about the rampant military industrial contamination in the 1970s but were largely ignored. The Pentagon, the Energy Department and the National Aeronautics and Space Administration (NASA) and the Coast Guard considered pollution on their property to be confidential matter. Leaders feared not only embarrassment from public disclosure but also that solving problems would divert money from projects they considered more worthwhile. Spending on military-environmental projects is causing private companies, some of them among the largest contractors for the *military industry,* to establish new divisions to compete for government contracts. Many of them are worth $100 million to $100 billion." This lack of communication and/or unwillingness to cooperate within the federal government has created a multimedia problem that has just begun to surface. The years of indifference and neglect have allowed pollutants/wastes to contaminate the environment significantly beyond what would have occurred had the responsible parties acted sooner.

4.1 Environmental Problems

Environmental problems result from the release of waste (gaseous, liquid, and solid) that are generated daily by industrial and commercial establishments as well as from households. The lack of consciousness regarding the conservation of materials, energy, and water has contributed to the wasteful habits of society. The rate of waste generation has been increasing in accordance with the increase in population and the improvement in living standards. With technological advances and changes in lifestyle, the composition of waste has likewise changed. Chemical compounds and products are being manufactured in new forms with different half-lives. It has been difficult to manage such compounds and products once they have been discarded. As a result, these wastes have caused many treatment, storage, and disposal problems.

Many environmental problems are caused by products that are either misplaced in use or discarded without proper concern for the environmental impacts. Essentially all products are potential wastes, and it is desirable to develop methods to reduce the waste impacts associated with products or to produce environmentally friendly products. Environmental agencies have been lax in promoting and automating tracking mechanisms that identify sources and feats of new products.

Solving problems, however, can sometimes create problems. For example, implementation of the Clean Air Act and Clean Water Act has generated billions of tons of sludge, wastewater, and residue that could cause soil contamination and underground water pollution problems. The increased concern over cross-media shifts of pollutants has yet to consistently translate into a symptomatic understanding of pollution problems, and viable changes.

As indicated above, environmental protection efforts have emphasized media-specific waste treatment and disposal after the waste has already been created. Many of the air pollutants that harm the environment come from "area or point sources" such as industrial complexes and waste disposal facilities. Therefore, they simply cannot be solely controlled by end-of-pipe solutions. Furthermore, these end-of-pipe controls that tend to shift pollutants from one media to another have often caused secondary pollution problems. Therefore, for air pollution control purposes, the environment must be perceived as a single integrated system and pollution problems must be viewed holistically. Air quality can hardly be improved with water and land pollution continuing to occur. Similarly, water quality cannot be improved if the land and air are polluted.

Many secondary pollution problems today can be traced in part to education, i.e., the lack of knowledge and understanding of cross-media principles for the identification and control of pollutants. Neither the Clean Air Act nor the Clean Water Act enacted in the early 1970s adequately addressed the cross-media nature of environmental pollutants. More environmental professionals now realize that pollution legislation is too fragmented and compartmentalized. Only proper education and training will address this situation and hopefully lead to more comprehensive legislation of a total environmental approach.

4.2 The Multimedia Approach

The environment is the most important component of life-support systems. It is comprised of air, water, soil, and biota through which elements and pollutants cycle. The cycle involves the physical, chemical, or biological processing of pollutants in the environment. It may be short, turning hazardous into non-hazardous substances as soon as they are released, or it may continue indefinitely with pollutants posing potential health risks over an extended period of time. Physical processes associated with pollutants cycling include leaching from the soil and groundwater, volatilization from water to land, or air deposition from air to land or water. Chemical processes include decomposition reaction of pollutants to products with properties that are possibly quite different from those of the original pollutants. Biological processes involve microorganisms that can break down pollutants and convert hazardous pollutants into less toxic forms. However, these microorganisms can also increase the toxicity of a pollutant, for example by changing mercury into methylmercury in soil.[3]

Although pollutants sometimes remain in one medium for a long time, they are most likely mobile. For example, settled pollutants in river settlements can be dislodged by microorganisms, flooding, or dredging. Displacement such as this gave rise to the PCB problem in New York's Hudson River. Pollutants placed in landfills have been transferred to air and water through volatilization and leaching. About 200 hazardous chemicals were found in the air water and soil at the Love Canal land disposal site in New York State.

The advantages of applying multimedia approaches lie in their ability:

1) To manage the transfer of pollutants
2) To avoid duplicating efforts or conflicting activities
3) To save resources by consolidation of environmental regulations, monitoring, database management, risk assessment, permit insurance, and field inspection

In recent years, the concept and goals of multimedia pollution prevention have been adopted by many regulatory and governmental agencies, industries, and the public in the United States and abroad. Multimedia efforts in the United States have been focused on the EPA's pollution prevention office, which helps coordinate pollution prevention activities across all EPA headquarters offices and regional offices. The current EPA philosophy recognizes that multimedia pollution prevention is best achieved through education and technology transfer rather than through regulatory imposition and mandatory approaches. But the progress of implementing these policies has been slow.

Recognition of the need for multimedia pollution prevention approaches has been extended from the government industry and the public to professional societies. The Air Pollution Control Association (APCA) was renamed as the Air and Waste Management Association (AWMA) to incorporate waste management. The American Society of Civil Engineers (ASCE) has established a multimedia management committee under the environmental engineering division. The American Institute of Chemical Engineers (AICHE) has reorganized its environmental division to include a section devoted to pollution prevention. The Water Pollution Control Federation (WPCF) has also adapted a set principle addressing pollution prevention.

4.3 Multimedia Application

Perhaps a meaningful understanding of the multimedia approach can be obtained by examining the production and ultimate disposal of a product or service.[3,4] A flow diagram representing the situation is depicted in Figure 4.1. Note that each of the ten steps in the overall process has potential inputs of mass and energy in may produce an environmental pollutant and/or substance or form of energy that may be used in a subsequent or later step. Traditional approaches to environmental management can provide some environmental relief, but a total systems approach is required if optimum improvements – in terms of pollutant/waste reduction – are to be achieved.

One should note that a product and/or service is usually conceived to meet a specific market need with little thought given to manufacturing parameters. At this stage of consideration, it may be possible to avoid some significant waste generation problems in future operations by answering a few simple questions:

1) What raw materials are used to manufacture the product?
2) Are any toxic or hazardous chemicals likely to be generated during manufacturing?
3) What performance regulatory specifications must the new product or service meet? Is extreme purity required?
4) How reliable will delivery manufacturing distribution processes be? Are all steps commercially proven? Does the company have experience with the operations required?
5) What types of waste are likely to be generated? What is their physical and chemical form? Are they hazardous? Does the company currently manage those wastes on-site or offsite?

Mycock et al. provide a detailed multimedia application involving a chemical plant.[5]

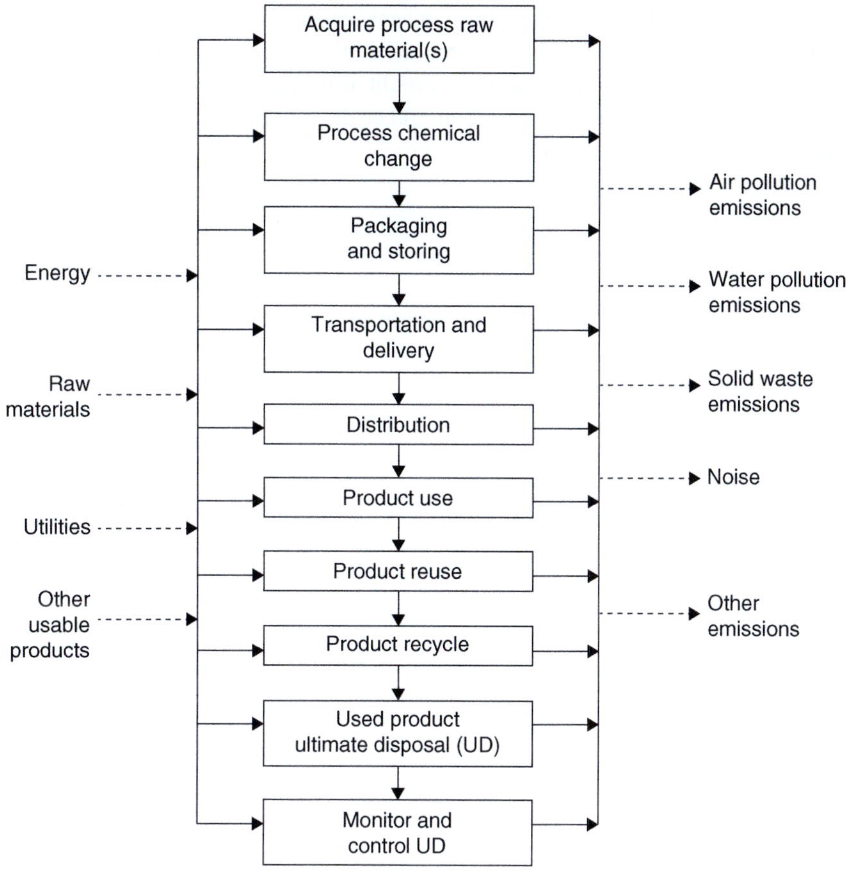

Figure 4.1 Overall multimedia flow diagram.

4.4 Education and Training

The role of environmental professionals in waste management and pollution control has been changing significantly in recent years. Many talented, dedicated environmental professionals in academia, government, industry, research institutions, and private practice need to cope with this change and extend their knowledge and experience from media-specific "end-of-pipe" treatment and disposal strategies to multimedia pollution prevention management. The importance of this extension and reorientation in education, however, is such that the effect cannot be further delayed. Many air pollution, water, pollution, and solid waste supervisors in government agencies spend their entire careers in just one function, because environmental quality supervisors usually work in only one of the media functions. Some may be reluctant to accept such activities. This is understandable, given the fact that such a reorientation requires time and energy to learn new concepts, and that time is a premium for them. Nevertheless, they must support such an education and training in order to have well-trained young professionals.

Successful implementation of multimedia programs will require well-trained environmental professionals who are fully prepared in the principles and practices of such programs. These programs need to develop a deep appreciation for the necessity of multimedia pollution prevention and all levels of society, which will require a high priority for educational and existing curricula of elementary and secondary education, colleges, and universities in training institutions. The use of computerized automation offers much hope.

Government agencies need to conduct a variety of activities to achieve three main educational objectives:

1) Ensure an adequate number of high-quality environmental professionals.
2) Encourage groups to undertake careers in environmental fields and to stimulate all institutions to participate more fully in developing environmental professionals.
3) Generate databases that can improve environmental literacy of the general public and especially the media.

These objectives are related to and reinforce one another. For example, improving general environmental literacy should help to expand the pool of environmental professionals by increasing awareness of the nature of technical careers. Conversely, steps taken to increase the number of environmental professionals should also help improve the activities of general groups and institutions. Developing an adequate human resource base should be the first priority in education. The training that environmental professionals receive should be of top quality.

References

1 **L. Theodore, R. Dupont, and K. Ganesan**, *Pollution Prevention: Sustainability, Industrial Ecology, and Green Engineering*, CRC Press Taylor & Francis Group, Boca Raton, FL, 2017.
2 **K. Schneider**, *A New Strategy*, New York Times, New York, NY, 1991.
3 **G. Burke, B. Singh, and L. Theodore**, *Handbook of Environmental Management and Technology*, 2nd Ed., John Wiley & Sons, Hoboken, NJ, 2000.
4 **L. Theodore**. *Personal Notes*. 1990.
5 **J. Mycock, J. McKenna, and L. Theodore**, *Air Pollution Control Engineering and Technology*, CRC Press Taylor & Francis Group, Boca Raton, FL, 2010.

5

Regulations

Environmental regulations in a general sense are laws that attempt to protect the environment, not only within the borders of the United States but also throughout the entire planet. These regulations are aimed at protecting the air, land, and water that are subjected to society's private and commercial interactions.

In the United States, environmental regulations can be dated back to 1898 with the River and Harbor's Appropriations Act. This law was directed at the removal of logs and other debris from the nations navigable waterways in order to maintain clear passage. The amount of environmental regulation has not only increased since then but also accelerated and broadened in scope. Environmental regulations within the United States have expanded the Federal Water Pollution Control Act of 1948 which was simply aimed at maintaining water quality, to the Clean Air Act Amendments of 1990 that mandates and enforces air pollution reduction, issues plans to meet air quality standards, and specifies compliance dates. This chapter deals specifically with air pollution legislation from the United States with major emphasis on the Clean Air Act Amendments of 1990.

Historically, California was the first state to take steps against air pollution as a result of deteriorating air quality, especially in the Los Angeles area. As it became apparent that other urban areas were experiencing a decline in air quality, it was realized that federal intervention would become necessary. Although air pollution legislation was promulgated between 1955 and 1970, it was not until the Clean Air Act of 1970, combined with the creation of the Environmental Protection Agency (EPA), that the framework for the current regulations was developed.

The legislative approval process for environmental law is the same for any law. The legislation must be approved by the United States Congress and sent to the president for approval and signature. If approved by the president, the legislation is signed into law. EPA is the branch of the federal government that is empowered to set and enforce environmental quality standards. The EPA is required by law to set national air and water quality standards and ensure that the standards are met by the states. It is the individual state's responsibility to ensure that the respective state will attain and remain in compliance with the quality standards.

5.1 Early Air Pollution Legislation

As mentioned above, the environmental regulations in the United States began in 1898 with the passage of the River and Harbors Appropriations Act. The federal government's involvement in air pollution control began in 1955 with Public Law 159. The law authorized federal funding for the US Public Health Service to initiate research into the nature and extent of the nation's air pollution problem. With the passage of the Clean Air Act of 1963, grants were authorized to state and local agencies to assist them in their own control programs. It also provided some limited authority to the federal government to take action to relieve interstate pollution problems. The basic federal control authority was expanded and strengthened by the Air Quality Act of 1967. One of its more significant measures gave citizens, for the first time, the statutory right to participate in the control process through public hearings. Details of these acts are described further in the subsections below.[1]

Optimizing Air Pollution Control Equipment Performance: Operation & Maintenance, First Edition. Jay Richardson and Louis Theodore.

5.1.1 Air Pollution Control Act of 1955

This Act, the first federally promulgated air pollution legislation, considers prevention and control of air pollution at its source to be primarily the responsibility of state and local governments. However, it offers federal financial assistance and leadership for the development of corporate federal state and local air pollution control programs.

The 1955 Act initiated the following:

1) Research on the effects of air pollution by the US Public Health Service.
2) Provision for technical assistance to the states by the federal government.
3) Training of individuals in the area of air pollution.
4) In-house and out-of-house research on air pollution control.

5.1.2 Clean Air Act of 1963

As with the Air Pollution Control Act of 1955, this Act directed the responsibility for air pollution toward the state and local governments. It did, however, reserve the right for the federal government to intervene in the case of interstate commerce. Specifically, the Act provided for:

1) Acceleration of the research and training program.
2) Grants to the state and local agencies for air pollution regulatory programs.
3) Development of air quality criteria.
4) Initially addressing air pollution control from federal facilities.
5) Federal authority to abate interstate air pollution.
6) Encouraging the automobile and fuel industries to prevent pollution.

5.1.3 Air Quality Act of 1967

By this time, the subject of national admission standards versus regional admission standards was undergoing debate in Congress. Although national emissions standards were not included in the Act, it did provide for a 2-year study of the concept. Also provided in the Act were:

1) Establishment of eight specific geographic areas on the basis of common climate, topography, and meteorology.
2) Designation of Air Quality Control Regions (AQCRs). These regions, either inter- or intrastate, were to undergo an evaluation of the air pollution problem.
3) Development of air quality criteria for specific pollutants, which caused indefinable health effects in humans.
4) Development and issuance of information on pollution control techniques. This would allow the federal government to make recommendations to the state and local agencies on the appropriate air pollution control technologies for specific areas.
5) A fixed timetable for state and local agencies to establish air quality standards based on the air quality criteria. Included in this was the opportunity for a public health hearing in adopting the standards.

The Act proved to be ineffective. The federal program was not implemented according to the time schedule. The procedure for determining AQCRs was so complicated that only 108 AQCRs had been designated by 1970. By 1970, a new legislation had been proposed.

5.2 Clean Air Act of 1970

The 1970 Act dealt with two major types of air pollution sources: stationary and mobile. Stationary sources with a fixed location (i.e., power plants) are a primary concern in this chapter. Some of the major provisions of the 1970 Clean Air Act (as they appeared at that time) included the following:

1) EPA, assisted by the states, is to designate AQCRs. These are basic geographic units in which air quality is evaluated and the control process takes place. Regional boundaries are based on considerations of climate, meteorology, topography, urbanization, and other factors affecting conditions in each area. A region can cover only part of one state, or it can

include portions of several states sharing a common air pollution problem. The country has been divided into about 250 regions. As pollution patterns change, or as more information about problems is gathered, the boundaries of some of the regions may change.

2) EPA is required by law to develop air quality criteria for the major pollutants: particulate matter, sulfur oxides, hydrocarbons, nitrogen oxides, carbon monoxide (CO), and so on. These criteria, which are issued in "criteria documents," give levels for which these pollutants – by themselves in and combination with other pollutants – are known to have adverse effects on public health or welfare. At the same time, EPA is to provide information on control technologies for each technology, the costs, and the economic feasibility of alternative control methods. EPA is required to review both criteria documents and control technology documents from time to time to revise them as new information become available.
3) A National Ambient Air Quality Standard (NAAQS) is the maximum level that will be permitted for a given pollutant. There are two kinds of such standards: primary and secondary. Primary standards are to be sufficiently stringent to protect the public health; secondary standards must protect the public welfare. EPA sets the standards after it issues the criteria document and a controlled technology document on the pollution in question. Both primary and secondary standards apply to all control regions. Each control region must satisfy the standards set forth. It is bound by federal enforcement. Radioactive effluents, however, are still regulated by the Nuclear Regulatory Commission. Ambient air quality standards are presented in Table 5.1.
4) Within nine months after the EPA issues the primary and secondary standards for a pollutant, each state must formulate an implementation plan to meet, maintain, and enforce those standards in each air quality control region within its jurisdiction. The states must hold public hearings on the plans, adopt them, and submit them for EPA approval. Each state plan must provide for the attainment of primary standards within three years of EPA's approval; secondary standards must be obtained within a reasonable time. If a state fails to submit a satisfactory plan, the EPA has the authority to write its own plan for that state which the state must then carry out.
5) The Act requires EPA to set standards of performance for new and modified stationary sources of pollution. The standards are distinct from the ambient air quality standards described above. They constitute direct emissions limitations for all major pollutants from specified types of sources, such as cement plants and municipal incinerators. All performance standards are applicable nationally to all pollution sources. However, they also apply to existing sources whenever modification, physical change, or change in method of operation results in increased emissions of old pollutants, or in new emissions of new pollutants. For all existing or modified sources in the specified categories, the states are required to set state performance standards under procedures to be established by the EPA. EPA will also prescribe procedures under which the states may choose to enforce the federal standards for new modified sources. These regulations have been presented in numerous issues of the Federal Register and can be found in the Code of Federal Regulations under Title 40, Chapter 1, Subchapter C, Part 60. When the standards are established and the source of pollution identified, a method can be used to determine how much reduction is required to reach the national standard. A rather simple mathematical equation (see Equation (5.1) was devised as a general estimate of the percentage reduction necessary

Table 5.1 National primary and secondary ambient air quality standards in effect in 1991.

Pollutant	Average Time	Primary Standard	Secondary Standard
SO_2	Annual mean	80 μg/m^3	
	24 hr	365 μg/m^3	
	3 hr		1300 μg/m^3
PM10	Annual mean	50 μg/m^3	Same as primary
		150 μg/m^3	Same as primary
CO	8 hr	9 ppm	No secondary
	1 hr	35 ppm	No secondary
O_3	1 hr	0.12 ppm	Same as primary
Nitrogen oxides	Annual mean	0.053 ppm	Same as primary
Lead	3 mo	1.5 ppm	Same as primary

throughout the control region. These facts must be known first about the gaseous pollutant: (i) the existing concentration of the pollutant, (ii) the air quality standard for the pollutant, and (iii) the background level of that pollutant in the region (sometimes quite difficult to determine).

$$\text{Percentage of emission reduction} = \left(\left(\mathrm{A}-\mathrm{C}\right)-\left(\mathrm{B}-\mathrm{C}\right)\right)/\left(\mathrm{A}-\mathrm{C}\right) \tag{5.1}$$

6) The EPA Administrator was charged with promulgation of standards for hazardous air pollutants regardless of whether they emanated from new or existing sources. Standards have since been promulgated for a number of hazardous materials including asbestos, beryllium, mercury, vinyl chloride, arsenic, and benzene. Enforcement of the standards is the responsibility of the EPA; however, at the states' option, implementation plans submitted to EPA may include regulatory control procedures for these materials. After EPA approval of the implementation plan, states would then be authorized to enforce the hazardous pollution standards within their jurisdiction. Source test methods were identified with the hazardous polluting standards. These test methods were the only approved means by which compliance could be determined. The regulations also specify reporting requirements for the source. Not only must operational data be maintained, but an application must be made to the EPA prior to any modification of existing sources. No new or modified source of a hazardous pollutant may start operation without proper notification to the EPA.
7) EPA may require states and individual sources to monitor polluted emissions, to keep records, and to submit public reports. All such records and reports are considered to be public information, with one exception: EPA may keep confidential any trade secrets or any other information whose public availability the manufacturer has shown to be a potential harm to his or her business. Emission data, however, are specifically exempt from such protection.
8) Once standards and implementation plans are in effect, the EPA is required to oversee state enforcement. Where widespread violations indicate that the state is failing to enforce the plan, EPA may step in and enforce it. Or EPA may enforce portions of the plan by issuing orders of compliance or by bringing civil actions in federal court for violations. EPA is also empowered to sue for immediate restraint of any pollution source that is endangering the health of persons if state or local authorities have failed to abate such pollution under their own regulations. Each major source of pollution within a control region must prepare to follow a detailed step-by-step schedule of measures it will bring it into accord with the implementation plan. Such individual timetables are referred to as compliance schedules. EPA requires the state to negotiate compliance scheduled with all major pollutant sources. Once negotiated, such schedules become legally enforceable and part of the state's implementation plan.
9) Any citizen may bring suit against any person or corporation alleged violating an emission standard or other limitation applicable under the Act. Citizens may also sue the administrator of the EPA for failure to perform an action required by the Act. In cases brought by citizen plaintiffs, the courts are empowered to award the cost of the litigation to such plaintiffs whenever the court determines such an award is appropriate.

5.3 Clean Air Act Amendments of 1977

The 1977 Amendments were similar in design to the Clean Air Act of 1970, but added new programs. These are now briefly introduced and treated in more detail in the subsections to follow. The Prevention of Significant Deterioration (PSD) program was developed for permitting major sources in areas meeting the air quality standards. The PSD program is designed to ensure that those new major stationary sources do not cause the area to violate the NAAQS. A Nonattainment New Source Review program was added to control permitting in areas that did not meet the NAAQS. This would allow continued economic development in these areas while attempting to achieve the air quality standards. In addition, the 1977 Amendments provided the foundation for EPA's Controlled Trading Program; the essential elements of which include:

1) Bubble policy (or bubble exemption under PSD)
2) Offsets policy (under nonattainment)
3) Banking and brokerage (under nonattainment)

5.3.1 Prevention of Significant Deterioration

Of all the federal laws placing environmental controls on industry (and, in particular, on new plants), perhaps the most confusing and restrictive are the limits imposed for the PSD of air quality. These limits apply to areas of the country that already are cleaner than required by ambient air quality standards. This regulatory framework evolved from judicial and administrative action under the 1970 Clean Air Act and subsequently was given full statutory foundation by the 1977 Clean Air Act Amendments.

Sources subject to PSD regulations (40CFR52.21) are major stationary sources and major modifications located in attainment or unclassified areas. A major stationary source is defined as any listed source with the potential to emit 100 tons/year (TPY) or more of any Clean Air Act pollutant. Sources not listed may also be classified as a major stationary source if the potential to emit exceeds 250 TPY. The "potential to emit" is defined as the maximum capacity to emit the pollutant under applicable emission standards and permit conditions (after the application of any air pollution control equipment) excluding secondary emissions. A "major modification" is defined as any physical or operational change of a major stationary source producing a "significant net emissions increase" of any Clean Air Act pollutant.

Continuous monitoring is required of all Clean Air Act pollutants with emissions greater than or equal to values for which there are NAAQS (except hydrocarbons). Continuous monitoring is also required for other Clean Air Act pollutants for which the EPA or state determines necessary. The EPA or the state may exempt any Clean Air Act pollutant from these monitoring requirements if the maximum air quality impact of the emissions' increase is less than the values in Table 5.2 or if the present concentration of the pollutant in the area that the new source would affect is less than the Table 5.2 values.

The EPA or the state may accept representative existing monitoring data collected within three years of the permit application to satisfy monitoring requirements. EPA regulations provide exemption from best available control technology (BACT) and ambient air impact analysis if the modification that would increase emissions is accompanied by other changes within the plant that would net zero increase in total commissions. BACT is the most stringent control technology after taking into account technical, environmental, energy, or economic considerations. This exemption is referred to as the "bubble" (described in further detail in this section) or the "no net increase" exemption.

A full PSD review would include a case-by-case determination of the controls required by BACT, an ambient air impact analysis to determine whether the source might violate applicable increments or air quality standards an assessment of the effective visibility, soils, and vegetation, submission of monitoring data, and a full public review.

The EPA regulations exempted smaller sources from the major elements of a PSD review, and, in particular, relieved those sources from compliance with BACT (though they still had to comply with applicable New Source Permit Standards [NSPS] as well as the requirements under the State Implementation Plan [SIP] program). Smaller sources were also exempted from conducting an ambient air impact analysis and submitting data supporting an ambient air quality analysis. Smaller sources, however, were not exempted from the program altogether. They remain subject to the statutory requirements to obtain preconstruction approval, including procedures for public review, and they still might be required, at EPA's request, to submit data supporting their application. Also, if emissions from a smaller source would affect a Class 1 area (national parks, wilderness areas, etc.) or if an applicable increment were already being violated, the full PSD requirements for ambient air impacts would apply.

Table 5.2 Significant monitoring levels.

Pollutant	mg/m^3	Maximum Averaging Time
CO	575	8 hr
NO_2	14	Annual
TSP	10	24 hr
SO_2	13	24 hr
Lead	0.1	3 mo
Mercury	0.25	24 hr
Beryllium	0.001	24 hr
Fluorides	0.25	24 hr
Vinyl chloride	15	24 hr
Total reduced sulfur	10	1 hr
H_2S	0.2	1 hr
Reduced sulfur compounds	10	1 hr

5.3.2 Nonattainment Areas

Those areas of the United States that failed to retain compliance with ambient air quality standards were considered Nonattainment Areas (NAs). The new plants could be constructed in NAs only if stringent conditions were met. Emissions had to be controlled to the greatest degree possible, and more than equivalent, offsetting omissions, reductions had to be obtained from other sources to ensure progress toward achievement of the standards. Specifically:

1) The new source must be equipped with pollution controls to ensure the lowest achievable emissions rate (LAER), which, in no case can be less stringent than any applicable NSPS.
2) All existing sources owned by an applicant in the same region must be in compliance with applicable SIP requirements or be under an approved schedule or an enforcement order to achieve such compliance.
3) The applicant must have sufficient offsets to more than make up for the emissions to be generated by the new source after application of LAER.
4) The emission offsets must provide a positive net air quality benefit in the affected area. LAER was deliberately a technology-forcing standard of control. The statute stated that LEAR must reflect: (i) the most stringent emission limitation contained in the implementation plan of any state for such category of sources, unless the applicant can demonstrate that such a limitation is not achievable, or (ii) the most stringent limitation achievable in practice within the industry category, whichever is more stringent. In no event could LAER be less stringent than any applicable NSPS. While the statutory language defined in BACT directed that "energy environmental and economic impacts in other costs" be taken into account, the comparable provision on the LAER provided no instruction that costs be considered. For existing sources emitting pollutants for which the area is a nonattainment one, reasonable available control technology (RACT) would be required. A control technology may be considered RACT if it results in the lowest emission limitation after considering reasonable, economic, and technical feasibility. EPA is in the process of defining RACT by industrial category.

5.3.3 Controlled Trading Program

Legislation enacted under the Clean Air Act Amendments of 1977 also provided the foundation for EPA's controlled trading program, essential elements of which include:

1) Bubble policy (or bubble exemption under PSD)
2) Offsets policy (under nonattainment)
3) Banking in brokerage (under nonattainment)

While these different policies vary broadly in form, their objective is essentially the same: to substitute flexible economic incentive systems for the current rigid, technology-based regulations that specify exactly how companies must comply. These market mechanisms could make regulating easier for EPA and less burdensome and costly for industry.

5.3.4 Bubble Policy

The bubble concept, introduced under PSD provisions of the Clean Air Act Amendments of 1977, was formally proposed as EPA policy on January 18, 1979, with the final policy statement being issued on December 11, 1979. The bubble policy allows a company to find the most efficient way to control the emissions of a plant as a whole rather than by meeting individual point source requirements. If it is found less expensive to tighten control of a pollutant at one point and relax control that another, this would be possible as long as the total pollution from the plant would not exceed the sum of the current limits on the individual point source pollution from the plant. Properly applied this approach should promote greater economic efficiency and increased technological innovation. There are some restrictions, however, in applying the bubble concept:

1) The bubble may only be used for pollutants in an area where the SIP has an approved schedule to meet air quality standards for that pollutant.
2) The alternatives must ensure that air quality standards will be met.
3) Emissions must be quantifiable, and trades among them must be even. Each emission point must have a specific admission limit, and that limit must be tied to enforceable testing techniques.
4) Only pollutants of the same type may be traded, i.e., particulates for particulates and hydrocarbons for hydrocarbons.

5) Control of hazardous pollutants cannot be relaxed through trades with less toxic pollutants.
6) Development of the bubble plan cannot delay enforcement of federal and state requirements.

Some additional considerations must be noted:

1) The bubble may cover more than one plant within the same area.
2) In some circumstances, states may consider trading open dust emissions for particulates (although EPA warns that this type of training will be difficult).
3) EPA may approve compliance state extensions in specific cases. For example, a source may obtain a delay in a compliance schedule to install a scrubber if such a delay would have been permissible without the bubble.

EPA will closely examine particle size distribution in particular emissions trades because finer particulates disperse more widely and remain in the air longer.

It will be the responsibility of the industry to suggest alternative control approaches and demonstrate satisfactorily that the proposal is equivalent in pollution reduction, enforceability, and environmental impact to existing individual process standards.

5.3.5 Offsets Policy

Offsets were EPA's first application of the concept that one source could meet its environmental protection obligations by getting another source to assume additional control actions. In NAs, pollution from a proposed new source, even one that controls its emissions to the lowest possible level would aggravate existing violations of ambient air quality standards and trigger the statutory prohibition. The offsets policy provided these new sources with an alternative. The source could proceed with construction plans, provided that:

1) The source would control emissions to the lowest achievable level.
2) Other sources owned by the applicant are in compliance or an approved compliance schedule.
3) Existing sources were persuaded to reduce emissions by an amount at least equal to the pollution that the new source would add.

5.3.6 Banking Brokerage Policy

EPA's banking policy is aimed at providing companies with incentives to find more offsets. Under the original offset policy, a firm shutting down or modifying a facility could apply the reduction in omissions to a new construction elsewhere in the region only if the changes were made simultaneously. However, with banking, a company can "deposit" the reduction for later use or sale. Such a policy will clearly establish that clean air or the right to use it has a direct economic value. Banking and trading other emission credits are addressed again in the 1990 Amendments.

5.4 Clean Air Act Amendments of 1990

New Clean Air Act Amendments were signed into law on November 15, 1990. This comprehensive law is expected to cost the industry billions of dollars annually. The Act is divided into 11 "Titles":

Title I – Provisions for Attainment and Maintenance of NAAQS
Title II – Provisions Relating to Mobile Sources
Title III – Hazardous Air Pollutants
Title IV – Acid Deposition Control
Title V – Permits
Title VI – Stratospheric Ozone Protection
Title VII – Provisions Relating to Enforcement
Title VIII – Miscellaneous Provisions
Title IX – Clean Air Research
Title X – Disadvantaged Business Concerns
Title XI – Clean Air Employment Transition Assistance

The EPA is required to develop regulations addressing all applicable sections of the Act. Many sections of the Act, however, have statutory trigger dates specified. It is clear, given the magnitude of the Act, that the burden on both the regulators and the regulated community will increase dramatically. Details on each of the Titles particularly Titles I through V are provided below.

5.4.1 Title I. Provisions for Attainment and Maintenance of National Ambient Air Quality Standards (NAAQS)

Several areas of the country have experienced exceedances in the ozone, CO, and PM_{10} air quality standards to varying degrees. To differentiate the severity of the problem, nonattainment classifications were developed. The more severe the problem the more requirements there are. Attainment dates are specified for the different classifications.

Ozone NAs have been divided into five categories: marginal, moderate, serious, severe, and extreme. CO and PM_{10} (particulate matter 10 μm and smaller) NAs are divided into two categories: moderate and serious.[4–6]

5.4.1.1 Ozone Nonattainment Areas

The NAAQS for ozone is 0.120 ppm. Several areas of the country have experienced exceedances of this standard to varying degrees. To differentiate the severity of the problem, no attainment classifications were developed (see Table 5.3). There are different attainment dates for different classifications. In addition, there are requirements for areas classified as ozone transport regions.

Added control of NO_x and VOC sources is mandated under the Act. The states must revise their SIPs and meet the requirements of the Act. All nonattainment area SIP revisions must address at least the following:

1) New source review of major NO_x sources.
2) Develop a comprehensive, accurate emissions inventory.
3) Correct or add all RACT requirements to the SIP.
4) Enhance the vehicle, I&M (inspection and maintenance) program.
5) Revise the construction and operating permit program.
6) Require stationary sources of NO_x and VOC to provide a certified statement of actual emissions.
7) The offset ratio is 1.1–1 for VOCs.

Additional requirements for moderate NAs are:

1) VOC reductions of 15% from baseline by November 1996.
2) Require existing VOC sources to comply with RACT requirements.
3) Require vapor recovery devices at gas stations selling more than 10,000 gallons per month.
4) The offset ratio is 1.15–1 for VOCs.

Serious ozone containment area requirements include all of the listed above plus:

1) The major source threshold is reduced to 50 TPY VOCs.
2) Monitoring of ambient ozone, NO_x, and VOCs must be enhanced.
3) VOC reductions of 3% per year per three years beyond the initial 15% reductions unless it can be demonstrated that a combined VOC/NO_x reduction strategy will result in equivalent ozone reductions.

Table 5.3 Ozone nonattainment classifications.

Classification	O_3 (ppm)
Marginal	0.121–0.138
Moderate	0.138–0.160
Serious	0.160–0.180
Severe	0.18–0.280
Extreme	Above 0.280

4) Upgrading the vehicle emissions testing program.
5) Institute a program for clean fuel usage in fleet cars.
6) Implement transportation controls if actual vehicle miles traveled are greater than projected beginning six years after enactment.
7) The offset ratio is 2.1–1 for VOCs.

There are two severe nonattainment categories. The "Severe II" classification is given to the Chicago, Houston, and New York metropolitan areas. These are more severe than "Severe I" metropolitan areas of Baltimore, Milwaukee, Muskegon, Philadelphia, and San Diego. SIP revisions for these areas must contain all requirements for serious areas and:

- The major source threshold is reduced to 25 TPY VOCs.
- The state must identify and adopt enforceable control strategies to reduce vehicle miles traveled.
- Employers of 100 persons or more will be required to increase vehicle occupancy during peak periods.
- The offset ratio is 1.3–1 for VOCs.

The only extreme ozone nonattainment area is the Los Angeles metropolitan area. In addition to the above requirements.

1) The major source threshold is reduced to 10 TPY VOCs.
2) Boilers must use clean fuels or advanced control technologies.
3) The offset ratio is 1.5–1 for VOCs.

In all ozone NAs, the requirements for VOCs may also apply to NO_x

5.4.1.2 Carbon Monoxide Nonattainment Areas

CO nonattainment problems are believed to be primarily due to automobile exhausts. It is no surprise that control measures are geared at reducing vehicle emissions.

Moderate areas have CO design values between 9.1 and 16.4 ppm. These areas must develop a comprehensive, accurate inventory of actual emission sources. Areas above 9.5 ppm will require oxygenated fuels to become available. Areas above 12.7 ppm must implement an enhanced vehicle I&M program and develop plans for reducing vehicle miles traveled. The SIP revision in those areas must also include a plan for specific annual emissions reductions. Areas with design values above 16 ppm must develop a clean fuel fleet program. All of the above apply to severe CO NAs, but the major stationary source threshold is reduced to 50 TPY of CO.

5.4.1.3 PM_{10} Nonattainment Areas

Areas that do not meet the standard PM_{10} will be designated as either moderate or serious. These areas will be required to develop a permitting program for major stationary sources and modifications of PM_{10}. The major source threshold is reduced to 70 TPY in a serious area. Additionally, sources of moderate areas must employ reasonably available control measures, while the best available control measures must be implemented in serious areas.

5.4.2 Title II. Provisions Relating to Mobile Sources

The mobile source provisions specify the means of reducing CO and hydrocarbon emissions from motor vehicles and refueling.

Tailpipe standards for the normal operation of cars and light trucks have been established for non-methane hydrocarbons, CO, and NO_x. Urban buses have been targeted for reduction in particulate matter. A standard for CO in cold weather has been established as well.

A clean fuels program is specified for all centrally fueled fleets of 10 or more vehicles and serious, severe, and extreme ozone NAs and serious CO NAs. In 1998, fleets in these areas must buy 30% of the vehicles that meet the standard; by 2001, they must buy 70%. In addition, 150,000 clean fuel cars are to be sold in California. This increased to 300,000 in 1999.

Fuel reformulation is also emphasized in this title. Oxygen content in motor vehicle fuel will increase especially in winter months. Gasoline will have a maximum Reid vapor pressure of 9 PSI during the summer months. A sulfur limit will be placed on diesel fuel. And as of January 1, 1996, lead was banned from using in motor vehicle fuel.

5.4.3 Title III. Hazardous Air Pollutants

The air toxics provisions in the Act will add to the National Emission Standards for Hazardous Air Pollutants (NESHAPs) that are currently regulated. Instead of 8, there will be 189 listed air toxics.

EPA will publish a list of specific source categories that emit the listed air toxics. For example, research and development laboratories will be a unique source category. Then a schedule will be published detailing proposed implementation dates for the new control requirements for each source category.

The current terminology associated with the new standard is maximum achievable control technology or MACT. The MACT standard will apply to major sources that are defined in the section as having the potential to emit 10 or more tons per year of any individual listed pollutant or 25 tons of any combination. The MACT standard will be based on the following: the emission limit will be at least stringent as the average control efficiency of the best controlled 12% of existing similar sources (for categories were 30 or more sources). The standards can be technology-based, health-based, or if necessary, work practice-based, or combined thereof. Existing sources would have three years after promulgation to comply with applicable MACT standards.

Eight years after MACT standards are established (except for those established two years after enactment), standards to protect against residential health and environmental risks remaining must be promulgated, if necessary. The standards would be triggered if more than one source in a category exceeds a maximum individual risk of cancer of one in 1 million (or 10^{-6}). These residual risk regulations would be based on current CAA language that specifies that standards must achieve an "ample margin of safety."

Standards to prevent accidental releases of toxic chemicals are required. EPA must establish a list of at least 100 chemicals and threshold quantities. All facilities with these chemicals on site in excess of the threshold quantities would be subject to the regulations which would include hazard assessments and risk management plans. An independent chemical safety board is established to investigate major accidents, conduct research, and promulgate regulations for accidental release reporting.[2,3]

A study of area source emissions and a strategy to reduce the cancer incidence from these emissions by 75% is required. Regulation of source categories accounting for 90% of the emissions of the 30 most hazardous area source pollutants is also required. Coke ovens can receive an extension of the residual risk standards until the year 2020 in exchange for compliance with stringent emission standards. Air toxics regulations of utilities will be based on the results of toxic emissions studies. A study of depositions to the Great Lakes, Lake Champlain, Chesapeake Bay, and coastal waters will determine whether additional regulation is needed. Standards are required for all types of municipal waste combustors with an exclusion for facilities that burn 30% or less municipal waste.

5.4.4 Title IV. Acid Rain

Acid rain has been widely publicized in years past. Sulfur dioxide and nitrogen oxides are the acid rain precursors of interest. This piece of legislation is geared toward utilities. The objective of this title is to reduce the effects of acid deposition by reducing annual SO_2 and NO_x emissions by 10 million and 2 million tons, respectively.

A 10-million-ton reduction in annual SO_2 emissions from 1980 levels will be sought primarily from utility sources. Annual SO_2 emissions will be capped at approximately 8.9 million tons by the year 2000. The SO_2 reductions will be met through a market-based system. Affected sources are allocated allowances based on required emissions reductions and past energy use. An allowance is worth 1 ton of SO_2 and is fully marketable. Sources must hold allowances equal to their level of ambitions of a $2,000/excess ton, penalty, and requirement to offset excess tons in future years. EPA will also hold special sales in auction allowances.

Reductions of SO_2 will be achieved in two phases. Phase I allowances are allocated to large units of 100 MW or greater that emit more than 2.5 lb/MMBTU in an amount equal to 2.5 lb/MMBTU times their 1985–1987 energy usage baseline. Phase I must be met by 1995 but units that install certain control technologies may postpone compliance until 1997 and may be eligible for bonus allowances. Units in Illinois, Indiana or Ohio or allotted upload pro rata share of additional 200,000 allowances annually during Phase I.

Phase 2 began in 2000. All utility units greater than 25 MW that emit at a rate of 1.2 lb/MMBTU will be allocated allowances at the rate times their baseline fuel consumption. Cleaner plants generally will be provided with 20% more allowances than would have been received based on their baseline consumption. Plants in 10 midwestern states that make Phase 1 reductions will be allocated 50,000 bonus allowances.

Utility NO_x reductions will help to achieve a 2-million-ton reduction from 1980 levels. Reductions will be accomplished through required EPA performance standards for certain existing plants in Phase I and others in Phase II. In addition, EPA will develop a revised NO_x New Source Performance Standard for utility boilers.

Incentives are available for clean coal technology, energy conservation, and renewable energy projects. Certain clean coal technology demonstration projects have been exempted from NSPS, NSR, and no attainment requirements. Units repowering with qualifying clean coal technologies receive a four-year extension for Phase II complaints. Such units may be exempt from NSR and NSPS requirements. Energy conservation and renewable energy demonstration projects may be allocated a portion of up to 300,000 incentive allowances.

5.4.5 Title V. Operating Permits

Within three years of enactment, the states must develop operating permit programs. The programs based on guidelines established by the EPA will be submitted to the EPA for approval.

Permits will apply to major sources covered under Title I as well as sources covered by other Titles of the Act. All sources subject to the program must submit permit applications to the state within one year of the date that the program received final approval from the EPA. The state must establish a schedule for acting on initial permit applications, which assures that at least a third of the submitted applications will be acted upon annually for three years.

Permits will be effective for up to five years. Permits must include all CAA requirements applicable to the source. They must also include a schedule of compliance and applicable monitoring and reporting requirements. Permit fees will be paid by the sources to cover the costs of the permitting program.

EPA must veto a permit if it does not comply with any applicable CAA requirements. The public may sue to compel the EPA to perform non-discretionary duty. Such cases are reviewable in the federal court of appeals if the EPA fails to veto a permit that does not comply with the Act.

Once issued, the permit replaces the otherwise applicable requirements, specifically identified in the permit, but the EPA may require that the permit is reopened for cause. A permit with a term of three or more years must be reopened if the new requirements applicable to the source or promulgated.

5.5 Other Considerations

5.5.1 Subsequent Key Clean Air Act Amendments Actions

CAAA action since the passage of the Amendments of 1990 include clean fuels, acid, rain, visibility in the CAAAC.[7] Information on these is detailed below.

1) EPA negotiated regulatory agreements on reformulated gasoline and oxygenated fuels. The resulting rules will reduce tailpipe emissions by requiring the use of oxygenated fuel in areas that are not in attainment for CO by November 1992 and, starting in 1995, require the use of reformulated gasoline and severe ozone NAs.
2) EPA proposed a clean air rule that will cut emissions of sulfur dioxide, a major contributor to acid rain, by 50%. This is a product of the Acid Rain Advisory Committee, a diverse group representing utilities, state regulators, environmental and consumer groups, the coal and gas industry, and others.
3) On September 18, 1991, EPA promulgated controls on Arizona's Navarro Generating Station. EPA also established a visibility transport commission to identify and evaluate sources and source regions that affect the visibility of the Grand Canyon National Park. In 1993, EPA and the national park service issued an interim report.
4) The CAAAC was established in November 1990 to provide high-level independent advice to EPA and policy issues related to the implementation of the 1990 Clean Air Act Amendments. The CAAAC is comprised of 50 senior representatives from state and local governments, academic institutions, unions, environmental and public interest groups, and industries.

5.5.2 Impact of the 1990 CAAA on Wastewater Treatment Plants

A wide range of residential commercial and industrial dischargers contribute VOC and toxic pollutants to publicly owned treatment works (POTWs). An even wider range of pollutants is potentially discharged to industrial wastewater treatment plants, depending on the specific type of industrial activity generating the wastewater. Limited information on the air emissions of VOCs and air toxics from industrial wastewater plants in POTWs is currently available; however, more extensive information on treatment plant wastewater and fluid quality is more readily available, in the literature particularly for POTWs due to the monitoring requirements of all the wastewater treatment plants under the National Pollutant Discharge Elimination Systems (NPDES) program. In addition to those pollutants commonly present in the influence of POTWS, byproducts of wastewater treatment processes are considered to be VOCs or toxic air pollutants can also potentially be emitted from POTWS. For example, chloroform is usually formed as a byproduct of wastewater chlorination.

The provisions of the Amendments dealing with a reduction of VOCs producing urban smog (Title I) and control of air toxics (Title III) significantly affect the water pollution control field in the areas of water quality and wastewater treatment.

Many POTW's located in ozone NAs have initiated studies to identify VOC and other toxic emissions. Based on the available literature, none of the POTWs have estimated the emissions of all 189 hazardous pollutants listed in the Amendments. It is possible that some of the large POTWs may fall into the category of a major source if all 189 pollutants are included in the calculations. In Los Angeles, the only extreme ozone nonattainment area, the local air quality agency adopted rule 1401, which regulates emissions of known or suspected carcinogenic air toxics based on cancer risk assessments. Local POTWs are covered under this rule.

Although air emissions for most POTWs do not pose significant risks to the general public, these regulations have already had a significant impact on the expansion plans and existing operating conditions of many POTWs and have led to the application of expensive off-gas control systems at some municipal and industrial wastewater treatment facilities. There are no published data that include chronic health risks to treatment plant personnel exposed to poorly ventilated confined spaces such as sewers and headworks buildings.

5.5.3 Impact of the 1990 CAAA on Hazardous Waste Incinerators

The impact of the 1990 Amendments on hazardous waste incinerators[8] is not clear at this time. Certainly, the Title I nonattainment provisions (ozone) and the Title III hazardous air pollutants provisions (air toxics) will have some effect. The emissions of VOCs and air toxics from not only the incinerator itself and ancillary equipment but also fugitive emissions can have a major impact on both the permitting process and operation of the incineration facility.

5.5.4 Cross-State Air Pollution Rule (CSAPR) and Good Neighbor

On July 6, 2011, the EPA finalized the CSAPR to address air pollution from upwind states that crosses state lines and affects air quality in downwind states, specifically targeting: SO_2 and NO_x. These emissions and the soot and smog they form can affect air quality and public health locally, regionally, and even in states hundreds of miles downwind. The transport of these pollutants across state borders, referred to as interstate air pollution transport, makes it difficult for downwind states to meet health-based air quality standards for PM2.5 and ozone.

The Clean Air Act's "good neighbor" provision requires EPA and states to address interstate transport of air pollution that affects downwind states' ability to attain and maintain NAAQS. Specifically, Clean Air Act section 110(a)(2)(D)(i)(I) requires each state in its SIP to prohibit emissions that will significantly contribute to the nonattainment of an NAAQS or interfere with maintenance of an NAAQS, in a downwind state. The Act requires EPA to backstop state actions by promulgating Federal Implementation Plans (FIPs) in the event that a state fails to submit or EPA disapproves of good neighbor SIPs. The Act is targeted toward power plants, but effects a large number of large industrial facilities as well. EPA sets a pollution limit (emission budget) for each of the states covered by CSAPR. Authorizations to emit pollution, known as allowances, are allocated to affected sources based on these state emissions budgets. The rule provides flexibility to affected sources, allowing sources in each state to determine their own compliance path. This includes adding or operating control technologies, upgrading or improving controls, switching fuels, and using allowances. Sources can buy and sell allowances and bank (save) allowances for future use as long as each source holds enough allowances to account for its emissions by the end of the compliance period.

References

1 **C. Meyers**, *Term Paper Submitted to L. Theodore as Part of Graduate Level Air Pollution Control Course*, Manhattan College, Bronx, NY, 1993.

2 **A. Martin**, *Term Paper Submitted to L. Theodore as Part of Graduate Level Air Pollution Control Course*, Manhattan College, Bronx, NY, 1993.

3 **L. Theodore, and R.R. Dupont**, *Environmental Risk Assessment: Principles and Calculations*, CRC Press/Taylor & Francis Group, Boca Raton, FL, 2012.

4 40CFR50, *National Primary and Secondary Ambient Air Quality Standards*, US Government Printing Office, Washington, DC, 1991.

5 USEPA, *Clean Air Act Amendments of 1990, Detailed Summary of Titles*, US Environmental Protection Agency, Washington, DC, 1990.

6 **L. Stander, and L. Theodore**, *Environmental Regulatory Calculations Handbook*, CRC Press/Taylor & Francis Group, Boca Raton, FL, 2012.

7 **J. Mycock, J. McKenna, and L. Theodore**, *Handbook for Air Pollution Control Engineering and Technology*, CRC Press/Taylor & Francis Group, Boca Raton, FL, 1995.

8 **J. Santoleri, J. Reynolds, and L. Theodore**, *Introduction to Hazardous Waste Incineration*, 2nd Ed., John Wiley & Sons, Hoboken, NJ, 2000.

6

Environmental and Health Risk

The rapid growth and expansion of the chemical process industry have been accompanied not only by a spontaneous rise of chemical emissions to the environment but also by human, material, and property losses because of fires, explosions, hazardous and toxic spills, equipment failures, other accidents, and business interruptions. Concern over the potential consequences of these continuous emissions and catastrophic accidents, particularly at chemical, petrochemical, and utility plants, has sparked interest at both the industrial and regulatory levels in obtaining a better understanding of the main subject of this chapter. This writing was undertaken in part as a result of this growing concern.

People face all kinds of risks every day, some voluntary and others involuntary. Therefore, risk plays a very important role in today's world. Earlier studies on cancer and the recent COVID-19 pandemic have caused a turning point in the world of risk because they opened the eyes of risk scientists and health professionals to the world of health risk assessments (HRAs).

The usual objective of HRA and the accompanying calculations is to evaluate the potential for adverse health effects from the release of chemicals into the environment. Unfortunately, the environment is very complex since there is a vast array of potential receptors present. The task of testing and evaluating each of the enormous number of chemicals on the market for their impact on human populations in ecosystems becomes extremely difficult. To further complicate the problem, health is a concept that has come to mean different things to different people. Some have defined it as: "… a state of complete physical, mental, and social well-being, and not merely the absence of disease or infirmary." Many other definitions and concepts have been proposed and appear in the literature.[1]

Since 1970, the field of HRA has received widespread attention within both the scientific and regulatory communities. It has also attracted the attention of the public. Properly conducted risk assessment plans and risk assessment calculations have received fairly broad acceptance in part because they put into perspective the terms toxic, health, hazard, and risk. *Toxicity* is an inherent property of all substances. It states that all chemical and physical agents can produce adverse health effects at some dose, or under some specific exposure conditions. In contrast, exposure to a chemical that has the capacity to produce a particular type of adverse effect represents a health *hazard*. *Risk* in a general sense, however, is the probability or likelihood that an adverse outcome will occur in a person or a group that is exposed to a particular concentration or dose of the hazardous agent. *Health risk* is a function of exposure and dose. Consequently, *HRA* is defined as the process or procedure used to estimate the likelihood that humans or ecological systems will be adversely affected by a chemical or physical agent under a specific set of conditions.

6.1 Risk Variables and Categories

6.1.1 Risk Variables

Placing a risk in perspective entails translating myriad technical risk analyses into concepts of risk that both the technical community and the general public can understand. The most effective technique for presenting risks in perspective is to contrast risks to other, similar risks. There are several comparison variables that affect acceptance of risk. Ten such variables include the following[2]:

Optimizing Air Pollution Control Equipment Performance: Operation & Maintenance, First Edition. Jay Richardson and Louis Theodore.

1) Voluntary versus involuntary
2) Delayed versus immediate
3) Natural versus man-made
4) Controllable versus uncontrollable
5) Known versus unknown
6) Ordinary versus catastrophic
7) Chronic versus acute
8) Necessary versus luxury
9) Occasional versus continuous
10) Old versus new

The public generally accepts voluntarily assumed risk more easily than an involuntarily imposed risk. Similarly, a naturally occurring risk is more easily accepted than a man-made risk. The more similar risks are with regard to these variables, the more meaningful it is to compare those risks.

6.1.2 Risk Categories

There are dozens of risk categories. Topping the list, for purposes of this book, are environmental and health risks. Two other important risks include financial risk and sports risk. The former topic is directly relevant to environmental risk and the latter is of interest to one of the authors.[3] Some other industrial risk categories (in alphabetical order) include:

1) Aerospace
2) Architecture
3) Chemical
4) Construction
5) Education
6) Energy
7) Governance
8) Medical
9) Pharmaceutical
10) Travel
11) Urban planning

However, irrespective of the risk, it is fair to say that the calculation of environmental and health risks has now become mandatory in environmental assessment and analysis studies/applications. More and more technical individuals are now required in this field as risk analysis is often performed using the best available data and information.

There are, of course, many other risk categories that the engineer, applied scientist, bureaucrat, society, and so on, are exposed to on a regular basis. Details of these "other" risks are available in the literature.[1]

6.2 Risk Assessment

The term risk assessment is not only used to describe the likelihood of an adverse response to a chemical or physical agent, but it has also been used to describe the likelihood of any unwanted event. These include risks such as explosions or injuries in the workplace; natural catastrophes; injury or death due to various voluntary activities such as skiing, sky diving, flying, and bungee jumping; diseases; death due to natural causes; and many others.[2]

Risk assessment of accidents serves a dual purpose. It estimates the probability that an accident will occur and assesses the severity of the consequences of an accident. Consequences may include damage to the surrounding environment, financial loss, injury, or loss of life. This chapter is also concerned with introducing the reader to the methods used to identify these hazards and the causes and consequences of accidents. Risk assessment of accidents (or hazard risk assessment, HZRA) provides an effective way to help ensure that a mishap either does not occur or reduces the likelihood of severe consequences as a result of the accident. The results of the HZRA allow concerned parties to take precautions to prevent an accident before it happens.

There are other classes of environmental health risks that do not pertain to chemicals but are an integral part of HZRA. For example, health problems can arise immediately/soon after a hazard, such as a hurricane or earthquake, that can leave local inhabitants without potable water for an extended period of time.

Environmental risk assessment may be broadly defined as a scientific enterprise in which facts and assumptions are used to estimate the potential for adverse effects. Risk management, as the term is used by the US Environmental Protection Agency (EPA) and other regulatory agencies, refers to a decision-making process that involves such considerations as risk assessment, technological feasibility, economic information about costs and benefits, statutory requirements, public concerns, and other factors. Risk communication is the exchange of information about risk.

Risk assessment may also be defined as the characterization of potential adverse effects to humans or to an ecosystem resulting from environmental hazards. Risk assessment supports risk management, including the set of choices centering on whether and how much to control future exposure to the suspected. Risk managers face the necessity of making difficult decisions involving uncertain science, potentially grave consequences to health or the environment, and large economic effects on industry and consumers. What risk assessment provides is an orderly, explicit, and consistent way to deal with scientific issues in evaluating whether a health problem or a hazard exists. This evaluation typically involves large uncertainties, because the available scientific data are limited, and the mechanisms for adverse health impacts or environmental damage are only imperfectly understood.

Risk assessment and risk management are two different processes, but they are intertwined. Risk assessment and risk management give a framework not only for setting regulatory priorities, but also for making decisions that cut across different environmental areas. Risk management refers to a decision-making process that involves such considerations as risk assessment, technology feasibility, economic information about costs and benefits, statutory requirements, public concerns, and other factors. Therefore, risk assessment supports risk management in that the choices on whether and how much to control future exposure to a suspected problem may be determined during the risk management process.[4,5]

From a risk management standpoint, whether dealing with a site-specific situation or a national standard, the deciding question is ultimately, "What degree of risk is acceptable?" In general, this does not mean a "zero risk" standard, but rather a concept of negligible risk. At what point is there really no significant health or environmental risk, and at what point is there an adequate safety margin to protect public health and the environment? In addition, some environmental statutes require consideration of benefits together with risks in making risk management decisions.

Thus, it should be noted that health risk addresses risks that arise from health and health-related problems. Chemicals are generally the culprit. Both the effect on and exposure to a receptor (in this case, generally a human) ultimately determine the risk to the individual for the health problem of concern. The risk can be described in either qualitative or quantitative terms, and there are various terms that may be used, e.g., 10 individuals will become sick, or 1×10^{-6} (one in a million) will die, or something as simple as "it is a major problem."

Another category of environmental risk is the aforementioned hazard risk. This class of risk is employed to describe risks associated with hazards or hazard-related problems, for example, accidents, negative events, and catastrophes. Unlike most health problems, these usually occur over a short period of time, say a few seconds or minutes. Both the probability and the consequence associated with the accident/event ultimately determine the hazard risk. Once again, the risk can be described in either qualitative or quantitative terms, and there are various terms that may be used.[1]

As noted earlier, once a risk has been calculated, one needs to gauge the estimated consequences (or opportunities if examining financial/economic scenarios) and evaluate and prioritize options for risk management or mitigation. These potentially strategic evaluations are usually fraught with uncertainties at numerous levels. Thus, the risk assessment process is normally followed by alternatives analyses; these options are usually based on decision-making procedures that are beyond the scope of this book. However, it is fair to say that there may be a full range of outcomes and consequences to various scenarios. It should also be noted that risk assessment is a dynamic process that can very definitely be a function of time. Much of this material is addressed later in the chapter.

6.3 Health Risk Assessment/Analysis

HRAs provide an orderly, explicit, and consistent way to deal with issues in evaluating whether a health problem exists and what the magnitude of the problem may be. This evaluation typically involves large uncertainties (to be discussed in a later section) because the available scientific data are limited and the mechanisms for adverse health impacts or environmental damage are only imperfectly understood.

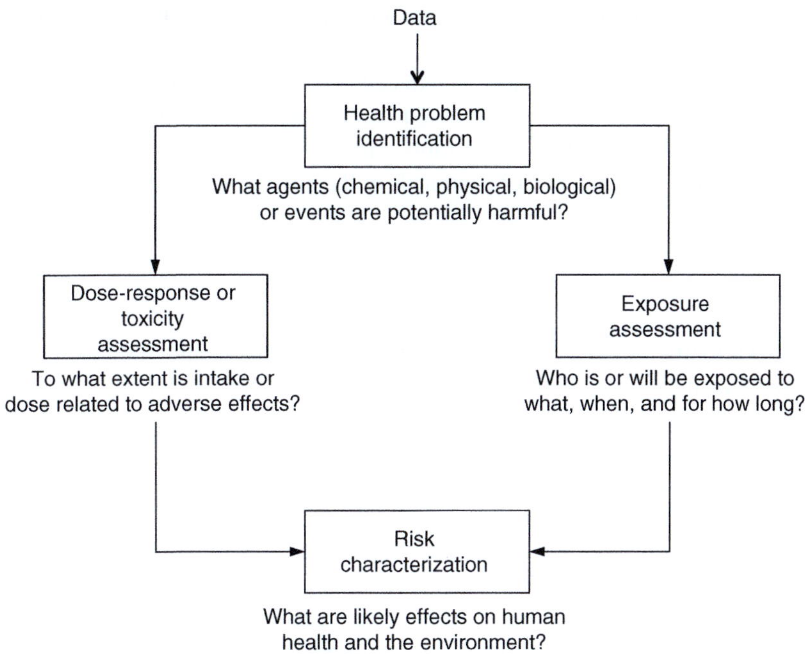

Figure 6.1 Health Risk Evaluation Process.

When one examines risk, how does one decide how safe is "safe," or how clean is "clean?" To begin with, one must look at both inputs of the risk equations, that is, both the toxicity of a pollutant and the extent of exposure. Information is required for both the current and the potential exposure, considering all possible exposure pathways. In addition to human health risks, one needs to look at the potential ecological or other environmental effects.

In recent years, several guidelines and handbooks have been published to help explain approaches for conducting HRAs. As discussed by a special National Academy of Sciences Committee which convened in 1983, most human or environmental health hazards can be evaluated by dividing the analysis into four parts: health problem identification, dose–response assessment or toxicity assessment, exposure assessment, and risk characterization (see Figure 6.1). The risk assessment might stop with the first step, health problem identification, if no adverse effect is identified or if an agency elects to take regulatory actions without further analysis.[4,5] Regarding identification, a health problem is defined as a toxic agent or a set of conditions that has the potential to cause adverse effects to human health or the environment. Health problem identification involves an evaluation of various forms of information in order to identify the different problems potentially caused by the toxic agent. Dose–response or toxicity assessment is also required in an overall assessment; responses and effects can vary widely since all chemicals and contaminants vary in their capacity to cause adverse effects. This step frequently requires that assumptions be made to extrapolate experimental results from animal tests to expected effects on exposed humans. Exposure assessment is the determination of the magnitude, frequency, duration, and routes of exposure of toxic agents to human populations and ecosystems. Finally, in health risk characterization, toxicology, *and* exposure data/information are combined to obtain a qualitative or quantitative expression of risk. An expanded presentation on each of the four HRA steps is provided in the next four subsections of Section 6.4.

6.4 Health Risk Assessments Components

6.4.1 Health Problem Identification

Health problem identification is defined as the process of determining whether human exposure to a chemical at some dose could cause an increase in the incidence of an adverse health condition (cancer, birth defect, etc.), or whether exposure to nonhumans, such as fish, birds, and other forms of wildlife, could cause adverse ecological effects. In other words, does exposure to a chemical have the potential to cause harm? It involves characterizing the quality, nature, and strength of the evidence of causation. It may not give a yes or no answer; however, it is intended to provide an assessment on which to base a decision as to whether a health problem has been identified. This identification characterizes the problem in terms of the agent and dose of the agent of concern. Since there are few hazardous chemicals or hazardous agents for which definitive exposure data in humans exist, the identification of health hazards is often characterized by the effects of health problems on laboratory test animals or other species and test systems.[2]

There are numerous methods available to identify the potential for chemicals to cause both adverse health conditions and significant effects on the environment. These can include, but are not limited to toxicology, epidemiology, molecular and atomic structural analysis, safety data sheets (SDSs), standardized mortality ratios (observed deaths or expected deaths), engineering approaches to problem-solving, analysis of the fate of chemicals in the environment, and evaluations of carcinogenic versus noncarcinogenic health hazards.

6.4.2 Toxicity and Dose–Response Assessment

Dose–response assessment is the process of characterizing the relationship between the dose of an agent administered or received and the incidence of an adverse health effect in exposed populations. This process considers such important factors as intensity of exposure, age, pattern of exposure, and other variables that might modify the response, such as sex and lifestyle. In effect, it involves the evaluation of the effects expected from various quantity/concentration levels of a particular chemical in the environment. Dose and response are therefore fundamental concepts that provide a relationship between the dosage of a toxic agent and the biological response. The magnitude of the biological response depends on the concentration of the contaminant/physical agent at the site of action, while the concentration of the contaminant at the active site depends on the dose. Thus, the dose and the response are causally related. Toxicity data exhibit a dose–response relationship if a mathematical model can be formulated to describe the response of the receptor and/or test organism in terms of the dose administered. The relation often takes the form of a percentage or number of receptors responding in a given manner either to a dose or to a specified range of concentrations over a given period of time. A dose–response assessment usually requires extrapolation from high to low doses and from animal to humans or one laboratory animal species to a wildlife species. A dose–response assessment should also describe and justify the methods of extrapolation used to predict incidence, and it should characterize the statistical and biological uncertainties in these methods. When possible, the uncertainties should be described numerically rather than quantitatively.

Once again, it is important to differentiate between the terms chronic and acute as they relate to toxicity. Chronic toxicity is caused by long-term or repeated exposure to low doses of the chemical, and the intensity is usually less than with acute exposures. Acute toxicity is caused by large doses of a chemical over short time periods, and is often characterized by the effects of health problems on laboratory test animals or other species and test systems.[2]

6.4.3 Exposure Assessment

As noted, a critical component of environmental HRA is exposure assessment. It is defined as the determination of the concentration of chemicals in time and space at the location of receptors and/or target populations. This description must therefore also include an identification of all major pathways for movement and transformation of a toxic material from a source to receptors. Ideally, concentrations should be identified as a function of time and location and should include all major transformation processes. The principal pathways generally considered in exposure assessments are atmospheric transport and surface and groundwater transport. Since atmospheric dispersion has received the bulk of treatment in the literature, a good part of the material to follow will address this topic.

The exposure assessment process consists of two basic methods for determining the concentration of a chemical to which receptor target populations are exposed:

1) The first is the direct measurement of the intensity, frequency, and duration of human or animal exposure to a pollutant currently present in the environment. This is a common practice in occupational settings.
2) In some situations, however, either concentrations are too low to be detected against the background, or direct measurement is too costly or difficult to implement. Under these circumstances, the second method is employed. It involves the use of mathematical models to estimate hypothetical exposures that might arise from the release of new chemicals into the environment.

In its most complete form, an exposure assessment should describe the magnitude, duration, timing, and route of exposure of the hazardous agent, along with the size, nature, and classes of the human, animal, aquatic, or wildlife populations exposed, and the uncertainties in all estimates. The exposure assessment can often be used to identify feasible prospective control options and predict the effects of available treatment technologies for controlling or limiting exposure.[2] However, the estimation of the likelihood of exposure to a chemical remains a difficult task. Attention in the past focused on too many overly conservative assumptions. This in turn resulted in an overestimation of the actual exposure risk posed to vulnerable receptors.

Obviously, without exposure(s), there are no risks. To experience adverse effects, one must first come into contact with the toxic agent(s). Exposure to chemicals can occur via inhalation of air (breathing), intake into the body via ingestion of water and food, or adsorption through the skin. These intake processes are followed by chemical distribution through the body via the bloodstream. After being absorbed and distributed, the chemical(s) may be metabolized and excreted, either as the parent compound or as their metabolites. The principal excretory organs are the kidney, liver, and lungs.

As noted earlier, the main pathways of exposure considered in human exposure assessments are via atmospheric, surface, and groundwater transport. However, the ingestion of toxic materials that have passed through the aquatic and terrestrial food chains, and dermal absorption are two other pathways of potentially significant human exposure.

The physical and chemical properties of the chemicals under study will dictate the primary route(s) by which exposure will occur. Naturally, the chemical under study should be analyzed for the primary route(s) of human exposure. There are instances where humans may be exposed to a compound via more than one route, for example, by inhalation and oral ingestion. Which is the most significant route of exposure? Assuming approximately equal exposure by both routes, it is recommended that the chemical exposure assessment should focus on the route posing the greater risk. For those situations where one route of exposure predominates over another, the dominant route should be considered. Once an exposure assessment determines the quantity of a chemical with which human populations may come in contact, the information can be combined with toxicity data to estimate potential health risks (as presented in the next subsection).[5]

The reader should once again note that two general types of potential health risk from chemical exposures exist. These are classified as follows:

1) *Chronic*: Risk related to continuous exposures over long periods of time, generally several months to a year. Concentrations of emitted chemicals are usually relatively low. This subject area falls in the general domain of HRA, and it is this subject that is addressed in this section. Thus, in contrast to the acute (short-term) exposures that predominate in HZRAs, chronic (long-term) exposures are the major concern in HRAs.
2) *Acute*: Risk related to exposures that occur for a relatively short period of time, generally from minutes to 1 to 2 days. Concentrations of emitted chemicals are usually high relative to their no-effect levels. In addition to inhalation, airborne substances might directly contact the skin, or liquids and sludges may be splashed on the skin or into the eyes, leading to adverse health effects in acute risk settings. This subject area falls, in a general sense, in the domain of HZRA and is addressed in the next section.

6.4.4 Health Risk Characterization

Health risk characterization is the process of estimating the incidence of a health effect under the various conditions of human or animal exposure described in an exposure assessment. It is performed by combining the aforementioned dose–response information and exposure assessment. From a receptor's perspective, the risk from exposure to any chemical also depends on the potency associated with the effects and the duration of the exposure. The summary effects of the uncertainties in the preceding steps should also be included in this analysis.

The quantitative estimate of the risk is of principal interest to the regulatory agency or risk manager making a decision. The risk manager must consider the results of the risk characterization when evaluating the economics, societal aspects, and various benefits of the assessment. Factors such as societal pressure, technical uncertainties, and severity of the potential impacts influence how the decision-makers respond to the risk assessment. As one might suppose, there is room for improvement in this step of the risk assessment process.[1,2]

A risk estimate indicates the likelihood of occurrence of the different types of health or environmental effects in exposed populations. Risk assessment should include both human health and environmental evaluations (e.g., impacts on ecosystems). Ecological impacts include actual and potential effects on plants and animals (other than domesticated species). The number produced from the risk characterization, often representing the probability of adverse health effects being caused, must be carefully interpreted.

6.5 Hazard Risk Assessment/Analysis

There are several steps in evaluating the risk of an accident (see Figure 6.2). If the system in question is a chemical plant, the specific steps to be followed in the risk evaluation process are listed here:

1) A brief description of the equipment and chemicals used in the plant is needed.
2) Any hazard in the system has to be identified. Hazards that may occur in a chemical plant include:
    a) Fire
    b) Toxic vapor releases
    c) Slippage
    d) Corrosion
    e) Explosions
    f) Rupture of pressurized vessels
    g) Runaway reactions
3) The event (or series of events) that will initiate an accident has (have) to be identified. An event could be a failure to follow correct safety procedures, improperly repaired equipment, or failure of a safety mechanism.
4) The probability that the accident will occur has to be determined. For example, if a chemical plant has a 10-yr lifespan, what is the probability that the temperature in a reactor will exceed the specified temperature range over that lifetime? The probability can be ranked qualitatively from low to high. A low probability means that it is unlikely for the event to occur in the lifetime of the plant. A medium probability suggests that there is a possibility that the event will occur. A high probability means that the event will likely occur during the lifetime of the plant.
5) The severity of the consequences of the accident must be determined.
6) If the probability of the accident and the severity of its consequences are low, then the hazard risk is usually deemed acceptable and the plant should be allowed to operate. If the probability of occurrence is too high or the damage to the surroundings is too great, then the hazard risk is usually unacceptable and the system needs to be modified to minimize these effects.

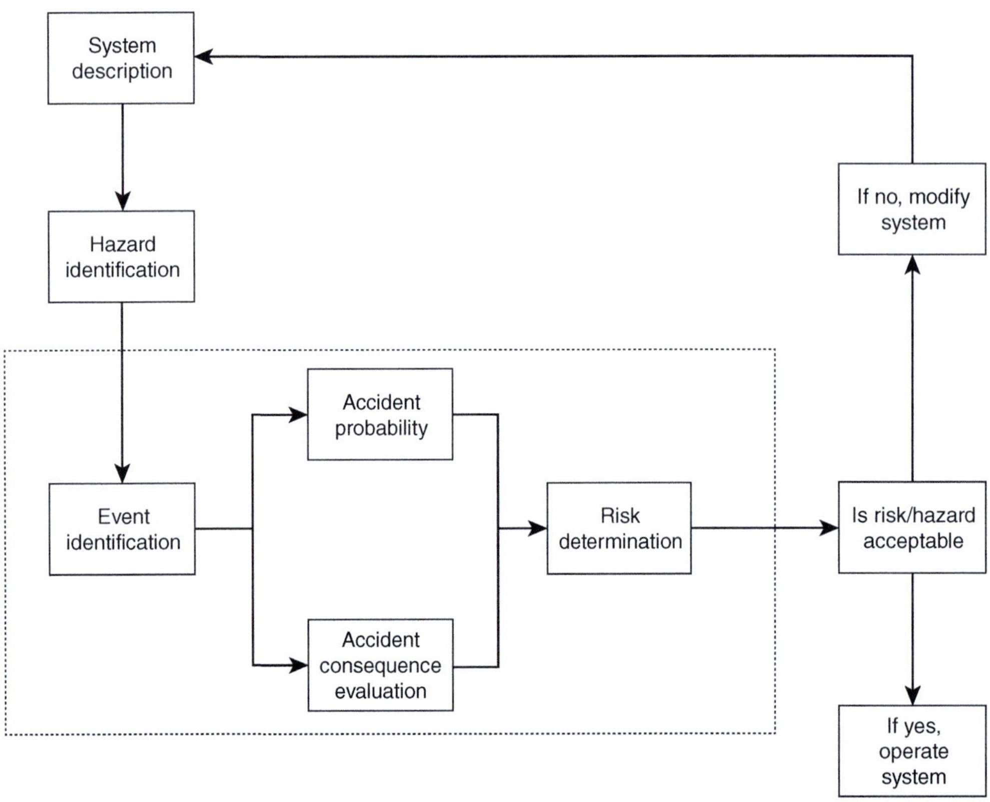

Figure 6.2 HZRA flowchart for a chemical plant.

As indicated in Figure 6.2, the heart of the HZRA approach is enclosed in the dashed box comprising Steps 3 through 6. The algorithm allows for re-evaluation of the process if the risk is deemed unacceptable (the process is repeated after system modification starting with Step 1).

Once again, it is important to note that an accident generally results from a sequence of events. Each individual event, therefore, represents an opportunity to reduce the frequency, consequence, and/or risk associated with the accident culminating from the individual events.

6.6 Risk Uncertainties/Limitations

The general subject of uncertainties/limitations is discussed in this section. The approach will examine the topic by briefly reviewing health risk and hazard risk concerns separately with their uncertainties/limitations highlighted. The accuracy of a measurement or a calculation refers to that value relative to the correct value while precision relates to the repeatability of that measurement or calculation and is expressed as the confidence interval around the mean of that value. Variability refers to the variation in the values that arise due to a calculation or measure while uncertainty, the concern at hand, refers to variations that occur due to the operation(s) or process itself. For example, in measuring the diameter of a heat exchanger tube in a condenser, accuracy refers to the value of the mean measurement compared to the true value of the tube diameter, precision refers to the range of values that arises from repeated measurements, while variability occurs from repeated measurements of the tube's diameter and uncertainty arises from the measurement of many tubes (in the population) of the condenser.

Although great controversy can surround the results of risk assessments, especially quantitative risk assessments, they are useful in particular applications. They can help establish priorities for regulatory action or intervention of any type. A consistent risk assessment performed across a range of substances can create a relative estimate of their health risks to humans. The limits of risk assessment can also be tested when government agencies (faced with the absence of other types of data and the need for action) must rely on risk assessment methods to establish health-based standards or guidelines to prevent human exposure to hazardous substances. Because of risk assessment shortcomings and the desire for greater specificity in measuring exposure, increasing interest is being shown in understanding pathologic changes at the molecular level with the hope that these investigations will lead to toxicological and epidemiological analyses of greater accuracy and sensitivity than are currently available.[6,7] In a general sense, problems in this area arise because of:

1) Uncertainty associated with available data
2) Uncertainty associated with governing equations
3) Concerns associated with assumed information
4) Concerns associated with limited and/or constrained governing equations
5) Concerns associated with overall analysis quality

It should be noted that there is no completely satisfactory way to generate accurate risk data since it is an inexact science fraught with uncertainties. At the very least, risk characterization should be checked against experience for reasonableness since the size and quality of the data employed do not permit an accurate quantitative estimate with a high degree of confidence. Careful documentation of all four parts of a risk assessment should also be maintained to prevent the practitioner from falling into traps that can influence the final results or pass the risk via a cross-media process onto another location or vulnerable population. The authors believe that the EPA and Occupational Safety and Health Administration (OSHA) have compounded the human health risk assessment uncertainty problem by some of their ambiguous and conflicting rules and regulations.

Finally, it should also be noted that less information is available on the similarities or differences in the degree of response of experimental animals as compared to humans to varying doses of a chemical. In these tests, the animals are, out of necessity, administered high doses of the chemical whereas humans are usually exposed to much lower levels. This makes it necessary to extrapolate from results *determined* at high doses to the results *expected* at low doses in humans. The validity of these extrapolations is, in most cases, not amenable to experimental verification. Thus, while the test species may serve as an approximate measure of the potential of a chemical to cause toxic effects in humans, attempts to quantify human risk on the basis of such studies remain subject to considerable scientific uncertainty. This uncertainty is particularly critical when, for example, attempts are made to predict carcinogenic responses in humans using data from tests in rats and mice.

References

1 **L. Theodore, and R.R. Dupont**, *Environmental Health Risk and Hazard Risk Assessment: Principles and Calculations*, CRC Press/Taylor & Francis Group, Boca Raton, FL, 2013.

2 **D. Paustenbach**, *The Risk Assessment of Environmental and Human Health Hazards: A Textbook of Case Studies*, John Wiley & Sons, Hoboken, NJ, 1989.

3 **L. Theodore**, *Basketball Coaching 101*, Theodore Tutorial, East Williston, NY, 2015.

4 **G. Burke, B. Singh, and L. Theodore**, *Handbook of Environmental Management and Technology*, 2nd Ed., John Wiley & Sons, Hoboken, NJ, 2000.

5 **L. Theodore, J. Reynolds, and K. Morris**, *Health, Safety and Accident Prevention: Industrial Applications*, Theodore Tutorials (originally published by USEPA, RTP, NC), East Williston, NY, 1996.

6 US Environmental Protection Agency, *Unfinished Business: A Comparative Assessment of Environmental Problems, Overview Report*. EPA-230-2-87-025a, Office of Policy, Planning and Evaluation, Washington, DC, 1987.

7 **P. Shields, and N. Hanes**, Molecular epidemiology and the genetics of environmental cancer, *JAMA J. Am. Med. Assoc.*, 266, 681–687, 1991.

7

Introduction to Air Pollution Control Equipment

In solving an air pollution emissions problem, an engineer must first carefully evaluate the system or process in order to select the most appropriate type(s) of control equipment(s) for the situation. After making a preliminary equipment selection, suitable vendors can be contacted for help in determining the capabilities and limitations of the technology and arriving at a final answer. An early and complete definition of the problem can help to reduce a poor decision that can lead to wasted pilot trials or costly installations.

Selecting an air pollution control device for cleaning a process gas stream can be a challenge. Some engineers, after trying to find shortcuts, employ quick estimates for both gas flow and collection efficiency that may be the entire extent of the collector specification – an inadequate approach. The end result will often be an ineffective installation that may have to be replaced or upgraded. Treating a gas stream, especially to reduce emissions, is usually not an income producer, but the costs – both capital and operating (see Chapter 23) – can be minimized, not by buying the least expensive collector but by thoroughly engineering the whole system as is normally done in the process design phase of a project.

Controlling the emission of pollutants from industrial and domestic sources is important in protecting the quality of air. Air pollutants can often exist as combinations of particulate matter and gases. Air cleaning devices have been reducing pollutant emissions from various sources for many years. Originally, air-cleaning equipment was used only if the contaminant was highly toxic or had some recovery value. Now with legislation including the Clean Air Act as well as individual state standards, control technology efficiency and effectiveness have been upgraded and more sources are regulated in order to meet the NAAQS. In addition, state and local air pollution agencies have adopted regulations that are in some cases more stringent than the federal emission standards.

Equipment used to control particulate emissions include gravity settlers (often referred to as settling chambers), mechanical collectors (cyclones), electrostatic precipitators (ESPs), scrubbers (gas–liquid contactors), and fabric filters (baghouses). Techniques used to control gaseous emissions are absorption, adsorption, combustion, catalytic and non-catalytic chemical reduction, and condensation. The applicability of a given technique depends on the physical and chemical properties of the pollutant and the exhaust stream. In addition, more than one technique may be capable of controlling emissions from a given source. For example, vapors generated from loading gasoline into tank trucks at large bulk terminals can be controlled by using several of the above gaseous control techniques. Most often, however, one control technique is used more frequently than others for a given source–pollutant combination. For example, absorption is commonly used to remove sulfur dioxide (SO_2) from boiler flue gas.

The material presented in this chapter regarding air pollution control equipment contains, at best, an overview of each control device. Equipment diagrams and figures, operation and maintenance procedures, and so on, have not been included in this development, but will be addressed in subsequent chapters in this Part. More details including predictive and design calculational procedures are available in the literature.[1–5]

7.1 Air Pollution Control Equipment for Particulates

As described above, the five major types of particulate air pollution control equipment are:

1) Gravity settlers
2) Cyclones

Optimizing Air Pollution Control Equipment Performance: Operation & Maintenance, First Edition. Jay Richardson and Louis Theodore.

3) Electrostatic precipitators
4) Scrubbers
5) Baghouses

Each of these devices is briefly described below.[2]

7.1.1 Gravity Settlers

Gravity settlers, or gravity settling chambers, have long been utilized industrially for the removal of solid and liquid waste materials from gaseous streams. Advantages accounting for their use are simple construction, low initial cost and maintenance, low-pressure losses, and simple disposal of waste materials. Gravity settlers are usually constructed in the form of a long, horizontal parallelepiped with suitable inlet and outlet ports. In its simplest form, the settler is an enlargement (large box) in the duct carrying the particle-laden gases: the contaminated gas stream enters at one end, the gas velocity decreases and larger particles settle out, and the cleaned gas exits from the other end.

The particles settle toward the collection surface at the bottom of the unit with a velocity at or near their settling velocity. One advantage of this device is that the external force leading to separation is provided by gravity (free by nature). Its use in industry is generally limited to the removal of large particles, i.e., those larger than 40 microns (or micrometers) and is useful with gas streams containing very high concentrations of particles. The gravity settler provides an initial collection device that can reduce particle loading into the final cleanup device(s).

7.1.2 Cyclones

Centrifugal separators, commonly referred to as cyclones, are widely used in industry for the removal of solid and liquid particles (or particulates) from gas streams. Typical applications are found in mining and metallurgical operations, the cement and plastics industries, pulp and paper mill operations, chemical and pharmaceutical processes, petroleum production (cat-cracking cyclones), and combustion operations (fly ash collection).

Particulates suspended in a moving gas stream possess inertia and momentum and are acted upon by gravity. Should the gas stream be forced to change direction, these properties can be utilized to promote centrifugal forces to act on the particles. In a conventional unit, the entire mass of the gas stream with the entrained particles enters the unit tangentially and is forced into a constrained vortex in the cylindrical portion of the cyclone. Upon entering the unit, a particle develops an angular velocity. Because of its greater inertia, it tends to move across the gas streamlines in a tangential rather than rotary direction; thus, it attains a net outward radial velocity. By virtue of its rotation with the carrier gas around the axis of the tube (main vortex) and its high density with respect to the gas, the entrained particles are forced toward the wall of the unit. Eventually, the particles may reach the outer wall, where they are carried by gravity and assisted by the downward movement of the outer vortex and/or secondary eddies toward the dust collector at the bottom of the unit. The flow vortex is reversed in the lower (conical) portion of the unit, leaving most of the entrained particles behind. The cleaned gas then passes up through the center of the unit (inner vortex) and out of the collector.

Multiple-cyclone collectors (multicones) are high-efficiency devices that consist of a number of small-diameter cyclones operating in parallel with a common gas inlet and outlet. The flow pattern differs from a conventional cyclone in that instead of bringing the gas in at the side to initiate the swirling action, the gas is brought in at the top of the collecting tube and the swirling action is then imparted by a stationary vane positioned in the path of the incoming gas. The diameters of the collecting tubes usually range from 6 in to 24 in. Properly designed units can be constructed and operated with a collection efficiency as high as 90% for particulates in the 5–10 micron range. The most serious problems encountered with these systems involve plugging and flow equalization.

One advantage cyclones have is that they can be constructed from abrasion- and heat-resistant materials and consequently have found wide applications in high-temperature operations such as boilers and fluidized bed combustors. One disadvantage of cyclones is that, to achieve relatively high (~90%) overall collection efficiencies it is usually necessary for there to be a high pressure drop across the cyclone which increases the energy consumption of the control system.

7.1.3 Electrostatic Precipitators

ESPs are satisfactory devices for removing small particles from moving gas streams at high collection efficiencies. They have been widely used in power plants for removing fly ash from the gases prior to discharge.

Two major types of high-voltage ESP configurations currently used are tubular and plate. Tubular precipitators consist of cylindrical collection tubes with discharge electrodes located along the axis of the cylinder. However, the vast majority of ESPs installed are the plate type. Particles are collected on a flat parallel collection surface spaced 8–18 in. apart, with a series of discharge electrodes located along the centerline of the adjacent plates. The electric field between the discharge electrodes and the collection surface imparts a static charge to the particles in the gas stream. The charged particles are attracted to the grounded collection electrodes and are separated out from the gas stream. The gas to be cleaned passes horizontally between the plates (horizontal flow type) or vertically up through the plates (vertical flow type). Collected particles are usually removed by periodic rapping of the tubes or plates.

Depending on the operating conditions and the required collection efficiency, the gas velocity in an industrial ESP is usually between 2.5 and 8.0 ft/s. A uniform gas distribution is of prime importance for precipitators, and it should be achieved with a minimum expenditure of pressure drop. This is not always easy, since gas velocities in the duct ahead of the precipitator may be 30–100 ft/s. It should be clear that the best operating condition for a precipitator will occur when the velocity distribution is uniform. When significant maldistribution occurs, the higher velocity in one collecting plate area will decrease efficiency more than a lower velocity at another plate area will increase the efficiency of that area. Both physical and Computational Fluid Dynamics (CFD) modeling have been used to assist in designing ESP entrance and exit ducts that produce uniform velocity distributions.

The maximum voltage at which a given field can be maintained depends on the properties of the gas and the dust being collected. These parameters may vary from one point to another within the precipitator, as well as with time. In order to keep each section working at high efficiency, a high degree of sectionalization is recommended. This means that many separate power supplies and controls will produce better performance in a precipitator of a given size than if there were only one or two independently controlled sections. This is particularly true if high efficiencies are required.

Additionally, multiple power supplies increase the reliability of the ESP since one power supply failure will not catastrophically affect ESP performance.

7.1.4 Scrubbers

Wet scrubbers have found widespread use in cleaning contaminated gas streams because of their ability to effectively remove both particulate and gaseous pollutants. Specifically, wet scrubbing involves a technique of bringing a contaminated gas stream into intimate contact with a liquid. Wet scrubbers include all the various types of gas absorption equipment (to be discussed later). The term "scrubber" will be restricted to those systems that utilize a liquid, usually water, to achieve or assist in the removal of particulate matter from a gas stream. The use of wet scrubbers to remove gaseous pollutants from contaminated streams is considered in the next section.

Another important design consideration for the wet scrubber (as well as absorbers) is concerned with suppressing the steam plume. Water-scrubber systems removing pollutants from high-temperature processes (i.e., combustion) can generate a supersaturated water vapor that becomes a visible white plume as it leaves the stack. Although not strictly an air pollution problem, such a plume may be objectionable for aesthetic reasons. Regardless, there are several ways to avoid or eliminate the steam plume.[6]

7.1.5 Baghouses

The basic filtration process may be conducted in many different types of fabric filters in which the physical arrangement of hardware and the method of removing collected material from the filter media will vary. The essential differences may be related, in general, to

1) Mode of operation
2) Cleaning mechanism
3) Type of fabric
4) Equipment

Gases to be cleaned can be either pushed or pulled through the baghouse. In the pressure system (push through), the gases may enter through the cleanout hopper in the bottom or through the top of the bags. In the suction type (pull-through), the dirty gases are forced through the inside of the bag and exit through the outside. The suction type of operation is generally preferred since any equipment/ductwork leaks will pull in clean outside air into the gas stream rather than leaking dust-laden gases into the atmosphere as can happen with pressurized operation. Additionally, with the suction (negative pressure) type of operation, the fan will be moving dust-free gas and not have to contend with erosion issues that could arise when passing dust-laden gas through the fan.

A wide variety of woven and felted fabrics are used in fabric filters. Clean felted fabrics are more efficient dust collectors than woven fabrics, but the efficiency of woven fabric filter bags will achieve near-equal filtration efficiency to felted fabrics after a dust layer accumulates on the surface. When a new woven fabric is placed in service, visible penetration of dust within the fabric may occur. This normally takes from a few hours to a few days for industrial applications, depending on the dust loadings and the nature of the particles. In many cases, new woven filter bags are prepared for operation by coating the bags with a powder, such as powdered limestone, prior to being exposed to the dirty gas stream. This pretreatment produces the initial dust layer needed for efficient filtering.

Baghouses are constructed as single units or as multi-compartment units. The single unit is generally used on small processes that are not in continuous operation, such as grinding and paint spraying processes. Multi-compartment units are used in continuous operating processes with large exhaust volumes such as electric melt steel furnaces and industrial boilers. In both cases, the bags are housed in a shell made of rigid metal material.

7.2 Air Pollution Control Equipment for Gaseous Pollutants

As described in the introduction to this chapter, the five generic types of gaseous control equipment include:

1) Absorbers
2) Adsorbers
3) Combustion units
4) Condensers
5) Catalytic and non-catalytic reduction

Each of these devices is briefly described below.[2]

7.2.1 Absorbers

Absorption is a mass transfer operation in which a gas is dissolved in a liquid. A contaminant (pollutant exhaust stream) contacts a liquid and diffuses (is transported) from the gas phase into the liquid phase. The absorption rate is enhanced by (i) high diffusion rates, (ii) high solubility of the contaminant, (iii) large liquid–gas contact area, and (iv) good mixing between liquid and gas phases (turbulence).

The liquid most often used for absorption is water because it is inexpensive, is readily available, and can dissolve a number of contaminants. Reagents can be added to the absorbing water to increase the removal efficiency of the system. Certain reagents merely increase the solubility of the contaminant in the water. Other reagents chemically react with the contaminant after it is absorbed. In reactive scrubbing, the absorption rate is much higher, so in some cases a smaller, economical system can be used. However, the reactions can form precipitates that have to be taken into account in the design of the absorber and peripheral equipment.

If a gaseous contaminant is very soluble, almost any of the wet scrubber designs will adequately remove this contaminant. However, if the contaminant is of low solubility, the packed tower or the plate tower[7,8] is more effective. Both of these devices provide increased mass transfer rates between phases and have relatively low-pressure drops. The packed tower, one of the most common gas absorption devices, consists of an empty shell filled with packing. The liquid flows down over the packing, exposing a large film area to the gas flowing up the tower. Plate towers consist of horizontal plates placed inside the tower. Gas passes up through the orifices in these plates while the liquid flows down across the plate, thereby providing increased mass transfer.[7,8]

7.2.2 Adsorbers

Adsorption is a mass transfer process that involves removing a gaseous contaminant by adhering to the surface of a solid. Adsorption can be classified as physical or chemical. In physical adsorption, a gas molecule adheres to the surface of the solid due to an imbalance of natural forces (electron distribution). In chemisorption, once the gas molecule adheres to the surface, it reacts chemically with it. The major distinction is that physical adsorption is readily reversible whereas chemisorption is not.

All solids physically adsorb gases to some extent. Certain solids, called adsorbents, have a high attraction for specific gases; they also have a large surface area that provides a high capacity for gas capture. By far the most important adsorbent for air pollution control is activated carbon. Because of its unique surface properties, activated carbon will preferentially adsorb hydrocarbon vapors and odorous organic compounds from an airstream. Additionally, activated carbon has been used to remove elemental and ionic mercury from flue gas streams at power plants. Most other adsorbents (molecular sieves, silica gel, and activated aluminas) will preferentially adsorb water vapor, which may render them useless for removing other contaminants.

For activated carbon, the amount of hydrocarbon vapors that can be adsorbed depends on the physical and chemical characteristics of the vapors, their concentration in the gas stream, system temperature, system pressure, humidity of the gas stream, and the molecular weight of the vapor. Physical adsorption is a reversible process; the adsorbed vapors can be released (desorbed) by increasing the temperature, decreasing the pressure, or using a combination of both. Vapors are normally desorbed by heating the adsorbent with steam.

Adsorption can be a very useful removal technique since it is capable of removing very small quantities (a few parts per million) of vapor from an airstream. The vapors are not destroyed; instead, they are stored on the adsorbent surface until they can be removed by desorption. The desorbed vapor stream is normally highly concentrated. It can be condensed and recycled, or incinerated as an ultimate disposal technique.

The most common adsorption system is the fixed bed adsorber. These systems consist of two or more adsorber beds operating on a timed adsorbing/desorbing cycle. One or more beds are adsorbing vapors, while the other bed(s) is being regenerated. If particulate matter or liquid droplets are present in the vapor-laden airstream, this stream is sent to pretreatment to remove them.

7.2.3 Combustion Units

Combustion is defined as a rapid, high-temperature gas-phase oxidation. Simply, the contaminant (an organic substance) is burned with air and converted to carbon dioxide and water vapor. The operation of any combustion source is governed by the three T's of combustion: temperature, turbulence, and time. For complete combustion to occur, each contaminant molecule must come in contact (turbulence) with oxygen at a sufficient temperature, while being maintained at this temperature for an adequate time. These three variables are dependent on each other. For example, if a higher temperature is used, less mixing of the contaminant and combustion air or shorter residence time may be required. If adequate turbulence cannot be provided, a higher temperature or longer residence time must be employed for complete combustion.

Combustion devices can be categorized as flares, thermal incinerators, or catalytic incinerators. Flares are direct combustion devices used to dispose of small quantities or emergency releases of combustible gases. Flares are normally elevated (from 100 ft to 400 ft) to protect the surroundings from the heat and flames. Flares are often designed for steam injection at the flare tip. The steam provides sufficient turbulence to ensure complete combustion; this prevents smoking. Flares are also very noisy, which can cause problems for adjacent neighborhoods, and are found, for the most part, at petroleum refineries.

Thermal incinerators are also called as afterburners, direct flame incinerators, or thermal oxidizers. These are devices in which the contaminant airstream passes around or through a burner and into a refractory-line residence chamber where oxidation occurs. To ensure complete combustion of the contaminant, thermal incinerators are designed to operate at a temperature of 700–800°C (1,300–1,500°F) and a residence time of 0.3–0.5 s. Ideally, as much fuel value as possible is supplied by the waste contaminant stream; this reduces the amount of auxiliary fuel needed to maintain the proper temperature.

In catalytic incineration, the contaminant-laden stream is heated and passed through a catalyst bed that promotes the oxidation reaction at a lower temperature. Catalytic incinerators normally operate at 370–480°C (700–900°F).

7.2.4 Condensers

Condensation is a process in which the volatile gases are removed from the contaminant stream and changed into a liquid. Condensation is usually achieved by reducing the temperature of a vapor mixture until the partial pressure of the condensable component equals its vapor pressure. Condensation requires low temperatures to liquefy most pure contaminant vapors. Condensation is affected by the composition of the contaminant gas stream. The presence of additional gases that do not condense at the same conditions – such as air – hinders condensation.

Condensers are normally used in combination with primary control devices. Condensers can be located upstream of (before) an incinerator, adsorber, or absorber. These condensers reduce the volume of vapors that the primary control equipment must handle. Therefore, the size and the cost of the primary control device can be reduced. Similarly, condensers can be used to remove water vapors from a process stream with a high moisture content upstream of a control system. A prime example is the use of condensers in rendering plants to remove moisture from the cooker exhaust gas. When used alone, refrigeration is required to achieve the low temperatures required for condensation. Refrigeration units are used to successfully control gasoline vapors at large gasoline dispensing terminals.

Condensers are classified as being either contact condensers or surface condensers. Contact condensers cool the vapor by spraying liquid directly on the vapor stream. These devices resemble a simple spray scrubber. Surface condensers are normally shell-and-tube heat exchangers. Coolant flows through the tubes while vapor is passed over and condenses on the outside of the tubes. In general, contact condensers are more flexible, simpler, and less expensive than surface condensers. However, surface condensers require much less water and produce nearly 20 times less wastewater that must be treated than contact condensers. Surface condensers also have an advantage in that they can directly recover valuable contaminant vapors.

7.2.5 Catalytic and Non-Catalytic Reduction

Catalytic and non-catalytic reduction equipment has been used, for the most part, in the electric utility and other industries for the destruction of nitrogen oxides in combustion flue gas. Both systems employ a chemical reaction between ammonia (NH_3) or ammonium hydroxide (NH_4OH) and the nitrogen oxides in the gas stream to produce nitrogen and water vapor. Selective non-catalytic reduction (SNCR) involves injecting ammonia or urea into the flue gas to yield nitrogen and water. The ammonia or urea must be injected into specific high-temperature zones in the upper furnace or convective pass for this method to be effective.

The other flue gas treatment method, selective catalytic reduction (SCR), involves injecting ammonia into the flue gas in the presence of a catalyst. SCR promotes the reactions by which oxides of nitrogen (NO_x) are converted to nitrogen and water at lower temperatures than required for SNCR. The SNCR process involves injecting ammonia or urea into boiler flue gas at specific temperatures. The ammonia or urea reacts with NO_x in the flue gas to produce N_2 and water. For the ammonia-based SNCR process, ammonia is injected into the flue gas where the temperature is 950 ± 30°C ($1{,}750 \pm 90$°F). In the urea-based SNCR process, an aqueous solution of urea is injected into the flue gas at one or more locations in the upper furnace or convective pass. The urea reacts with NO_x in the flue gas to form nitrogen, water, and carbon dioxide (CO_2). SCR involves injecting ammonia or ammonium hydroxide into boiler flue gases in the presence of a catalyst to reduce NO_x to N_2 and water.

7.3 Hybrid Systems

Hybrid systems in the air pollution field are defined as those types of control devices that involve combinations of control mechanisms – for example, fabric filtration combined with electrostatic precipitation. Unfortunately, the term hybrid system has come to mean different things to different people. The two most prevalent definitions employed today for hybrid systems are:

1) Two or more different air pollution control equipment connected in series, e.g., a baghouse followed by an absorber or a dry scrubber followed by a baghouse.

2) An air pollution control system that utilizes two or more collection mechanisms simultaneously to enhance pollution capture, e.g., an ionizing wet scrubber (IWS), will be discussed shortly.

The two major hybrid systems found in practice today include IWSs and dry scrubbers. These are briefly described below.

7.3.1 Ionizing Wet Scrubbers

The IWS is an innovative development in the technology of the removal of particulate and gaseous contaminants from a gas stream. These devices have been incorporated into commercial incineration facilities[9,10] after a quench tower. In the IWS, high-voltage ionization in the charge section places a static electric charge on the particles in the gas stream, which then passes through a crossflow packed-bed scrubber. The packing is normally polypropylene: in the form of circular-wound spirals and gear-like wheel configurations, providing a large surface area. Particles with sizes of 3 microns or larger are trapped by inertial impaction within the bed. Smaller charged particles pass close to the surface of either the packing material or a scrubbing water droplet. An opposite charge on that surface is induced by the charged particle, which is then attracted to an ion attached to the surface. All collected particles are eventually washed out of the scrubber. The scrubbing water also can function to absorb gaseous pollutants, especially when enhanced with an alkali such as soda ash or caustic soda.

According to Celicote (the IWS vendor), the collection efficiency of the two-stage IWS is greater than that of a baghouse or a conventional ESP for particles in the 0.2–0.6 micron range. For 0.8 micron and above, the ESP is as effective as the IWS.[4] Scrubbing water can include caustic soda or soda ash when needed for efficient adsorption of acid gases.

7.3.2 Dry Scrubbers

The success of fabric filters in removing fine particles from flue gas streams has encouraged the use of combined dry-scrubbing/fabric filter systems for the dual purpose of removing both particulates and acid gases simultaneously. Dry scrubbers offer potential advantages over their wet counterparts, especially in the areas of energy savings, zero liquid discharge, and capital costs. Furthermore, the dry-scrubbing process design is relatively simple, and the product is a dry waste rather than a wet sludge.

There are three major types of dry scrubber systems: spray drying, the circulating dry scrubber (CDS), and dry injection. The first two processes are often referred to as wet–dry systems. When compared to the conventional wet scrubber, they use just enough water to lower the temperature of the gas to the desired outlet temperature. This outlet temperature is generally 20–40°F above the adiabatic saturation temperature. The third process has been referred to as a dry–dry system because no liquid scrubbing is involved. The spray-drying and the semi-dry systems are predominantly used in utility and industrial applications where high acid gas removal efficiencies are needed.

The method of operation of the spray dryer gas cleaning system is relatively simple, requiring only two major items: a spray dryer similar to those used in the chemical food-processing and mineral-preparation industries, and a baghouse or ESP to collect the entrained solids. In the spray dryer, the sorbent solution, or slurry, is atomized into fine droplets and injected into the incoming flue gas stream to increase the liquid–gas interface and to promote the mass transfer of the SO_2 (or other acid gases) from the gas to the slurry droplets where it is absorbed. Simultaneously, the thermal energy of the gas evaporates the water in the droplets to produce a dry powdered mixture of alkali salts and some unreacted alkali. Because the slurry contains just enough water to lower the temperature of the flue gas to a point above the adiabatic saturation temperature, the flue gas is not saturated and contains no liquid carryover. After leaving the spray dryer, the solid bearings gas passes through a fabric filter (or ESP), where the dry product is collected and where a percentage of unreacted alkali reacts with the remaining acid gases for further removal. The advantage of using a fabric filter over an ESP is that the dust layer on the filter bags contains some of the unreacted alkali. As the cooled gas passes through the dust layer, additional acid gases are removed, in some cases equaling the acid gas removal in the spray dryer. With an ESP the only additional acid gas removal is when the particles are still in the gas stream before their deposit on the collection plates. The cleaned gas is then discharged through the ductwork to an induced draft (ID) fan and to the stack.

Dry-injection processes generally involve the pneumatic introduction of a dry, powdery alkaline material, usually a sodium-based sorbent (sodium bicarbonate, trona, or nahcolite), into the flue gas stream with subsequent fabric filter collection. The injection point in such processes can vary from the boiler-furnace area all the way to the flue gas entrance to the baghouse, depending on operating conditions and design criteria. This process has the advantage of being very simple, mechanically, with almost no increase in the pressure drop of the system. The disadvantage that the dry-injection process has is that gas–solid mass transfer is much slower than gas–liquid mass transfer. Therefore, the acid gas removal for dry injection is much lower than that achieved with spray dryer or CDS systems.

Additional details on hybrid systems are found in Chapter 17.

7.4 Factors in Selecting and Comparing Equipment

There are a number of factors to be considered prior to selecting a particular air pollution control system. In general, they can be grouped into three categories: environmental, engineering, and economic. Additional details are found in Part III, Chapter 26.

References

1 **L. Theodore**, *USEPA Training Manual, Air Pollution Control Equipment for Particulates*, RTP, NC, 1981.
2 **L. Theodore**, *USEPA Training Manual, Air Pollution Control Equipment for Gaseous Pollutants*, RTP, NC, 1982.
3 **L. Theodore, and R. Allen**, *Air Pollution Control Equipment, A Theodore Tutorial*, Theodore Tutorials (Originally Published by the USEPA/APTI, RTP, NC), East Williston, NY, 1993.
4 **M.K. Theodore, and L. Theodore**, *Introduction to Environmental Management*, 2nd Ed., CRC Press/Taylor & Francis Group, Boca Raton, FL, 2021.
5 **L. Theodore**, *Air Pollution Control Equipment Calculations*, John Wiley & Sons, Hoboken, NJ, 2008.
6 **L. Theodore**, *Personal Notes*, East Williston, NY, 1986.
7 **L. Theodore**, and F. Ricci, *Mass Transfer Operations for the Practicing Engineer*, McGraw-Hall, New York, NY, 2010.
8 **L. Theodore**, *Chemical Engineering: The Essential Reference*, McGraw-Hall, New York, NY, 2014.
9 **J. Santoleri, J. Reynolds, and L. Theodore**, *Introduction to Hazardous Waste Incineration*, 2nd Ed., John Wiley & Sons, Hoboken, NJ, 2000.
10 **L. Theodore**, *Hazardous Waste Incinerators, A Theodore Tutorial*, Theodore Tutorials (Originally Published by the USEPA/APTI, RTP, NC), East Williston, NY, 1989.

8

Introduction to Operation, Maintenance, and Inspection

This chapter is concerned primarily with operation, maintenance, and inspection (OMI) issues as they apply to air pollution control equipment. However, part of the presentation to follow will address OMI from a heat exchanger perspective; these issues obviously vary depending on the type of other air pollution control device under consideration. Because of the general nature of the material to follow, the reader should note that this can also be applied to most air pollution control equipment.[1,2]

Chapter contents include equipment description, personnel, installation procedures, operation, maintenance, inspection, improving operation and performance, special equipment, and record keeping. **Note**: the bulk of material for this chapter has been drawn from the original work of Connery[3] and subsequent publications based on his work.[4]

8.1 The Need for an Operation and Maintenance Program

Operation and maintenance problems have plagued users since the installation of the first pollution control device. However, because of the proliferation of equipment installed as a result of the Clean Air Act Amendments of 1970, the operation and maintenance problem has magnified intensely. The number and complexity of these problems have increased at a nearly exponential rate since the early 1970s. The 1950s and 1960s initially involved installations designed for the low to medium collection efficiency range. During the 1970s, designs of higher efficiency were developed that contained more and more component parts. The time normally required for the training of field technicians and engineers on the problems (and solutions) with these newer installations for both gas and particulate control was not available. Although there has been a concentrated effort in recent years to better understand air pollution control equipment, the result – unfortunately – is likened to placing the cart before the horse.

There are several reasons for an operation and maintenance program. The most important are:

1) Continuously meeting present emission regulatory control codes and improving air quality
2) Prolonging control equipment life
3) Maintaining productivity (reducing production interruptions) of the process unit served by the control device
4) Prolonging process life
5) Reducing operating costs
6) Promoting better public relations
7) Avoiding community alienation
8) Promoting better relations with regulatory officials

Thus, the proper operation of the control device requires a planned program of operator's training, equipment know-how, preventive maintenance, and adequate record keeping.

Optimizing Air Pollution Control Equipment Performance: Operation & Maintenance, First Edition. Jay Richardson and Louis Theodore.

8.2 System Description

It is important that operating personnel be provided with detailed descriptions of the unit operations and processes of the system(s) that the air pollution control equipment serves. Guidance on operating and controlling the air pollution control device is provided later in this text. The purpose of this section is to assist the reader in understanding the construction drawings, the purpose and function of the plant, and to simply state the engineer's concept of operation and control of the air pollution control process(es).

A detailed description of the operation, control, and relationship to other plant units of each control device and its components should be made available. Photographs and/or schematic diagrams should be included to supplement the verbal descriptions of routine, alternate, and emergency operations. The user should also be aware of all applicable manuals. Information concerning applicable safety features; fail-safe features; operations problems, causes, and suggested cures; a startup guidance should also be available.

Suitable texts, reports, and references for further study by plant operators should be provided, with appropriate consideration given to the complexity of the control units/processes and the expertise of the plant personnel.

Process descriptions can be divided into two major categories: a general division containing information generally applicable to all similar control processes, and a specific section containing operating information pertinent to the specific plant in question. An example of the information required is as follows:

1) General
   a) Unit description
      i) Basic principles
      ii) Operational features
      iii) Design efficiency
      iv) Performance test results
   b) Relationship to adjacent units
   c) Classification and control
   d) Major components
   e) Common operating problems
      i) Pumps
         1) Pumps will not start
         2) Reduced rate of discharge
         3) High power requirements
         4) Excess noise
         5) Other
      ii) Plugging
      iii) Others
   f) Startup
      i) Equipment inspection
      ii) Startup procedures
      iii) Controlling startup
2) Specific plant operation
   a) Normal operation
      i) General
      ii) Schematics
   b) Alternate operation
      i) General description of alternate operation modes
      ii) Information for alternate operating conditions
      iii) References to construction drawings, equipment shop drawings, and construction specifications
   c) Emergency operations and fail-safe features
      i) Warning devices
      ii) Standby power
      iii) Pump capacities
      iv) Piping configuration

8.3 Personnel

Regardless of the care that goes into the design and construction of an air pollution control facility, without qualified personnel in adequate numbers to operate the process(es), the full capabilities of the control facility cannot be realized. A well-thought-out staff requirement will assist the system's management as they seek funds for staffing their facility. The personnel/workforce requirements recommended should also be compatible with existing federal and state guidelines, if required and applicable.

Up-to-date training for operators and maintenance personnel should be stressed as being of critical importance in the proper functioning of the control facility. The purpose is to protect the huge investment in plant equipment from damage or deterioration and to reduce the quantity of pollutants discharged into the atmosphere.

To prepare personnel/workforce recommendations adequately, a task analysis of each job within the control system should be made. This analysis will provide details on the skills and qualifications required for each position. There are specialized methods and techniques and professionals available for the determination of personnel/workforce and training requirements. Persons responsible for this task should consider using personnel with specialized skills in personnel/workforce factors to determine personnel and training needs.

The qualifications for all types of personnel should be listed. Job descriptions for all personnel should be given. This description might include the following for an operator: follow shift standard procedures, keep daily operating records, check all meters and recorders for proper operation, and so on. The following is a list of some of the types of personnel commonly employed for the operation and maintenance of air pollution control systems:

1) Superintendent
2) Assistant superintendent
3) Clerk
4) Operations supervisor
5) Shift foreperson
6) Operator II
7) Operator I
8) Automotive equipment operator
9) Maintenance supervisor
10) Mechanical maintenance foreperson
11) Maintenance mechanic II
12) Maintenance mechanic I
13) Maintenance helper
14) Electrician II
15) Electrician I
16) Laborer
17) Painter
18) Storekeeper
19) Custodian
20) Chemist
21) Laboratory technician

The following outline provides a job description qualifications profile.

1) Formal education
2) General requirements
3) General educational development
   a) Reasoning
   b) Mathematical
   c) Language
4) Special vocational preparation
5) Aptitudes as related to general working conditions
   a) Intelligence
   b) Verbal

    c) Numerical
    d) Form perception
    e) Spatial perception
    f) Clerical perception
    g) Motor coordination
    h) Finger dexterity
    i) Manual dexterity
    j) Eye-hand-foot coordination
    k) Color discrimination
6) Interests
7) Temperament
8) Physical demands
9) Working conditions

Complete and accurate job descriptions are sometimes difficult to prepare. Good job descriptions should include but are not limited to the following:

1) List items, equipment, or processes that an individual must operate
2) State if monitoring of gauges or meters is required
3) Discuss interpreting any meter or gauge readings for process control actions
4) List any logs or records to be maintained
5) Outline of any maintenance duties required
6) State any other title that the individual might carry
7) Discuss decision-making requirements
8) State responsibilities and authority given to the individual on the job being described
9) List any report or budget functions that must be performed
10) Discuss any supervisory or inspection functions
11) Discuss training requirements, if applicable

The pertinent details of any existing state certification program should be discussed, with emphasis on how they apply to the control system at hand. This discussion should include a copy of the rules and regulations of the state certifications board.

Only properly trained personnel should be expected to perform satisfactory inspections, repairs, and preventative maintenance tasks. Properly trained personnel should possess a thorough knowledge of the functions and operations of their equipment and the procedures for servicing it. A good maintenance management program must consider the limitations of plant operators and maintenance or factory representatives to perform certain required maintenance functions. Outside contractors should be brought in for specialized tasks outside the knowledge base of in-house staff.

Job titles, job descriptions, and qualifications for maintenance personnel should be provided. Any general information regarding maintenance personnel or particular information on possible sources of maintenance help should be available.

8.4 Installation Procedures

The preparation of equipment for installation begins on delivery of the unit from the manufacturer. On delivery of the unit, the general condition should be noted to determine any damage sustained during transit and prior to receipt. Any dents or cracks should be reported to the manufacturer prior to attempting to install the unit. Any flanged or threaded connections are usually blanked with suitable pipe plugs. These closures are necessary to avoid the entry of debris into the unit during shipping and handling and should remain in place until actual piping connections are made (if applicable). Sufficient clearance is required for at least inspection of the unit or in-place maintenance.

Equipment must be supported on structures of sufficient rigidity and strength to avoid imposing excessive strains due to settling. Additional considerations may be required for rotating or vibrating equipment. Units should be carefully leveled and squared to ensure proper drainage, venting, and alignment with attached piping. On occasions, some of these units are purposely aligned to facilitate venting and drainage, and the alignment with piping then becomes the prime concern.

Installation and arrangement drawings and manuals should be carefully considered and reviewed *before, during,* and *after* installation.

8.5 Operation

The maximum allowable working pressures and temperatures are normally indicated on the unit's nameplate or in the design data sheet. These excursion values should not be exceeded. Special precautions should be taken if any individual part of the unit is designed for a maximum temperature lower than the unit as a whole. In addition, maintaining an adequate flow of various media may be required at all times. Equipment is typically designed for a particular fluid throughput. Generally, a reasonable overload can be tolerated without causing damage. If the equipment is operated at excessive flow rates, problems such as erosion or destructive vibration could result. Erosion could occur at normally acceptable flow rates if other conditions such as entrained liquids or particulates in a gas stream or abrasive solids are present. Evidence of erosion should be investigated to determine the root cause. Vibration can be propagated by problems other than flow overloads such as improper design, fluid maldistribution, or corrosion-erosion of internal flow-directing devices such as baffles. Considerable studies and research have been conducted to develop a reliable vibration analysis procedure to predict or correct damaging vibration.

Most equipment should be warmed up slowly and uniformly; the higher the temperature range, the slower the warm-up should be. This is generally accomplished by introducing any applicable coolant or heated fluid and increasing the flow rate to the design level and gradually adding any other streams. Equipment is usually cooled down by shutting off the feed stream first and then the heating or cooling stream. Equipment containing flammable, corrosive, or high-freezing-point fluids should be thoroughly drained for any prolonged outages.

8.6 Maintenance and Inspection

Recommended maintenance of all equipment requires regular inspection to ensure the mechanical integrity of the unit and a level of performance consistent with the original design criteria. A brief general inspection should be performed on a regular basis while the unit is operating. Vibratory disturbance, leaking gasketed joints, excessive pressure drops, decreased operating efficiency, and any intermixing of fluids are all signs that a thorough inspection and maintenance procedure are required.

Complete inspection requires a shutdown of the unit for access to internals plus testing and cleaning. Scheduling can be determined only from experience and general inspections. Potential causes of deterioration include general corrosion, intergranular corrosion, stress cracking, galvanic corrosion, impingement, erosion attach, and the lack of a formal maintenance-and-inspection program.

Fouling occurs because of the deposition of foreign material on the interior or exterior of the equipment. Evidence of fouling during operation is increased pressure drop and a general decrease in performance. Fouling can be so severe that it can lead to mechanical damage to the equipment. The nature of the deposited fouling determines the method of cleaning that should be utilized. Cleaning is generally performed using chemicals, steam, or other suitable fluids. Mechanical cleaning is occasionally performed but requires the utmost care to avoid damaging the equipment. Proper maintenance also requires testing of the unit to check its integrity.

8.7 Improving Operation and Performance

Within the constraints of the existing system, *improving operation and performance* generally refers to maintaining or improving operation and original (or consistent) performance. Several factors previously mentioned are critical to the design and performance of a unit: operating pressure, amount of non-condensable gases in vapor stream, temperatures, flow rate, fouling resistance, and mechanical soundness. Any pressure drop, if applicable, should be minimized. Deaerators or similar devices should be operational where necessary to remove gases in solution with liquids. Proper and regular venting of equipment and leakproof gasketed joints in vacuum systems are all necessary to prevent problems. Flow rates and temperatures (if applicable) should be checked regularly to ensure that they are in accordance with the original design and performance criteria. The importance of this can be illustrated simply by comparing the winter and summer performance of equipment using a cooling tower or river water [5,6]. Decreased performance due to fouling will generally be exhibited by a gradual decrease in operating efficiency and should be corrected as soon as possible. Mechanical malfunctions can also be gradual but will eventually lead to a near-total or total inability for the unit to perform.

Fouling and mechanical soundness can be controlled only by regular and complete maintenance. In some cases, fouling is much worse than originally predicted and requires frequent cleaning regardless of the precautions taken in the original design. These cases require special action to alleviate the problems associated with fouling. Most equipment manufacturers provide designs for alternate conditions as a guide to estimating the cost of improving efficiency via alternate configurations or process fluids.

8.8 Special Tools and Equipment

A review of the work to be performed by the maintenance personnel will aid in developing a list of the tools and equipment required for the facility. The tools and equipment should be maintained in good working order and be available when required. Where some items may be required in more than one location, this should be taken into account. It is suggested that maintenance tools and supplies not be placed any more than 100–200 ft from the point or points of use. Tool boards with specialized or frequently used tools should be located with appropriate equipment where required.

Control facilities large enough to warrant a central tool room should maintain a complete tool inventory. Tools should be issued under a tool check control system. All tools should be regularly checked for their condition in terms of personnel safety and equipment protection.

Equipment manufacturers' lubrication specifications should be readily available. An interchangeable lubricant chart should be provided. This should contain information on the use of color-coded lubrication tags for all equipment. Sample forms for recording quantities of lubricants consumed and in stock should be included. A sample lubrication route should be outlined to assist maintenance supervisors in developing lubrication routes for the control facility. Those responsible for preventative maintenance should also be responsible for lubrication. Their duties should include the following:

1) Conduct lubrication studies
2) Prepare lubrication specifications
3) Establish schedules
4) Train lubricators
5) Standardize application methods
6) Maintain consumption and inventory records
7) Establish proper handling and storage
8) Investigate new lubricants; evaluate and revise specifications as necessary
9) Standardize lubricants whenever possible to eliminate stocks of identical material under various trade names

8.9 Records

An important factor in any efficient operation of an air pollution control system is the maintenance of accurate operational, performance, and financial records. Without a record of past operational performance, it is impossible to identify trends in any process. Operating cost records are also essential if meaningful budgets are to be prepared. Accurate records permit plant operating personnel and management to maintain control of their facility.

The objective of this section is to describe the records and reports that should be maintained. The importance of keeping neat, well-written, and accurate records is stressed. The suggested types of records that should be maintained include:

1) Monthly operating report
2) Daily log, by shift, of process operations
3) Operating cost records

Other records, such as those pertaining to laboratory, maintenance, and safety, should also be kept.

Operators' worksheets should be maintained. These sheets are temporary until data are transferred to the daily operating log, or they may be kept if they apply to a remote location. A daily operating log should also be maintained. Information on this daily log can include the following:

1) Routine operating duties
2) Unusual conditions (operational and maintenance)
3) Accidents to personnel
4) Complaints
5) Power consumption
6) Plant visitors

Physical plant records should be available for reference at the plant and should include:

1) Plant operation and maintenance manual
2) As-build engineering drawings
3) Copy of construction specifications
4) Equipment suppliers' manuals
5) Piping and wiring diagrams
6) Data cards on all equipment
7) Construction photographs

Operating costs should also be recorded. The major categories of operating costs are labor, utilities, chemicals, and supplies. Labor is usually broken down into operation, administration, and maintenance. Utilities include electricity, fuel oil, telephone, internet, gas, and water. Chemicals should be limited to those used in the control processes (e.g., absorption with chemical reaction). Supplies include chemicals, cleaning materials, maintenance supplies, and other expendable items.

Finally, records that reflect such things as training individuals have received and employee turnover rate are valuable. A record of emergency conditions affecting the control system should also be maintained. A system for maintaining these records should be developed and available to management.

References

1 **L. Theodore**, *Heat Transfer Applications for the Practicing Engineer*, John Wiley & Sons, Hoboken, NJ, 2011.

2 **A.M. Flynn, T. Akashige, and L. Theodore**, *Kern's Process Heat Transfer*, 2nd Ed., Scrivener-Wiley, Beverly, MA, 2019.

3 **W. Connery et al.**, *Energy and the Environment*, Proceedings of the 3rd National Conference, AICHE, NY, 1975, 276–282.

4 **W. Connery**, Condensers, in *Air Pollution Control Equipment, A Theodore Tutorial*, (originally published by USEPA/APTI, RTP, NC) Theodore, L. and Buonicore, A. J. Eds, Theodore Tutorials, East Williston, NY, 1992.

5 **L. Theodore, and R.R. Dupont**, *Environmental Health Risk and Hazard Risk Assessment: Principles and Calculations*, CRC Press/Taylor & Francis Group, Boca Raton, FL, 2013.

6 **L. Theodore, and R.R. Dupont**, *Chemical Process Industries: Environmental Health Risk Calculations*, CRC Press/Taylor & Francis Group, Boca Raton, FL, 2023.

Part II

Air Pollution Control Equipment

As noted repeatedly in Part I, this part is the heart of the book. All the important air pollution control equipment in use today receive extensive treatment from an operation, maintenance, inspection, and optimization performance perspective. Following a brief introduction, most chapters include:

- Description of Control Device
- Design Considerations
- Inspection Procedures
- Operation
- Maintenance
- Improving Operation and Performance
- Recent Developments
- Conclusions
- References

Optimizing Air Pollution Control Equipment Performance: Operation & Maintenance, First Edition. Jay Richardson and Louis Theodore.

9

Absorbers

Gas absorption, as applied to the control of air pollution, is concerned with the removal of one or more pollutants from a contaminated gas stream by treatment with a liquid. A necessary condition is the solubility of these pollutants in the absorbing liquid. The rate of transfer of the soluble constituents from the gas to the liquid phase is determined by diffusional processes[1, 2] occurring on each side of the gas–liquid interface.

Consider, for example, the process taking place when a mixture of air and sulfur dioxide is brought into contact with water. The SO_2 is soluble in water, and those molecules that come into contact with the water surface dissolve fairly rapidly. However, the SO_2 molecules initially dispersed throughout the gas phase, can reach the water surface only by diffusing through the air, which is essentially insoluble in the water. When the SO_2 at the water surface has dissolved, it is distributed throughout the water phase by a second diffusional process. Consequently, the rate of absorption is determined by the rates of diffusion in both the gas and liquid phases.[1]

Equilibrium is another extremely important factor to be considered that affects the operation of absorption systems. The rate at which the pollutant will diffuse into an absorbent liquid will depend on the departure from the equilibrium that is maintained. The rate at which equilibrium is established is then essentially dependent on the rate of diffusion of both the pollutant through the non-absorbed gas and through the absorbing liquid. Equilibrium concepts and relationships are considered in a later section.

The rate at which the pollutant mass is transferred from one phase to another depends also on a so-called mass transfer or rate coefficient which relates the quantity of mass being transferred with the driving force. As can be expected, this transfer process ceases upon the attainment of equilibrium.

Gas absorption can also be viewed on a molecular scale as a mass transfer or diffusional operation characterized by a transfer of one substance through another. The mass transfer process may be viewed as occurring as a result of a concentration difference driving force, for example, the diffusing substance moving from a place of relatively high to one of a relatively low concentration. The rate at which this mass is transferred depends to a great extent on the diffusional characteristics of both the diffusing substance and the media involved.

9.1 Description of Control Device

The principal types of gas absorption equipment may be classified as follows:

1) Packed columns (continuous operation)
2) Plate columns (staged operation)
3) Miscellaneous

Of the three categories, the packed column is by far the most commonly used for the absorption of gaseous pollutants. The first type of unit is discussed below.

Packed columns are used for the continuous contact between a liquid and a gas. The countercurrent packed column (see Figure 9.1) is the most common type of unit encountered in gaseous pollutant control for the removal of an undesirable gas, vapor, or odor. This type of column has found widespread application in both the chemical and pollution control industries. The gas stream containing the pollutant moves upward through the packed bed against a downward-flowing, absorbing, or reacting liquid that is injected at the top of the packing. This results in the highest possible transfer/control efficiency. Since

Optimizing Air Pollution Control Equipment Performance: Operation & Maintenance, First Edition. Jay Richardson and Louis Theodore.

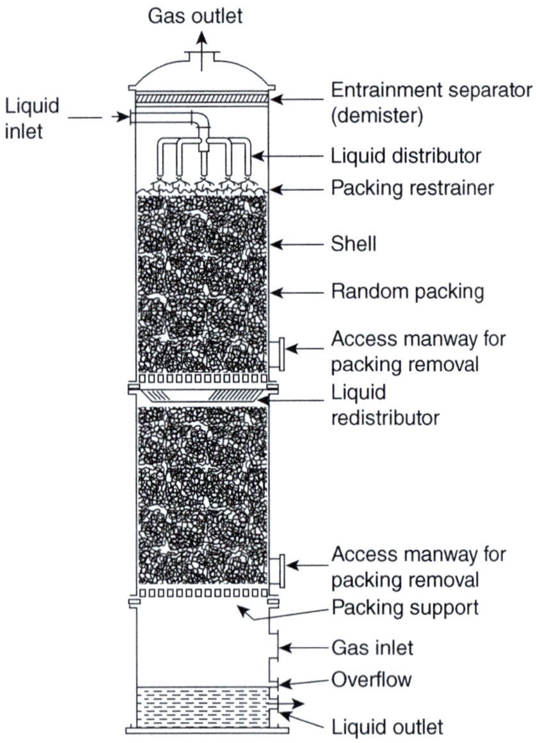

Figure 9.1 Diagram of countercurrent packed column adsorber.

the pollutant concentration in the gas stream decreases as it rises through the column, there is constantly fresher liquid available for contact. This provides a maximum average driving force for the transfer process throughout the packed bed.[1]

Liquid distribution plays an important role in the efficient operation of a packed column. A good packing from a process viewpoint can be reduced in effectiveness by poor liquid distribution across the top of its upper surface. Poor distribution reduces the effective wetted packing area and promotes liquid channeling. The final selection of the mechanism of distributing the liquid across the packing depends on the size of the column, type of packing, tendency of packing to divert the liquid to column walls, and the materials of construction. For stacked packing, the liquid usually has little tendency to cross-distribute and thus moves down the column fairly uniformly in the cross-sectional area that it enters. In the dumped condition, most flow profiles follow a conical distribution down the column, with the apex of the cone at the liquid impingement point. To produce efficient use of the packed bed for well-distributed liquid flow and reduced channeling of gas and liquid, the impingement of the liquid onto the bed must be as uniform as possible. The liquid flowing down through the packing and on the inside wall of the column should be redistributed after a bed depth of approximately 3 column diameters for Raschig rings and 5–10 column diameters for the other packing (check literature for details of packing). As a guide, Raschig rings usually have a maximum of 10–15 ft. of packing per section, while other packing can use a maximum of 12–20 ft. As a rule of thumb, however, the liquid should be redistributed every 10 ft. of packed height. The redistribution process brings the liquid off the wall and outer portions of the column and directs it toward the center area of the column. As noted earlier, redistribution is seldom necessary for stacked bed packings, as the liquid flows essentially in vertical streams.

Cross-flow-packed scrubbers are particularly successful when the process air stream requires both gas absorption and particulate removal. The cross-flow scrubber operates by allowing the gas to flow horizontally across the scrubber. The scrubbing liquid is introduced at the top of the scrubber and drains vertically downward through the packing. Contact between the gas and liquid therefore occurs at a right angle. In general, cross-flow scrubbers operate at lower pressure drops and liquid recycle flow rates than countercurrent packed-bed absorbers. Vendors claim that cross-flow scrubbers operate with a liquid rate and pressure drop of approximately 60% less than a comparable countercurrent packed tower. In addition, the cross-flow scrubber generally operates at a much higher gas-to-liquid ratio than does the countercurrent packed-bed absorber.

9.2 Design Considerations

Air pollution control equipment design has had a reputation as being a mystical art rather than theoretically predictable designed devices. As air pollution equipment has grown out of its infancy, much of the mysticism has been replaced by the acceptance of specific design parameters for sizing and predicting performance. As these design concepts have become more widely accepted, greater emphasis is now being placed on methods of reducing operating cost and improving performance. Both cost reduction and improved performance are directly related to the proper operation and maintenance of the device, two topics that receive attention later in this chapter. The material to follow in this section keys on the factors that impact the design of this unit.

Basic gas absorption principles are common to all packed scrubbers and wet collectors used for gas absorption, regardless of their operating modes. The gas to be absorbed into the liquid must be soluble or be capable of reacting chemically with the scrubbing liquid. The degree of solubility depends on the type of gas, its concentration (mole fraction or vapor pressure), the system pressure, the type of absorbing liquid, and the liquid temperature. Any liquid has a limited amount of gas which it can absorb before it becomes saturated. This is known as the aforementioned term: *equilibrium*. If the liquid reaches equilibrium within a packed scrubber, it cannot absorb additional gas. The liquid is recirculated from a reservoir (sump) to the liquid distributor in most packed scrubbers. This liquid cannot be recycled indefinitely or it will reach saturation. Equilibrium or saturation will be reached first at the gas outlet of the scrubber because it is here that the contaminant has the lowest concentration in the air stream. To prevent equilibrium from occurring and drastically reducing the removal efficiency, fresh liquid is introduced and contaminated liquid is withdrawn from the system continuously, or by introducing chemicals to react with the absorbed gaseous pollutant.

In addition to equilibrium considerations, one must consider how the gas can be absorbed into the liquid most effectively. Since it is the liquid that is absorbing the gas, a means of extending the liquid surface must be provided. This is the premise for the packed absorber itself. By providing a packing medium over which the liquid is spread, a greater area of contact can be achieved. One could interpret, therefore, that the area of gas–liquid contact is the area of the packing. If this were true, the relative efficiency of two packings would be proportional to the ratio of their areas. Yet, ½-in. Raschig rings having twice the surface area are *only* 70% more efficient than 1-in. Raschig rings. Therefore, there must be more than the area of the packing relating to performance. This additional factor, liquid surface renewal, has a greater effect with the use of thermoplastic packing than with ceramic packing.

In industrial applications, the packed scrubber (absorber) is used predominantly for the control of gaseous pollutants where efficiencies in the excess of 90–95% are required. There are three modes of operation of packed scrubbers: countercurrent, co-current, and cross-flow. Extensive detailed design calculations and accompanying illustrative examples are available in the literature.[1–3]

9.3 Installation Procedures

Most equipment is shipped by truck or rail and should be inspected immediately upon receipt for any signs of damage to the exterior and interior of the gas absorber or other equipment. Specific attention should be given to checking any protruding parts, points that come in contact with the shipping media, or areas where lifting lugs, tie-down lugs, ladder clips, or other exterior components are attached, and around all internal components. If damage is observed, it should be reported to the carrier and the supplier promptly. Any damage should be repaired before placing the equipment into service. When unloading large scrubbers and setting on foundations, one should always use proper rigging and handling procedures as recommended by the supplier. Some helpful suggestions for handling are as follows[4]:

1) Exercise care to prevent dropping, abrasion, and impact.
2) Always use padding where necessary to prevent abrasion and impact.
3) Always provide sufficient support to prevent undue deflection and distribute pressure over a large surface area.
4) Use lifting lugs, keeping pull on the lugs' centerline, in a radial direction only.
5) Always use recommended guidelines to control the absorber when moving to prevent striking objects.
6) Never place concentrated stresses at any one point.
7) Do not roll or drag the absorber over anything that will cause damage.
8) Never roll over a fitting, lug, or other component.
9) Do not lift or pull the absorber by fittings or other components except lifting lugs.

Scrubbers with lifting lugs can be lifted into a vertical position using a crane. When setting the absorber in place, make certain that the area has been swept clean. Set the unit down gently to avoid damage. Most absorbers must have full bottom support. The foundation (particularly on roofs) must have sufficient strength to support the absorber when full of liquid media without sagging or deflecting. The absorber should be well anchored by securing with all the hold-down lugs before proceeding with the piping on scrubbers with bottom drain nozzles, and a drain recess must be provided in the installation pad which does not allow the weight of the vessel to rest on the drain nozzle. After the absorber has been placed in position, it should be inspected again to ensure that no damage has been incurred during installation.

Once the absorber is positioned and anchored, the absorber internals must be installed. The packing support plates and entrainment separator support plates are often shipped separately to prevent breakage. These plates should be carefully installed through access doors and subsequently oriented properly. If the support plates were factory installed, check to ensure that they are positioned properly and not damaged. The packing material, which is normally shipped separately, can then be placed in the absorber. If thermoplastic packing is used, it can be randomly dumped into the absorber through the gas outlet or the upper access door, making sure that bridging of the packing does not occur. Do not tamp down the packing, because this will cause a higher pressure drop. After filling to the proper height, the packing should be leveled. This can be easily done by hand or with a long-handled tool such as a rake. If ceramic packing is used, the tower should first be filled with water and then the packing material carefully and slowly added to prevent breakage. After filling to the proper packing height, the water should be emptied from the absorber and the packing leveled.

The liquid distributor can now be installed through the access provided. If a weir-type distributor is provided, it will normally be shipped loose and must be placed in the absorber on the internal supports provided. These weir boxes must be placed apart at specified distances and must be leveled to prevent maldistribution of the liquid. If spray-type distributors are provided, these are normally assembled and installed in the scrubber prior to shipment. Often, the spray nozzles are shipped loose and must be installed in the field. If the spray nozzles are not installed, it is recommended that the piping and fittings be inspected to ensure they are clean and are at least hand-tight. If possible, the lines should be blown out with air. Install the spray nozzles by screwing them in hand-tight. Then, using a strap wrench, tighten them one or two turns more. On thermoplastic threads, it is suggested that a single layer of Teflon ribbon thread seal be used. Some spray distributors are designed so that they can be completely removed from the scrubber. If this type is provided, care must be taken when reinstalling the spray distributors so that the spray nozzles are pointed in the proper direction. The entrainment separator should now be installed above the liquid distributor as recommended by the equipment manufacturer. Once the absorber internals are properly installed, the recycle piping (if required) and all piping and instrument connections to the absorber can be made. All connections should be properly aligned and supported to prevent abnormal stresses on the shell nozzles.

The overflow and drain connection should be barometrically sealed or a U-trap provided to allow the liquid to flow out properly through these connections. Inlet and outlet duct connections should be made, taking care not to force the flanges to match up with the absorber. If a fan or blower is connected to the absorber or duct, it is suggested that a flexible rubber joint be used to prevent vibrations from affecting the scrubber's operation.

9.3.1 Pre-Startup Check

After installation is completed, it is recommended that each component of the system be checked prior to the actual startup of the system. Procedures include:

1) The scrubber sump should be filled with fresh water and then drained to remove any dirt or foreign material.
2) Remove all foreign matter from the duct system and fan housing.
3) Turn the fan impeller by hand to assure that it runs freely and does not contact the fan inlet or housing.
4) Check the fan belt tension for tightness and adjust if necessary.
5) Bump the fan start switch to ensure proper rotation of the fan impeller.
6) Check the recycle pump and motor alignment.
7) If the pump is provided with a packed stuffing box, check to ensure that the packing and seal cage have been installed.
8) Turn the pump shaft by hand to be sure that all rotating parts run freely.
9) Disconnect the pump coupling and bump the pump start switch to ensure proper rotation of the impeller. Some pumps will be severely damaged if operating in the wrong rotation. After proper rotation is achieved, reconnect the coupling.
10) Check to ensure that oil has been added for lubrication of the pump bearings.
11) If a mechanical seal rather than a packed stuffing box is provided on the pump, it may require lubrication or continuous freshwater flushing. If so, adjust the water rate as recommended by the pump and seal manufacturer.
12) Close the absorber drain valve and fill the sump with fresh water until it begins to flow out of the overflow connection.
13) If required, adjust the continuous water-makeup rate to the absorber as recommended by the manufacturer.
14) Open all valves in the recycle piping and start the recycle pump. The liquid level in the sump will fall slightly due to the liquid being held up in the packing and piping. This is normal, and the liquid level will again rise to the overflow position after a few minutes of operation.
15) Set the recycle liquid flow rate by using the throttling valve in the recycle piping.
16) Check for leaks in the recycle piping and tighten flanges where leaks occur.
17) Visually observe the spray pattern or liquid distribution in the absorber to make certain that the liquid is evenly distributed. Adjust if necessary to achieve even distribution.

18) Close all absorber access doors, and open all dampers in the system except for the inlet damper on the fan suction, which should be closed. Start the fan and slowly open the inlet damper to the design flow rate.
19) The system volumetric flow rate should be adjusted to its designed capacity. This can be accomplished by adjusting the dampers in the system.
20) Check the fan for excessive noise or vibration, which may indicate misalignment. This should be adjusted if required.
21) Check all instrumentation and record all system functions under these clean conditions.
22) After allowing the recycle pump and fan to operate for approximately 2 hours, shut down the system and drain the sump.
23) Recheck the spray nozzles and strainers for plugging, and clean if required.

9.3.2 Startup

Once the above pre-startup check has been completed, a routine startup of the system can be initiated.

1) Close the sump drain valve and fill the sump with water to the overflow position.
2) Open the makeup water valve, noting the rate of makeup has been preset during the initial system check.
3) Start the pump; the throttling valve has already been preset for the proper flow rate. Allow the pump to run for two to three minutes, then start the fan. Always start the pump before initiating the fan. On system shutdown, always stop the fan before the pump.
4) Initiate the chemical feed system, making sure that the chemical feed tank is filled.
5) Check all liquid flow rates, gas flow rates, pressures, and the pressure drop across the absorber.

9.4 Operation

After startup, the system's liquid and gas flow rates should again be checked to ensure that they are operating within the design parameters of the system. After approximately two weeks of operation, the system should be shut down and inspected for any possible nozzle pluggage or settling of the packing which frequently occurs during the first two weeks of operation. If the packing has settled, the packing should be topped off to the appropriate design level. During normal operation, daily checks should be made of the recycle liquid flow to the liquid distributor and the liquid makeup rate. Also, if a chemical feed system is employed, daily checks of the chemical solution should be made to ensure an adequate supply of chemical. Daily logs should also be kept of any instrumentation readings, such as pressure indicators, temperature indicators, flow indicators, and all other operations pertinent to the specific system. Should any drastic variance or excursion(s) be noted in these readings an investigation should be conducted to indicate the cause of these abnormal readings. The following list is a guide for any abnormalities that may be encountered during the operation of the equipment.[4]

1) If the static pressure drop across the scrubber continually increases over a long period of time, this could indicate one of the following:
   a) The liquid flow rate to the liquid distributor has increased and should be checked.
   b) The packing in the irrigated bed could be partially plugged due to solids deposition and may require cleaning.
   c) The entrainment separator could be partially plugged and may require cleaning.
   d) The packing support plate at the bottom of the packed section could be blinding and causing increased pressure drop, and will require cleaning.
   e) The packing could be settling due to corrosion or to solids deposition, again requiring cleaning or additional packing.
   f) The airflow rate through the absorber could have been increased by a change in the damper setting, which may need adjustment.
2) The pressure drop across the absorber begins to decrease either slowly or rapidly. This could be caused by the following possibilities:
   a) The liquid flow rate to the distributor has decreased and should be adjusted accordingly.
   b) The airflow rate to the scrubber has decreased due to a change in the fan characteristics or damper settings.
   c) Partial plugging of the spray or liquid distributor, causing channeling through the scrubber, could be occurring. The liquid distributor should be inspected to ensure that it is totally operable.
   d) The packing support plate could have been damaged and fallen into the bottom of the absorber, allowing the packing to fall to the bottom and produce a lower pressure drop. This should be checked.

3) A pressure or flow change in the recycle liquid, usually indicating a lower liquid flow rate. This may be caused by the following:
   a) A plugged strainer or filter in the recycle piping, which may require cleaning.
   b) Plugged spray nozzles, which may require cleaning.
   c) The piping may be becoming partially plugged with solids and need cleaning.
   d) The liquid level in the sump could have decreased, causing pump cavitation.
   e) The pump impeller could have been worn excessively.
   f) A valve in either the suction or discharge side of the pump could have been inadvertently closed.
4) A high liquid flow indicates the following:
   a) A break in the internal distributor piping.
   b) A spray nozzle that has been inadvertently "uninstalled."
   c) A spray nozzle that may have come loose or eroded away, creating a low-pressure drop.
   d) A change in the throttling valve setting on the discharge side of the pump, allowing larger liquid flow. Reset to the proper conditions.
5) Excessive entrainment carryover from the outlet of the absorber. This could be caused by the following:
   a) A partially plugged entrainment separator, causing channeling and re-entrainment of the collected liquid droplets.
   b) The airflow rate to the absorber could have increased above the design capability, causing re-entrainment.
   c) If a packed-type entrainment separator is used, the packing may not be level, causing channeling and re-entrainment of moisture.
   d) If a packed entrainment separator is used, and a sudden surge of air through the absorber occurred, this could have caused the packing to be carried out of the absorber or the packing being blown aside, creating an open area "hole" through which the air passes.
   e) The entrainment support plate could be damaged and have dropped, causing channeling through the separator.
   f) The velocity through the absorber has decreased to a point that absorption does not effectively take place, producing low removal efficiencies.
   g) Conceivably during wintertime operation, some moisture due to condensation may be observed and considered a problem with the entrainment separator. However, it should be recognized that this is not uncommon, since the air discharging off of the absorber is saturated with water vapor, and any difference or lowering of the temperature will cause condensation to occur.
6) If a reading indicates a low airflow or no airflow across the absorber, the following may be the cause:
   a) The packing in the absorber may be plugged, causing a restriction to the airflow.
   b) The liquid flow rate to the absorber could have been increased inadvertently, again causing greater restriction and pressure drop, creating a lower gas-flow rate.
   c) The fan belts have worn or loosened, reducing the airflow to the equipment.
   d) The fan impeller could be partially corroded, reducing fan efficiency.
   e) The ductwork to or from the absorber could be partially plugged with solids and may need cleaning.
   f) A damper in the system has been inadvertently closed or the setting changed.
   g) A break or leak in the duct could have occurred due to corrosion.
7) Should the airflow through the absorber be increased or has suddenly increased, it may be due to the following:
   a) A sudden opening of a damper in the system.
   b) Low liquid flow rate to the absorber.
   c) The packing has suddenly been damaged and has fallen to the bottom of the absorber.
8) A sudden decrease in the absorber efficiency could indicate the following:
   a) The liquid makeup rate of the absorber has been inadvertently shut off or throttled to a low level, decreasing the absorber efficiency.
   b) If a chemical feed system is employed, the system may have run out of the chemical required.
      i) This could indicate a malfunction of pH probes (if employed), requiring replacement.
   c) The set point of the pH control may have to be adjusted to allow more chemical feed.
   d) The set point on the pH control may have to be adjusted to allow more chemical feed.
   e) A problem may exist with the chemical metering pump, control valve, or line pluggage.
   f) The liquid flow rate to the absorber may be too low for effective removal.
   g) Pluggage or solids buildup in the absorber liquid distributor or packing may have caused channeling.[5]

The foregoing list indicates only some of the common operational problems encountered with absorber systems. There are a number of related problems. Excessive moisture may be blown from either the fan or the blower if it is on the suction side of the system. Excessive moisture droplets being emitted could be caused by condensation in the stack or in the fan. Improper condensate drainage and trapping or line pluggage from the fan could be the problem. Scrubber liquid levels can also be too high. This can occur by having too large an inlet continuous freshwater makeup with the overflow piping being too restrictive, or with the overflow and drain piping trap, or barometric seal being broken.

Generally, an absorber system is overdesigned for the capacity it is to handle. On occasion, lower volumetric flow rates are introduced through the system below the design condition. These lower volumetric flow rates usually increase gas absorption. However, if too low a gas input is used, creating a very low static pressure drop through the absorber, channeling can occur and can lower the absorber's performance and entrainment separator's performance. Caution should be taken to use the manufacturer's minimum gas-flow-rate recommendation to prevent this problem from occurring.[4]

9.5 Maintenance

Normal preventative maintenance requires only periodic checks of the fan, pumps, chemical feed system, piping, duct, and absorber liquid distributor. The normally irrigated packing may never have to be cleaned during the life of the absorber as long as the absorber has been properly designed and operated. However, should the absorber run dry or the entering air stream contain unexpectedly high solids loadings, a heavy formation of solids, crystallized salts, or other foreign matter may accumulate on the packing, and this *must* be removed. In most cases, removal can be accomplished by recirculating, for a short period of time, a chemical solution into which the solids will dissolve or react. Sometimes, chemical removal will not remove the buildup and it may be necessary to use high-pressure water, hot water, or atmospheric steam. Prior to using any chemical, hot water, or steam, the absorber manufacturer should be consulted to verify the resistance of the internals and shell to the cleaning medium. In very few instances, such as CaF_2 deposition or extremely heavy solids deposition, it may be necessary to either remove the packing medium from the absorber for cleaning or replace the packing.

The normally unirrigated entrainment separator must be periodically flushed with sprays to prevent buildup and eventual plugging. The intervals between routine washing must be determined by experience, as the collection of solid materials is a function of the specific operation conditions. The following is a suggested maintenance checklist to ensure proper operation and unexpected problems with the absorber system.

1) Pump maintenance
   a) Check the bearing lubrication to make sure that the proper amount of oil is present. (weekly)
   b) If a packing stuffing box is provided, make sure that the drip rate is not excessive or that the packing is lubricated. (weekly)
   c) Check pump and motor bearings for unusual noise or heat. If this is noticed, it could indicate excessive wear on the bearings, which may result in excessive shaft runout, requiring frequent repacking of the stuffing box. (weekly)
   d) Inspect gasketing at the suction and discharge connections for leaks. (weekly)
   e) Check shaft alignment and levelness of the baseplate. (monthly)
   f) Look for crystallization of solution or solids in the packing, which could cause scouring of the shaft. (monthly)
   g) Check for water flush to the mechanical seal (if provided), and adjust if necessary. (weekly)
2) Fan maintenance
   a) Check the fan motor and fan bearings for unusual vibration noise or heat. (weekly)
   b) Check the bearing lubrication to ensure that the proper amount of oil is present. (weekly)
   c) Inspect the belts for wear and replace them if necessary. Also, check the belts for proper tension. Adjust if necessary. (monthly)
   d) Check all fan bolts for looseness and tighten if required. (monthly)
   e) Inspect the fan impeller and blades for any solids' buildup or erosion. Clean or replace the impeller if excessive buildup or erosion has occurred. Such buildup can unbalance the impeller and cause premature failure. (monthly)
   f) Check the fan housing drain to make sure it is not plugged so that it will permit proper condensate drainage. (monthly)

3) Absorber maintenance
   a) Check the pressure drop across the absorber. Increased pressure drop will usually indicate that plugging of the packing is occurring. (weekly)
   b) Open the absorber access door provided for inspection of the liquid distributor (access doors should never be opened while the fan is in operation). Observe the liquid distribution. Clean if required. (monthly to quarterly)
   c) Inspect the absorber sump for solids buildup and flush if necessary. (semi-annually)
   d) During the inspection of the liquid distributor, check the packing level for possible settling. Add packing to the proper level if necessary. (quarterly to semi-annually)
   e) Check the entrainment separator for solids or crystallization buildup and clean with sprays if required. (quarterly to semi-annually)
   f) Inspect the absorber packing for solids buildup and clean if required. (semi-annually)
   g) Open all access openings and inspect the tower internals for corrosion or breakage. (semi-annually)
   h) During the inspection of the tower internals, also inspect the absorber shell for any corrosion and repair if necessary. (semi-annually)
4) System maintenance
   a) Check and clean all strainers or filters in the piping system. (weekly to monthly)
   b) Inspect the recycle piping, duct and absorber flanged connections, access doors, and flexible connections for leaks, and tighten or replace gaskets if necessary. (monthly to quarterly)
   c) Check all system bypass or control dampers and cycle each one a few times to ensure proper operation. (weekly)
   d) Check all piping to and from the absorber, including the recycle piping, for erosion or plugging and clean or replace if required. (quarterly)
   e) If a chemical feed system is employed, check the metering pump or valves and the pH probes for proper operation. It is not unusual to replace the pH probe elements frequently if solids buildup occurs. (weekly)
   f) Inspect the duct for solids buildup due to settling. This may require periodic flushing to prevent excessive restriction to airflow. (semi-annually)
   g) Inspect the duct, fan, and absorber for any external corrosive attack or obvious deterioration. (semi-annually)
   h) Inspect structural supports, duct, and pipe hangers for any deterioration due to corrosion. (semi-annually)
   i) Frequent checks of any heat tracing or freeze protection during the winter season, if applicable, should be conducted. (weekly)
   j) All instrumentation should be inspected, checked, and calibrated to ensure proper function and control. (monthly)

The reader should note the periodic time intervals indicated for maintenance will vary depending on the specific system's operating conditions, the equipment manufacturer's recommendation, and the regular outages of the plant.

9.6 Improving Operation and Performance

The basic design parameters associated with absorbers have been discussed, primarily packed absorbers, and the installation, operation, and maintenance of this equipment. Much of the information presented is also applicable to other absorber types, such as spray towers, baffle towers, tray towers, and wet cyclones.

There are options available that should be considered when purchasing a packed absorber or any absorber to reduce maintenance costs and allow the equipment to be more operationally reliable between scheduled maintenance shutdowns. Any options generally increase the initial cost of the equipment, and for this reason, are often not considered. However, some of these items can pay for themselves in a short time by reducing maintenance cost. One such item is the provision of adequate access openings to all internal portions of the scrubber. The areas that should be considered are the sump section of the absorber; the lower packed section for removal of the packing and the internal packing support plate, if required; the liquid distributor for spray nozzle cleaning or removal of the internal liquid distributor, and access to the entrainment separator for inspection and removal of the packing and support plates, if necessary. The access openings should be large enough for accessibility and total removal of the internal components of the absorber. To reduce the initial cost of equipment, some manufacturers provide only two or three access doors. In many cases, it may be advantageous to consider clear, see-through access openings, especially in the liquid distributor region, for external observation of the absorber of the liquid distributor to ensure its proper operation without equipment shutdown.

The liquid distributor, especially spray-type distributors, is one area where maintenance is required on a periodic basis. Many manufacturers provide these as an integral part of the absorber shell construction where spray-type distributors are used. Should the spray nozzles need cleaning or removal with this type of spray distributor, a worker must physically enter the absorber for removal of the spray nozzles for cleaning. This can be a continual source of irritation. Since most absorbers are handling corrosive-type gases, the absorber must be shut down and totally purged with air to allow entry. In addition, in most plants, it is necessary that a worker wear a rubber suit and mask for safety when entering a vessel. The need for entry into the absorber for maintenance of a spray-type distributor can be totally eliminated by designing the distributor so that it can be totally removed from the absorber.

The entrainment separator is another unit requiring continuous maintenance. Because it is generally an unirrigated section, solids and/or salt deposition, which is not uncommon, needs to be eliminated periodically by washing. To clean the entrainment separator, it is necessary to shut down the absorber, open the equipment, and physically flush out the separator. Equipment shutdown for this maintenance can be eliminated by installation of a spray wash-down flush header above the entrainment separator which can provide intermittent flushing of the separator media. This eliminates equipment shutdown and maintenance cost. If this is provided, it should be recognized that if the fan is operating, the liquid will be carried over through the outlet of the system during the flush cycle. To prevent this, the fan should be shut down during this flushing cycle.

The absorber can be designed to provide future additional packing height requirements should greater future gas absorption capability be required. This can be accomplished in two ways: the shell of the absorber can be flanged to allow a future stub section to be inserted, or the tower can be initially designed with a void space above the packing between the liquid distributors so that future packing heights can be added without shell modification. Although the initial cost of this may seem high, it could preclude the need to field-modify the absorber or to purchase a new absorber for greater efficiency, if any air pollution laws are tightened.

Another source of periodic maintenance relates to the strainers or filters in the recycle piping, and the recycle pumps. To prevent total system shutdown during cleaning or maintenance of these items, dual strainers or pumps may be considered, valved separately, so that the system can be maintained in a fully operational mode during strainer cleaning or pump maintenance checks.

As discussed earlier in the chapter, instrumentation used for the purpose of providing operation data is useful in determining areas where problems or maintenance may be required. The type of instrumentation will depend totally on the system requirements. Areas of specific interest would be a pressure drop across the absorber, both the wetted bed and entrainment separator; low liquid recycle flow and/or pressure; and, liquid makeup rates to the system.

9.7 Recent Developments

The basic design of absorbers has remained relatively unchanged since first used in the early part of the twentieth century. Some modest equipment changes, new types of devices, and unique packing have appeared in the last 20 years, but all have essentially employed the same capture mechanisms used in the past. One area that has recently received some attention is hybrid systems – equipment that in some cases operate at higher efficiency and more economically than conventional devices. Tighter regulations and a greater concern for environmental control by society have placed increased emphasis on the development and application of these systems. Hybrid systems are defined as those types of control devices that involve combinations of control mechanisms – for example, fabric filtration combined with electrostatic precipitators, ionizing wet scrubbers, and dry scrubbers.

The advantages of cross-flow scrubbers continue to be of interest. The cross-flow mode has some attractive features over the countercurrent design while affording lower pressure drop. Lower pressure drop results in the cross-flow unit where the liquid is at right angles to the airflow, resulting in less pressure drop when compared to the countercurrent mode, where the gas and liquid flow in opposite directions.

Of the various types of gas absorption devices mentioned, packed columns, and plate columns are the most commonly used industrial practice. Although packed columns are used more often in air pollution control, both have their special areas of usefulness, and relative advantages and disadvantages of each are worth considering.[5]

9.8 Conclusions

This chapter presents a practical rather than theoretical approach to absorber operation and maintenance. It should be recognized that an absorber system used for air pollution control is an operating system, as is any production operating equipment producing a product. High maintenance costs associated with absorber systems are often due to misapplication of the equipment for a specific purpose. Air pollution control systems and absorber systems are considered nuisance areas required to prevent air pollution emissions into the atmosphere. They are nonproductive items required by the manufacturer of an item. It is an added cost to produce this item. It is for this reason that maintenance is often neglected on air pollution equipment in preference to operating equipment producing a product. If the air pollution control system is considered as a process system needed to produce a product and preventative scheduled maintenance is provided, in the long run, operating and replacement costs will be lowered.

Air pollution control regulatory agencies are becoming more cognizant of the fact that preventative maintenance on air pollution control equipment has been neglected. Because of this, many states are requiring continuous stack emissions monitoring to prevent excessive gaseous pollutants from being emitted to the atmosphere due to the absorber or air pollution system becoming ineffective due to lack of maintenance. With the greater emphasis being placed on preventative maintenance of absorbers, it is hoped that the information presented in this chapter will be helpful to the user in obtaining a better understanding of absorber operation and maintenance for greater equipment reliability and lower operating costs.

References

1 **L. Theodore**, *Air Pollution Control Equipment Calculations*, John Wiley & Sons, Hoboken, NJ, 2008.
2 **L. Theodore**, *Chemical Engineering: The Essential Reference*, McGraw Hill, New York, NY, 2012.
3 **L. Theodore**, *Engineering Calculations Sizing Packed Tower Absorbers Without Data*, pp. 18–19, CEP, 2005.
4 **Celicote Co., Inc.**, *Installation, Operation, and Maintenance Manual*, Berea, OH, 1975.
5 **L. Theodore, and F. Ricci**, *Mass Transfer Operations for the Practicing Engineer*, John Wiley & Sons, Hoboken, NJ, 2010.

10

Adsorbers

During adsorption, one or more gaseous components are removed from an effluent gas stream by adhering to the surface of a solid. The gas molecules being removed are referred to as the *adsorbate*, while the solid doing the absorbing is called the *absorbent*. Adsorbents are highly porous particles and adsorption occurs primarily on the internal surface of the particles.

The attractive forces that hold the gas to the surface of the solid are the same that cause vapors to condense (van der Waals forces). All gas–solid interfaces exhibit this attraction, some more than others. Adsorption systems use materials that are highly attracted to each other to separate these gases from the nonadsorbing components of an air stream. For air pollution control purposes, adsorption is occasionally not a final control process. The containment gas is merely stored on the surface of the adsorbent. After it becomes saturated with adsorbate, the adsorbent must either be disposed of and replaced, or the vapors must be desorbed. It is important to note that desorbed vapors are highly concentrated and may be recovered more easily and more economically than before the adsorption step.

The adsorption process is classified as either physical or chemical. The basic difference between physical and chemical adsorption is the manner in which the gas molecule is bonded to the adsorbent. In physical adsorption, the gas molecule is bonded to the solid surface by weak forces of intermolecular cohesion, i.e. van der Wall forces. The chemical nature of the adsorbed gas remains unchanged; therefore, physical adsorption is a readily reversible process. In chemical adsorption, a much stronger bond is formed between the gas molecule and the adsorbent. A sharing or exchange of electrons takes place – as happens in a chemical bond. Chemical adsorption is not easily reversible; in fact, it is usually very difficult.

The forces active in physical adsorption are essentially electrostatic in nature. These forces are present in all states of matter: gas, liquid, and solid. They are the same forces of attraction that cause gases to condense and real gases to deviate from ideal behavior. Physical adsorption is sometimes referred to as the aforementioned van der Waals adsorption. The electrostatic effect that produces the van der Waals forces depends on the polarity of both the gas and solid molecules. Molecules in any state are either polar or nonpolar, depending on their chemical structure. Polar substances are those that exhibit a separation of positive and negative charges within the compound. This separation of positive and negative charges is referred to as a *permanent dipole*. Water is a prime example of a polar substance. Nonpolar substances have both their positive and negative charges in one center, so they have no permanent charge. Most organic compounds, because of their symmetry, are nonpolar (1,2).

Chemical adsorption, or chemisorption, results from the chemical interaction between a gas and a solid. The gas is helped to the surface of the adsorbate by the formation of a chemical bond. Adsorbents used in chemisorption can be either pure substances or chemicals deposited on a carrier material. One example is using pure iron oxide chips to adsorb H_2S. Another example is using activated carbon which has been impregnated with sulfur to remove mercury vapor.

10.1 Description of Control Device

Although it is probable that all solids adsorb gases to some extent, adsorption as a rule is not very pronounced unless an adsorbent possesses a large surface area for a given mass. For this reason, such adsorbents as charcoals, activated alumina, silica gel, and molecular sieves are particularly effective as adsorbing agents. These substances have a very porous structure

Optimizing Air Pollution Control Equipment Performance: Operation & Maintenance, First Edition. Jay Richardson and Louis Theodore.

and with their large, exposed surfaces can adsorb appreciable volumes of various gases. The first three of the above adsorbents are amorphous in nature with a nonuniform internal structure. Molecular sieves, however, are crystalline and possess an internal structure of regularly spaced cavities with interconnecting pores of definite size. Of the above adsorbents, activated carbon is the primary nonpolar adsorbent. It is possible to manufacture other adsorbing materials having nonpolar surfaces, but since their surface areas are much less than that of activated carbon, they are not used commercially. Polar adsorbents will preferentially adsorb any water vapor that may be present in a gas stream. Since moisture is present in most pollutant airstreams, the use of polar adsorbents is severely limited for an air pollution application.

Activated carbon can be produced from a variety of feedstocks such as wood, coal, coconut, nutshells, and petroleum-based products. The activation process takes place in two steps. First, the feedstock is carbonized. Carbonization involves heating the material (usually in the absence of air) to a temperature high enough (600°C) to drive off all volatile material. Thus, carbon is essentially all that is left. To increase the surface area the carbon is then "activated" by using steam, air, or carbon dioxide at higher temperatures. These gases attack the carbon and increase the pore structure. The temperatures involved, the amount of oxygen present, and the type of feedstock all greatly affect the adsorption qualities of the carbon. Manufacturers vary these parameters to produce activated carbon suitable for specific purposes. In the sales literature, the activity and retentivity of carbons are based on their ability to adsorb a standard solvent such as carbon tetrachloride (CCL_4).

Because of its nonpolar surface, activated carbon is used to control the emission of organic solvents, odors, toxic gases, and gasoline vapors. Carbons used in gas phase adsorption systems are manufactured in granular form, usually ranging from 4×6 to 4×20 mesh in size. (A 4×6 mesh is one that will pass the carbon through a 4-wire-per-inch Tyler mesh screen, but will be captured by a 6-wire-per-inch screen.) The bulk density of the packed bed can range from $0.5\,g/cm^3$ to $0.8\,g/cm^3$ (from $30\,lb/ft^3$ to $50\,lb/ft^3$) depending on the internal porosity of the carbon.[1–6]

Activated alumina (hydrated aluminum oxide) is produced by special heat treatment which is precipitated from native aluminas or bauxite. It is available in either granule or pellet form and is used principally for drying wet gas steam.

The manufacture of silica gel consists of the neutralization of sodium silicate by mixing with dilute mineral acid, washing the gel formed to rid it of salts produced during the neutralization reaction, followed by drying, roasting, and grading processes. The name "gel" arises from the jellylike form of the material present during one stage of its production. It is generally used in granular form, although bead forms are also available. Silica gel also finds primary use in gas drying, although it has other applications (for example, gas desulfurization and purification).

Molecular sieves are dehydrated zeolites (i.e., aluminosilicates) in which atoms are arranged in a definite pattern. The fundamental building block of the molecular sieve is a tetrahedron of four oxygen anions surrounding a smaller silicon or aluminum cation. The cations serve to make up the positive charge deficit in the alumina tetrahedra. Each of the four oxygen anions is shared with another silica or alumina tetrahedron to extend the crystal lattice in three dimensions. The resulting crystal is unusual in that it is honeycombed with relatively large cavities, each cavity connected with six adjacent ones through apertures or pores. These pores serve as the adsorption surface within a molecular sieve.[1] Molecular sieves will effectively remove acetylene from olefins, and ethylene or propylene from sautéed hydrocarbons. Some of the basic types and pores of molecular sieves are available in the literature.[1] Although these adsorbents are the most commonly used, others are available. These adsorbents and others are listed with their applications in Table 10.1.

10.2 Design Considerations

After determining the particular adsorbent necessary to control the emissions, an applicable adsorber system can be selected. Two major categories of adsorbers that are presently manufactured for industrial and commercial use are air purification devices (nonregenerative) and solvent recovery devices (regenerative).

Air purification applications are effective where the pollutants are emitted at very low concentrations (below 100 ppm) but need to be controlled because of their highly malodorous or toxic nature. These systems are cost-effective only at low concentrations because of their nonregenerative design. In solvent recovery applications, activated carbon may be used to economically recover and reclaim solvents from process exhaust streams at concentrations in the range of 100–1,000 ppm. Activated-carbon regenerative systems, whether for pollution control or solvent recovery, are designated as carbon-resorb systems. The air pollution control device that is required for a particular application will depend upon the concentration range of the volatile organic compound to be controlled.[3]

The basic function of control devices applicable in the low-concentration range is to purify both recirculated indoor air and ventilation air brought in from outdoors so that it will be odorless and safe for breathing. The control devices used for

Table 10.1 Common adsorbents and their applications.

Adsorbent	Application
Activated carbon	Solvent recovery, elimination of odors, purification of gases
Alumina	Drying of gases, air, and liquids
Bauxite	Treatment of petroleum fractions; drying of gases and liquids
Bone char	Decolorizing of sugar solutions
Decolorizing carbons	Decolorizing of oils, fats, and waxes; deodorizing of domestic water
Fuller's earth	Refining of lube oils and vegetable and animal oils, fats, and waxes
Magnesia	Treatment of gasoline and solvents; removal of metallic impurities from caustic solutions
Molecular sieves	Selective removal of contaminants from hydrocarbons
Silica gel	Drying and purification of gases
Strontium sulfate	Removal of iron from caustic solutions

this purpose are designed as air purification systems. Air purification is generally carried out in a closed system, i.e., the carbon-purified air is recirculated into the occupied spaces for reuse by the occupants. There are several configurations of air purification devices, such as radial flow canisters, panels, and pleated carbon beds; all are designed to treat contaminated air in low concentrations.

To attain the most effective and economical service from air purification systems utilizing carbon adsorbent, various parameters need to be optimized. These parameters are flow rate, velocity, carbon mesh size, and bed thickness. They determine the resistance to flow, the adsorptive capacity, and the carbon bed area. Another important factor is the determination of the optimum service time at which the carbon bed should be replaced. Pollutant concentration is also an important factor because of its effect on capacity and service time.

The equipment necessary to constitute a complete adsorption system must be designed and sized adequately to achieve optimum emissions reduction efficiencies and maximum solvent reclamation and recovery. The sizing and overall design of the adsorption system depend on the properties and characteristics of both the inlet gas stream and the adsorbent. Design of stationary-bed activated-carbon systems must include consideration of the following operating parameters.

1) Gas stream
   a) Adsorbate concentration
   b) Temperature
   c) Temperature rises during adsorption
   d) Pressure
   e) Flow rate
   f) Presence of adsorbent contaminant material
   g) Gas density at operation temperature and pressure
   h) Gas viscosity at operating temperature and pressure
2) Adsorbent
   a) Adsorption capacity as used on-stream
   b) Temperature rise during adsorption
   c) Isothermal or adiabatic operation
   d) Adsorbent life in the presence of particulate matter
   e) Possibility of catalytic effect causing adverse chemical reaction in the gas stream and of formation of solid polymerizates on the adsorbent bed, with consequent deterioration.

The adsorbent is generally used in a fixed bed configuration in which the contaminated air is passed. Depending on the concentration and market value, the contaminant is either recovered or discarded when the loaded adsorbent requires regeneration. The evaluation of the dynamic capacity of an adsorbent bed is essential when determining adsorption rates.

10.3 Installation Procedures

After the control equipment design has been completed, the installation becomes an important factor in the future operation and maintenance procedures of the air pollution control system. Improper installation may contribute to excessive energy consumption and unnecessary losses of volatile organic materials to the environment and to the workplace. This may result in deterioration of the air resource and the water quality and may present excessive employee exposure to toxic and hazardous chemicals. Because of these consequences, everything possible must be done to ensure correct operation and ease of maintenance of all air pollution control equipment.

10.3.1 Location

Whether the adsorber is a nonregenerative odor-control device or a multiple-bed carbon adsorption system, it is desirable to locate the control equipment as close as possible to the emissions source. For optimum results, most carbon adsorber manufacturers recommend duct runs less than 100 ft. It is also important to select a location that will provide accessibility for the observation, operation, and maintenance of the equipment.

Regeneration adsorption systems designed for the reclamation and recovery of solvents have precise operation requirements and should be located in a restricted area. Security of the operation must be provided. The security may be as elaborate as the total enclosure of the equipment or as simple as a handrail with restricted-entry signs.

If the equipment, especially steam-stripping adsorbers, are to be located in an unheated area (outdoors or on the roof) where temperatures may be below freezing, adequate insulation must be provided on all water, steam condensate, and compressed-air piping and reservoirs.

Another important aspect of carbon adsorption system installation, and one that is often overlooked, is the preparation of the floor. To improve cleanup operations, bare concrete floors should be sealed and coated with a material such as epoxy paint that is impervious to the particular solvent being used in the process. The use of floor tile beneath the equipment is usually not suitable because of the ability of many organic solvents to penetrate the tile and dissolve the adhesives.

10.3.2 Noise Abatement

Most of the objectionable noise from a carbon adsorber originates with the air-handling portion of the system. The fan seldom generates enough noise to damage hearing; however, the location of the fan might prove to be objectionable to adjacent employees and/or neighbors. Exhaust ducts will carry sound from the system into adjacent workplaces and may produce uncomfortable working conditions in a remote area. Nuisance levels of noise may be projected toward neighboring residences by rooftop fans.[7]

A loud, high-pressure fan may be located in an area of already high noise levels within a factory and may not be a significant contributor to the resultant sound pressure levels. However, if a fan is located in a relatively quiet area or connected by ductwork into quiet areas, then even low-sound-level fan installations may be objectionable.

In fan installations where resultant sound levels are excessive or bothersome, acoustical treatment of the facility may be appropriate. To minimize noise levels, the fan and motor assembly should be mounted on isolation materials such as springs or rubber-in-shear pads. This will produce the added benefit of increased bearing life since the vibrational energy is absorbed in the springs and rubber rather than in the bearings. Isolation of the blower housing from the inlet and outlet ducts with flexible connections will further reduce noise transmission through the ductwork into adjacent areas. The application of adsorption materials along the length of adjacent ductwork, especially around transitions and elbows, will significantly reduce noise losses from the system. Some installations may require a complete acoustic enclosure around the blower assemblies in order to reduce radiated sounds from the fan. It is important to remember when dealing with sound enclosures that sound travels in a straight line from the source and will travel through even the smallest opening. Therefore, the enclosure must be sealed to eliminate noise leakage. Auxiliary cooling of the motor may be required in these cases and can be accomplished by running chilled water, available on the adsorber condenser, through finned coils within the enclosure to carry away the heat. The amount of cooling required depends upon the size and efficiency of the blower.[8]

10.3.3 Exhaust Duct Installation

Sound exhaust system design must be followed by good installation practices. The purpose of the duct, of course, is to convey the materials from the emission source to the air pollution control process. The duct velocity must be adequate to convey particulate matter without settling out in an unwanted location within the ductwork while minimizing airflow to reduce unnecessary evaporation of volatile materials at the source. If evaporated volatile solvents enter the duct airstream at a warm temperature and subsequently cool while traveling to the adsorber, condensation of the solvents may occur. Care must be taken to eliminate traps in the exhaust system where moisture and solvents may condense and collect. Moisture condensation is especially troublesome on the discharge side of the adsorber in processes using the steam-strip cycle for regeneration. While the carbon beds are drying, moisture is driven out of the beds and condensation will occur in the outlet stack. All trap areas in the stack must be drained periodically to remove water from the duct; otherwise, premature rusting of ductwork can occur.

Particulate matter that is entrained in the airstream must be removed prior to entry into the adsorption beds. A suitable air filter must be installed in the airstream to remove this matter or severe plugging of the carbon will occur.

Clean out doors should be provided throughout the duct system for ease of maintenance on internal exhaust duct cleaning. On multiple-emissions-source exhaust systems, blast gate dampers should be installed for balancing airflows in ducts as operating conditions change. It is extremely important to eliminate leaks, especially on the pressure side of the fan. The exhaust duct connections and seams should be either welded or soldered.

When the outlet from the adsorber is discharged through the roof, care must be taken to eliminate the possibility of blowing exhaust gases into an adjacent air intake downwind from the adsorber. An adequate rain hood should be installed on the stack to eliminate rainfall from entering the system. Some carbon adsorber manufacturers suggest that the outlet from the adsorber be discharged directly into the workplace. However, extensive monitoring of the effluent stream must accompany this practice.

In all inlet and outlet ducts to the adsorption equipment, openings in the ductwork should be provided to accomplish clean out of the duct, and monitoring and sampling of the airstream. Suitable pressure gauges should be provided across filters and carbon beds to indicate the free flow of the airstream. Airflow switches are typically installed in the airstream to sound a warning when the flow is lost. A thermocouple is usually installed in the inlet duct to the adsorber beds so that the incoming temperature may be periodically recorded on the outlet duct. It is a good idea to consult a handbook on ventilation design and practice prior to installing the exhaust system and components.

10.3.4 Cooling-Water Supply

An adequate supply of cooling water is necessary to condense the vapors of water and solvent while steam-stripping the carbon bed. The demand for this cooling water exists only while the steam-strip cycle is operational, and the installation should call for an automatic shutoff valve to interrupt the flow at all other times. The water inlet and outlet temperatures should be provided at the appropriate location. If water is being delivered at a temperature below 60°F, the incoming supply line should be insulated to eliminate condensate dripping. The outlet water temperature may reach 120°F on some heat exchangers where a high-boiling point solvent is being condensed.

Where flammables are being adsorbed or where a cool carbon adsorber bed is required for efficient adsorption, cooling-water coils may also be installed within the carbon beds. Installation of controls such as a flow control valve or a globe valve for regulation of flow on outlet piping with thermometer wells on the inlet and outlet is appropriate on all cooling-water lines. If a connection is made to a potable water supply, adequate backflow-prevention devices must be installed in the supply to eliminate the possibility of contamination of drinking water.

10.3.5 Steam Supply

On steam-stripping regenerative carbon adsorption systems, adequate steam at about 3–15 psig must be available to strip beds of adsorbed solvents. Where a high-pressure steam supply is available and a reduction of pressure is required, more precise control has to be obtained when a double pressure-reducing valve arrangement is used. For example, a first-stage reduction will lower steam pressure from 100 psig to 30 psig, then a second-stage reduction will lower steam pressure from 30 psig to the desired tank pressure of about 5 psig. Steam traps should be installed where required, as the adsorption process stripping cycle requires dry steam. Do not discharge steam traps into the same closed container as the water from the decanter, because the steam pressure during blowout will retard decanter flow. All steam piping must be properly insulated, not only to conserve steam, but also to reduce the burn hazard. The carbon tanks themselves may also be insulated, for the same reasons. A steam-supply cutoff valve should be provided in a readily accessible location in case of emergency. Do not use the shutoff valve to regulate steam flow because of severe eroding of the valve components by restricted steam flow.

10.3.6 Compressed-Air Supply

Dry, clean, oil-free compressed air is required to operate controls and dampers on carbon adsorption units. A shutoff valve should be installed in adjacent locations and a filter, regulator, lubricator, and pressure gauge should be provided to ensure correct delivery pressure. Prior to assembly, the shafts on the air cylinders should be lightly oiled to eliminate damage to the seals.

10.3.7 Electrical Supply

Provide necessary electrical service to the system according to the manufacturer's recommendations and in accordance with the National Electrical Code. Special considerations should be given to installations using flammable solvents.

10.3.8 Internal Solvent and Water Piping

The principle of operation of the decanter is based on the differing densities of the liquids being separated. The greater the differences in the density, the easier the separation of the liquids involved. In a process containing one solvent such as PCE, the heavier solvent will flow from the bottom of the decanter and the lighter water will flow from the outlet near the top of the cylinder. It is important that the water discharge and the solvent discharge be totally unrestricted in their flow from the decanter. Any back pressure acting on neither line from restricted flow or improper venting may cause solvent flow into the water discharge pipe or water to flow into the solvent discharge pipe. The same back-pressure condition within the decanter will produce erratic flow conditions and contamination of both wastewater and reclaimed solvent. The various water drain lines should be kept separated so that all flows are independent and will not interfere with other flow conditions. It is strongly recommended that a suitable recording flowmeter be installed on the solvent recovery piping. A sight glass on the reclaimed solvent tank is also desired with level controls on the solvent or water storage tanks to reduce overfills and spills. In all piping of water, solvent, compressed air, etc., unions and valves should be installed for ease of shutoff, disassembly, and repair.

Appropriate fire extinguishers should be provided for controlling local electrical and chemical fires, as well as flame arresters inside ductwork transporting flammable solvents. Follow all established safety precautions and use common sense in all installation procedures. A job well done during installation will return many times the effort in both operation and maintenance time savings in the future.

10.4 Operation

Utilizing adsorption as an air pollution control measure on sources emitting volatile organic hydrocarbons has proven to be extremely effective if proper design is applied and rigid operating procedures are established and followed. It is always desirable to check over the process being controlled to determine that normal operation is being experienced. Of course, the blower must be running with the fan turning in the correct direction. Three-phase motors that are running backward may be corrected by changing two of the phase wires to the motor. Initial startups will probably require a system balance of the exhaust ducts. If multiple processes are being exhausted into the same system, adjustments of individual slide dampers may be required to obtain the correct airflow in each branch duct. Airflow switches are inexpensive and should be placed in each branch to sense airflow and allow the process to operate when the airflow is adequate. Airflow velocities below 100 ft/min through the carbon beds will provide adequate retention time for solvents in the airstream to be adsorbed on the charcoal. Excessive flows will reduce carbon efficiencies and allow volatiles to escape into the atmosphere. Excessive airflows are also detrimental to process operations where unnecessary solvents are evaporated and lost from process tanks and delivered to the carbon beds, posing additional loads on the system. Additional expenses to replace conditioned factory air that is wasted can be significant, and every effort should be made to minimize airflow from the manufacturing plant.

Prior to startup, the prefilter ahead of the carbon tanks, should be checked to verify that it is properly secured to the housing. This filter should be closely monitored during the first few days of operation. It may tend to accumulate dirt or debris initially from the construction and installation activities, causing the filter to become ineffective and restricting solvent entry into the carbon beds. After about three days of operation, replace or clean the dirty filter bag and observe the manometer gauge that registers air pressure between the blower and the carbon bed. There is no exact guideline for pressure differential that is applicable to all devices, as the duct restriction of specific installations will yield different gauge readings. Each system should stabilize and provide consistent manometer readings. However, during the first two or three days of operation, the carbon beds will be settling, so the readings will not reflect final operating conditions. With both carbon tanks set on the adsorption cycle, one can place a mark on the manometer dial face to indicate a normal reading. Any changes in the reading can be used to help maintain the correct airflow to the adsorber. A change of 1″ WC pressure on this gauge is significant, and the problem should be corrected. If the reading drops by 1″ WC or more, check the following points ahead of the blower:

1) Dirty filter bag
2) Improper slide damper setting
3) Obstruction in the duct

If the reading increases by 1″ WC or more, check the following:

1) Appropriate beds should be in the adsorption cycle when the reading is noted
2) The adsorber tank dampers must be open
3) There must be no restriction in the exhaust duct
4) The filter bag may be loose or missing

A black discharge may be noticed coming from the adsorber exhaust upon startup. This is due to the carbon dust contained in all new charcoals and should subside after a couple of cycles. The duration of the adsorption cycle may be preset by manually set timers within the control panel of the unit.

The precise time from startup to saturation depends on a number of factors. No simple, universal formula can be given to cover every situation. The type of solvent, airstream temperature, airflow rates, variations in concentration, and total amount of solvent evaporated from the process per unit of time must be considered. The initial adsorbtion cycle should be sufficiently long to completely saturate the carbon beds with solvent. All subsequent cycles should then be shortened because total regeneration cannot be effectively achieved.

On startup, the initial cycle time and timer settings can be estimated by using the experimental data for carbon efficiencies contained in Table 10.2.

The inlet and the outlet solvent concentrations during actual operation may be continuously monitored with an organic vapor analyzer (OVA) to determine the breakpoint on the adsorption cycle. Timers can then be adjusted to discontinue adsorption and begin regeneration. A desirable feature of activated carbon in the control of solvent emissions is its ability to recover the adsorbed solvents on steam regeneration. Regeneration may be accomplished by passing hot gases through the carbon bed. Low-pressure steam at 3–15 psig is the usual source of heat and is sufficient to remove most solvents. Normally, the flow of steam passes in a direction *opposite* to the flow of gases during adsorption. With this arrangement, the steam passes upward through the carbon. The steam flow through the bed is typically only one-fifth to one-tenth of the air velocity and is too low to initiate boiling or cratering of the bed. This countercurrent flow is an advantage in regeneration because a solvent gradient exists across the adsorbent bed and, depending on the concentration of adsorbate and bed depth, the inlet side of the bed may be saturated before the outlet reaches the breakpoint. Thus, with countercurrent regeneration, the solvent, driven out of the adsorbent from the outlet side by the incoming steam, will in turn start to remove vapor at the inlet before it becomes heated because it is already saturated. This results in lower steam consumption and more efficient operation. Note that steam consumption per pound of solvent recovered varies with strip time and with the particular solvent adsorbed.

After the solvent is steam-stripped, the carbon beds are not only hot but also saturated with water. Cooling and drying are accomplished by opening the bed to the incoming airstream, allowing the entrapped water to evaporate and adsorb the heat of the carbon and to be swept away to the atmosphere by the airstream.

Table 10.2 Physical properties of common VOCs.

	Boiling Point (°F)	Molecular Weight	Soluble in Water	Flammable	LEL[a] (vol %)	Carbon Adsorption Efficiency[b]
Acetone	133	58.1	Yes	Yes	0.15	8
Benzene	176	78.1	No	Yes	1.4	6
Butyl acetate	259	116.2	No	Yes	1.7	8
Butyl alcohol	241	74.1	Yes	Yes	1.7	8
Carbon tetrachloride	170	153.8	No	No	—	10
Ethyl acetate	171	88.1	Yes	Yes	2.2	8
Ethyl alcohol	165	46.1	Yes	Yes	3.3	8
Heptane	209	100.2	No	Yes	1	6
Hexane	156	86.2	No	Yes	1.36	6
Isobutyl alcohol	241	74.1	Yes	Yes	1.68	8
Isopropyl alcohol	205	60.1	Yes	Yes	2.5	8
Methyl alcohol	153	32	Yes	Yes	6	7
Methylene chloride	104	84.9	Yes	No	—	10
Methylethyl ketone	174	72.1	Yes	Yes	1.81	8
Methylisobutyl ketone	237	100.2	Yes	Yes	1.4	7
Perchloroethylene	250	165.8	No	No	—	20
Toluene	231	92.1	No	Yes	1.27	7
Trichloroethane	189	131.4	No	No	—	15
Trichlorofluoroethane	117.6	186.3	No	No	—	8
Naphtha	208	—	No	No	0.81	7
Xylene	292	106.2	No	Yes	1	10

[a] LEL: Lower explosive limit. The lowest concentration value of a vapor that will support propagation of flame upward through a cylindrical tube.
[b] Carbon adsorption efficiency: efficiencies are based on 200 cfm laden air per 100 lb of carbon per hour at concentrations above 15 ppm.

The solvent vapors and the steam that emerges from the tank during stripping are then condensed in a water-cooled condenser. The cooling required for condensation may typically be 5 gal/min at an inlet temperature of 60°F. The temperature rise in the condenser will be approximately 40°F, making the outlet water temperature about 100°F. High cooling-water temperatures or low water-flow rates will slow the condensation process, thereby increasing the steam-strip cycle time and adding unnecessary costs to the process. The final step in the regeneration cycle is the separation of the steam and solvent condensate by using a gravity-type decanter that is adequately vented to eliminate "air locks," properly leveled and drained into an open container to reduce back pressure, and precharged with a sufficient amount of solvent to prevent initial steam condensate from entering the solvent discharge pipe that can contaminate the solvent storage tank.

It is important to sample both the wastewater discharge and the recovered solvent periodically to maintain the necessary effluent quality. Solvent solubility in the wastewater stream for the chlorinated solvents is quite low (0.015/100 g for PCE). Any additional solvent discharges into municipal sewer systems or to local streams should be eliminated. The integrity of the recovered solvent can be determined through tests for the moisture content and acidity. Moisture content should be less than the solubility of water in the solvent. Acid acceptance tests will reflect losses of solvent inhibitors that have been added to reduce the breakdown of the solvent.

10.4.1 Multiple Compound Solvents

The adsorption phenomenon becomes somewhat more complex if the gases or vapors to be adsorbed consist of several compounds. Carbon adsorption of the various components in a mixture is not uniform, and generally, these components are adsorbed in an approximately inverse relationship to their relative volatilities. Hence, when air containing a mixture of organic vapors is passed through an activated-carbon bed, the vapors are equally adsorbed at the beginning; but as the amount of the higher-boiling constituent retained in the bed increases, the more-volatile vapor revaporizes. After the breakpoint is reached, the exit vapor consists largely of the more-volatile material. At this stage, the higher-boiling component has displaced the lower-boiling component, and this is repeated for each additional component. Two or more VOCs in the airstream, as a general rule, will have the following effects:

1) The adsorption of organic compounds having higher molecular weights will tend to displace those having lower molecular weights. Lighter compounds will tend to be separated or partitioned from the heavier compounds and will pass through the bed at a faster rate. This will increase the mass-transfer zone (MTZ) and may require additional carbon bed depth, or shorter operating cycles.
2) Carbon retentivity may be reduced.
3) Efficiencies of any given system will tend to be lower on a multiple organic application.
4) The LEL of the mixture will vary directly with the LEL of the individual components. Safety considerations may dictate more or less dilution of air to reduce the flammability potential.

Decanter operation will also be affected by multiple organics in the system. Where multiple solvents are involved, the organic layer will consist of a mixture of the compounds roughly paralleling their feed stream concentrations. Some compound mixtures of immiscible organics may possess a variety of densities resulting in a multi-layer decanter separation requiring multiple decantation for recovery. The recovered organics may also be a compound mixture and their value as a reclaimed solvent will depend upon the following questions:

1) Can the mixture be reused as decanted?
2) Is it possible to reconstitute the mixture for reuse by solvent additions?
3) If mixtures cannot be reused, is it possible to sell them to a central refiner who would fractionate and resell them as used solvent?
4) If the mixture cannot be made suitable for reuse, can it be used as a fuel for the generation of thermal energy?
5) Is the recovered mixture of high-enough value to consider on-site distillation equipment?

Water-soluble organics usually present a special problem to steam regeneration systems. Because of water contact in the decanter, the water-soluble materials will be carried away in the wastewater stream. If losses of these organics are detrimental to the process or to water quality, alternative stripping mechanisms should be considered. Complex distillation columns may also be employed to recover the water solubles and return them to the source.[9]

The solvent corrosion factor should be considered in the selection of the materials of construction for the adsorption equipment. Corrosion should be prevented or retarded if long-term operation of the equipment is expected. Many volatile organic compounds are not particularly corrosive and may be handled in conventional carbon steel tanks. Some solvents are subject to hydrolysis and/or other chemical reactions and to the formation of corrosive by-products. This must be dealt with by the use of coatings of base metals such as stainless steel or other materials that have a high resistance to oxidation. Most carbon steel carbon adsorption systems will employ some type of protective coating to isolate the activated carbon from direct contact with the interior of the adsorption vessel. This preventive measure is necessary to reduce galvanic action between the carbon and the tank wall which will result in excessive pitting and corrosion in the tank. The cycle time and/or manner in which the system is to be utilized will also be a factor. Systems that are to be regenerated only once or twice a day will tend to be dryer than those that cycle hourly and may be satisfactorily designed using internal coatings rather than expensive base materials. More expensive materials for construction should be provided only in those inaccessible areas such as vapor lines, small valves, drain connections, and damper housings, where the satisfactory applications of coatings would be difficult. Where corrosive compounds are utilized and where systems are to be cycled on a frequent basis, the more resistant types of construction materials should be considered.

All hydrocarbon materials that may be introduced into the adsorption system are characterized as either aliphatic or aromatic. Aromatic compounds exhibit a special type of unsaturation having to do with resonance and the stable nucleus provided by the six-carbon benzene-ring configuration. Aromatics of the simpler type will resist the formation of by-products and may be used with mild steel. Many of the aliphatic compounds have saturated alkanes or paraffins, and do not usually react with aqueous solutions of acids, bases, or oxidizing agents and therefore are compatible with mild steel constructions.

The major families of corrosive hydrocarbons are the halogenated ketones or aldehydes and the esters. Esters are the product of a reaction between an acid and an alcohol. Such reactions are reversible, and the hydrolysis of an ester will yield the acid and alcohol from which they were produced. In use, most esters involve the formation of acetic acid and will require processing in stainless steel equipment.

The halogenated compounds represent a saturated hydrocarbon in which a hydrogen atom is replaced by a halide. The halogen atoms are easily displaced from their associated carbon atom, and the formation of other compounds – often acidic – can be expected. Many halogenated materials are stabilized by the addition of corrosion retardants and inhibitors, and can generally be processed in carbon steel equipment with an impervious surface coating. Monel, Hastelloy, or titanium are sometimes used in connection with difficult halogenated compounds.

Ketones and aldehydes are both characterized by the presence of a carbonyl group. This carbon/oxygen group is subject to chemical reactions that may produce a number of corrosive by-products requiring handling in the equipment of stainless steel construction. The ketones from the major category of "reactive organics," and the presence of the carbonyl group, with its ability to undergo chemical reaction, cause the problem in carbon adsorption systems. Often these reactions are exothermic, and these reactive organic operations will require additional hardware and modified control sequences to control maximum temperature within the beds. Live steam injection or permanently installed cooling coils are effective in controlling temperatures that may approach the lower explosive limit of the solvent.

Undesirable contaminants in the airstream may also produce detrimental effects if not removed prior to their introduction into carbon beds. Contaminants may be broken down into three categories: particulates, entrained liquids, and high-boilers. Almost all industrial applications will require solid-media-type filtration systems for the removal of airborne dust, lint, and general dirt in a particle size down to 3–5 μm. Solid media filters of the type generally available are made of cloth or fiberglass and are usually satisfactory for this application. Automatically operated filters of the moving-media type, controlled by pressure drop across the filter material, usually offer a satisfactory solution for excessive particulate contamination.

In other applications, where fine particles of resin or other solids used in the coating process have diameters of 1 μm or less, special filtration systems will be required. Electrostatic precipitators, for instance, may find application in this area.

Entrained liquid is also a form of contaminant. There are a number of mist eliminators available, most of which are designed to be mounted in the solvent airstream.

Problems with high-boilers have been mentioned previously and are listed in Table 10.3. There are volatile organics that have boiling points in excess of 500°F, such as resins, plasticizers, and/or compounds that react chemically on the carbon to form solid or polymerization products that will not be removed during steam desorption.

Some flammable materials will require special considerations in carbon adsorption applications. When solvents are being adsorbed that have low flash points, as indicated in Table 10.3, extreme care must be taken to ensure against carbon bed fires. It was previously mentioned that the phenomenon of adsorption is exothermic or "heat-liberating." If the carbon beds are not allowed to stay cool during adsorption and if sufficient oxygen reaches the flammable solvent in the bed, the potential exists for ignition, and a fire in the carbon bed may occur. Purging with an inert gas such as nitrogen rather than air will eliminate sufficient oxygen from entering the unit and reduce potential fire hazards. Carbon beds may also be cooled by direct water injection. The water will serve as a heat sink until sufficient flammable solvents displace the water molecules on the carbon beds during normal adsorption operation.

Table 10.3 VOCs not suitable for carbon adsorption.

	Reactive Compounds
Organic Acids	
	Aldehydes
	Ketones (some)
	Monomers (some)
High-Boilers	
	Plasticizers
	Resins
	Long-chain hydrocarbons: C_{14} and up
	Phenols
	Glycols
	Amines

10.5 Maintenance

Effective air pollution control utilizing carbon adsorption must be accompanied by a routine maintenance program. The program should provide for scheduled inspections of all equipment components as well as necessary monitoring of operating parameters to ensure correct operation and optimum performance of the control equipment. Monitoring of the inlet and outlet gas stream, quality of reclaimed solvent, wastewater quality, etc., will provide valuable information about the performance of the carbon adsorber in this section.

Routine maintenance of air pollution control equipment can also be important because equipment failure can be expensive in terms of lost production, lost solvents, degradation of air resources, and potential health effects on employees. Each component must operate properly to ensure steady, efficient output, and desired results from the system.

Establishing an equipment maintenance program need not be elaborate or complicated. The actual work in routine inspection and servicing may, in large part, be performed by shop personnel operating the control equipment. Of course, expensive repairs of rebuilding should be performed by skilled and trained maintenance personnel. System components of the carbon adsorber that require routine maintenance fall into four major categories: air handling, adsorbing, stripping, and reclaiming.

The function of air-handling apparatus is to collect, transport, and deliver particulate-free, solvent-laden air to the adsorber. Any leaks in the duct works on the suction side of the fan will introduce excessive ambient air into the system, resulting in a reduction of solvent concentration and poor adsorber efficiency. Air duct leakage on the pressure side of the fan will discharge unwanted solvent vapors into the workplace. Leakage checks should be performed periodically, especially at flexible connections, at joints in the ductwork, on the fan and filter housing, and around the adsorber bed dampers. Accurate collection velocity data should be established by routinely checking the capture velocity at the source using a vane-type velocity meter or hot-wire thermal anemometer. Correct operation of flow-indicating devices within ducts can be verified by mechanically stopping the flow to the duct or by running off the fan motor. Ventilation system imbalance may also occur from time to time and may require periodic adjustments to dampers to rebalance the system. The particulate filter bag installed in-line ahead of the adsorber beds should be equipped with a differential pressure gauge and/or switch to indicate dirty or plugged filter media. A bag should be changed or cleaned when the differential pressure increases by 1″ WC or more. Observance of the operation of the disk dampers with the adsorber tanks should occur periodically to determine the correct seating and opening of the valves.

Maintenance inspections on the adsorption cycle equipment are somewhat more complex than for the air-handling apparatus. The integrity of the activated carbon must be maintained to ensure efficient removal of solvent vapors from the airstream. As the carbon particles erode with time and the capillaries become plugged with contaminants and polymers, the granules gradually lose their ability to adsorb and retain solvent molecules. Carbon adsorptability and retentivity should be tested regularly by opening up the bed. Laboratory analysis will reveal the effectiveness of the carbon bed. Most manufacturers of activated charcoal will perform adsorptability and retentivity tests for their customers. If the carbon fails the tests, all the adsorbent should be removed from the system and be regenerated or replaced.

Maintenance requirements on the desorption-cycle equipment involve primarily the steam supply, valving, and timer controls. Steam pressure on the carbon tanks should be regulated to minimize steam-stripping pressure (3–10 psi). Lower steam pressure will require too much run time to adequately strip the beds, while pressure that is too high will tend to fluidize the bed and create excessive erosion of the carbon granules. Steam traps must be operative or water will be carried into the carbon beds and retard proper stripping action. Periodically, the steam pressure relief valves located in the supply line and in the main carbon tanks should be checked for correct pressure settings by increasing steam pressure until the valves "pop off." Steam leaks around gaskets and operating dampers should be corrected by replacing the gaskets and seals. Gasket materials that are in contact with the solvent vapors must withstand that particular solvent's chemical properties. Often, the gasket material supplied with the system may not be suitable for the solvent flowing to the adsorber and an alternate material may be required. Leaks around the carbon tanks may create additional problems of corrosion around the leaks. The appearance of corrosion throughout the system may indicate a loss of inhibitors in the solvent that has been lost through numerous steam-stripping cycles. Makeup inhibitors or makeup solvents may alleviate this condition. Boiler feedwater treatment may require some modification if carryover chemicals are introduced into the carbon beds and create corrosion problems.

The apparatus utilized to reclaim the solvent requires little maintenance. The automatic cooling-water valve should be checked for proper opening and closing operation. Automatic mechanical valve shafts and other mechanisms should be lightly oiled. The condenser, in time, may become inefficient because of excessive buildup of solubles from the cooling water. Acidizing of the water jacket or tubes may be required to renew condenser efficiency. Inadequate separation of water and solvent in the decanter may indicate a plugged vent line. All vent lines and drain lines must be unrestricted for the correct operation of the system.

In general, normal maintenance procedures should be followed in routine cleaning of electrical contacts, lubrication of all bearings, compressed-air components, air cylinder shafts, replacement of obviously broken or worn parts, and house-keeping practices around the adsorber. In the final analysis, common sense is the best maintenance tool available in view of the fact that a large percentage of carbon adsorber equipment failures can be traced to neglect, improper operation, or just plain abuse.

10.6 Improving Operation and Performance

Optimizing the performance of air pollution control equipment such as carbon adsorbers involves consideration and monitoring of the following aspects:

1) Operation of manufacturing process controls to minimize solvent emissions
2) Quality of the solvent/air inlet stream
3) Characteristics of the inlet stream such as concentration, temperature, and flow
4) Duration of the adsorption cycle; saturation, and working bed capacities
5) Quality and quantity of available steam for regeneration
6) Duration of steam-strip cycle
7) Saturation and retentivity of the carbon
8) Quantity and quality of cooling water
9) Effectiveness of the water/solvent separator
10) Quality of reclaimed solvent
11) Quality wastewater
12) Quality of exhaust stream from the adsorber bed

1) *Operation of the manufacturing process.* Prior to applying operating control to end-of-the-line air pollution control equipment such as a carbon adsorber, consideration must be given to process controls that will minimize solvent losses to the airstream and/or workplace. Listed below are some important considerations:
   a) Consider substitution of the solvent by one of lower volatility or environmental impact.
   b) Minimize local exhaust ventilation from the process.
   c) Utilize adequate freeboard height to reduce losses.
   d) Provide canopies, hoods, and enclosures over tanks if possible, and minimize all openings to process.
   e) Provide a parts-drying chamber within the process if possible.
   f) Cover tanks when not in use.
   g) Minimize the travel rate of parts into and out of the cleaning tank vapor zone.
   h) Solvent spraying should be performed in the vapor zone, preferably with a gentle flush rather than an atomized spray.
   i) Allow parts to drain properly prior to exiting the cleaning tank.
   j) Maintain a cold-air blanket above tanks with either chilled-water coils or direct-expansion refrigeration coils.
   k) Do not clean porous or adsorbent materials.
   l) Do not agitate the solvent bath with compressed air.
   m) Use compressed-air flow-off as a last resort.
   n) Do not direct ventilating fans on process.
   o) Establish and maintain filling, draining, startup, shutdown, operations, and maintenance procedures.

After the application of appropriate design parameters to the cleaning process, consideration can be given to end-of-the-line controls.

2) *Quality of solvent/air inlet stream.* The large surface areas initially offered by the activated carbon granules will gradually be lost in the adsorption/desorption process. The effective life of the carbon depends upon the quality of the incoming airstream flowing into the beds. Airborne contaminants must be removed from the vapor-laden stream

prior to introduction to the adsorbent to eliminate "plugging" of capillaries within the granules. Periodic and routine maintenance on any bag filter will ensure longer life for the carbon beds. Some solvents in the carbon beds may decompose or polymerize when contacting the adsorbent. Polymerization will significantly lower the adsorption capacity of the adsorbent and also reduce its regenerating capabilities by conventional means. These polymerizing solvents must be avoided.

3) *Characteristics of inlet stream.* In many instances, exhaust ventilation design is concerned primarily with contaminant control at the point of emission to protect worker health and reduce employee exposures to hazardous chemicals. As a result, many exhaust systems have been designed to satisfy only these requirements, with little regard for system efficiencies. Earlier discussions of process design characteristics indicated a desire to minimize exhausted air losses and excessive carryout of solvent vapors. Lower airflow will not only improve the cleaning process efficiency but will reduce fan cost, ventilation system cost, and power consumption, and will reduce previously conditioned room air losses to the atmosphere. Cost comparisons of annual operating and maintenance costs and costs to replace exhausted air from a building are shown in Figure 10.1. The costs shown are general in nature and will vary considerably from industry to industry and as the materials of construction vary to suit the process design criteria. Annual conditioned-air loss through exhaust ventilation reflects the cost associated with year-round comfort-air-conditioned factories. The operating costs are based on a 2-shift, 5-day operation. In addition to minimizing airflow, attempts should be made to present a cool gas to the adsorber beds. An increase in temperature results in a corresponding decrease in the quantity of gas adsorbed. The effect of temperature on gas adsorption is shown in Figure 10.2.[2] It would therefore be an advantage in

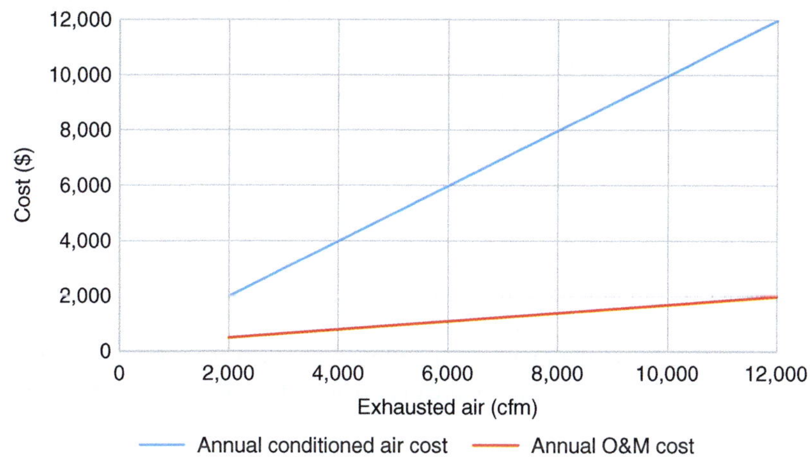

Figure 10.1 Representative annual fan operating cost.

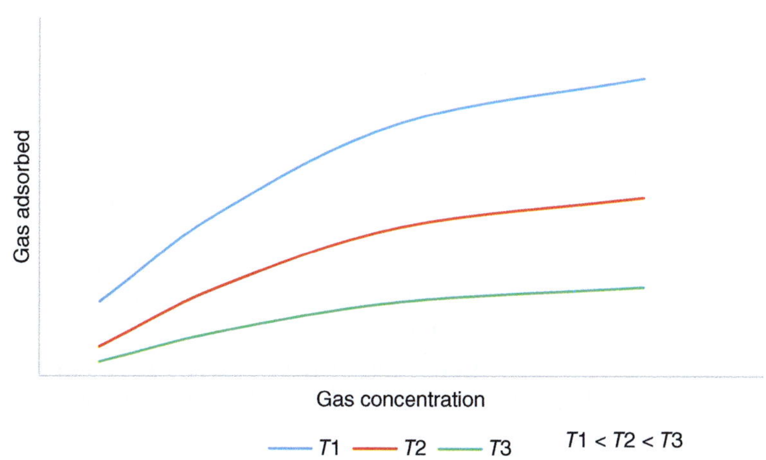

Figure 10.2 Effect of temperature on gas adsorption.

promoting adsorption to reduce mechanically the inlet gas temperature to the carbon beds if adequate cooling is not achieved in the transporting ductwork from the process to the adsorber. By minimizing the airflow, reducing the temperature, and thereby raising the solvent concentration, maximum utilization of the adsorptive process can be realized.

4) *Duration of adsorption cycle.* Working bed capacities vary considerably, depending upon the particular solvent being reclaimed and its regeneration characteristics. To maximize the performance of the carbon adsorber, the duration of the adsorption cycle should be extended to just below the breakpoint of the beds. Breakthrough can be determined using OVAs simultaneously on the inlet and outlet streams of the adsorber bed. Breakthrough history can then be initiated only when absolutely necessary.
5) *Steam quality.* Low steam pressure and wet steam will reduce the effectiveness of the stripping operation. Steam traps must be continually checked for correct operation. Because of the intermittent steam demand, condensate in the supply lines must be removed prior to stripping. Automatic steam traps should ensure the availability of dry steam for the desorption of the beds.
6) *Duration of the steam-strip cycle.* To provide efficient recovery of solvents, steam stripping must not be excessively long, to try to recover too much of the "heel," or too short, requiring more frequent regeneration.[1] Desirable strip-cycle times can be assured by continuous monitoring of reclaimed solvent quantities that flow from the decanter.
7) *Saturation and retentivity of carbon.* Ensure good operation and performance of the carbon beds by periodically testing the charcoal for its ability to adsorb solvent and to retain the solvent after the manufacturing process flow is discontinued. Adsorbent manufacturers will assist in this test. The adsorptability and retentivity of the carbon will be affected by airborne contaminants that enter the beds and plug the capillaries in the granule. These two conditions have reduced the efficiency of the carbon bed and retentivity tests indicate a necessity to perform extensive regeneration.
8) *Cooling-water quality.* Monitor cooling-water inlet and outlet temperatures. The steam-strip cycle is typically operated on a timer mode and cooling water is required to condense all the vapors presented to the heat exchanger in the same given amount of time. Inadequate cooling water will result in solvent losses to the airstream. The inlet and outlet temperatures will vary depending upon the particular solvent being reclaimed. The cooling-water supply must be sufficient to condense vapors of steam as well as solvent vapors that are presented to the condenser for the duration of the steam-strip cycle.
9) *Effectiveness of the water/solvent separator.* Vent properly, provide the unrestricted flow of solvent and water discharging from the decanter, and check the quality of both solvent and wastewater.
10) *Quality of reclaimed solvent.* Laboratory analysis is required to determine losses of inhibitors, decomposition of solvent, acidity, or contamination.
11) *Quantity of wastewater.* Wastewater analysis can also detect the presence of pollutants in the steam. Because of the rapid separation of water and solvent, gross amounts of nonsoluble solvents can be detected by observation. Dissolved contaminants must be detected through laboratory analysis.
12) *Quality of exhaust stream.* The final analysis required to determine the efficient operation of the adsorber is in the exhaust stream. By using an OVA the concentration of organic solvent in the airstream can be determined. The total quantity of solvent losses in pounds per unit of time should be recorded to establish a historical file on the system. Deviations from the file data could then be noticed, rectified, and carbon adsorber efficiencies maintained.

10.7 Monitoring

Improvements in the operation and performance of carbon adsorption systems can be obtained through a well-defined program of operation and maintenance, a program that encompasses the total manufacturing process, beginning with the solvent used and the design and operation of the process and its exhaust system thus should then be extended to the operation of end-of-the-line equipment. To ensure proper operation and maintenance, a formal monitoring program should be established. Implementation of this program will provide an accurate data file of such parameters as inlet concentrations, outlet concentrations, solvent removal efficiency, cycle times, and solvent quantities reclaimed.

Two methods recognized for monitoring of organic emissions are the source test screening method, which involves the OVA instrument, and the source testing verification method, which specifies an integrated bag as the sampling device with a gas chromatograph or infrared spectrophotometer analysis.

The OVA is one of the most popular instruments employed for organic gas stream concentration measurement. Both portable and bench-top OVAs are available and both have proven to be reliable for carbon adsorption monitoring.

The OVA works on the principle of hydrogen flame ionization for the detection and measurement of organic vapors. The instrument measures organic vapor concentrations by producing a response to an unknown sample, which can be related to the gas of known composition to which the instrument has previously been calibrated. An electric field drives the charged carbon ions which are produced by an incinerated solvent to a collector electrode, which in turn generates a current that is measured with a logarithmic electrometer preamplifier. The output signal is proportional to the log of the input or the ionization current. An amplifier conditions this signal and sends a readout to the external monitor. Units are also available that will produce a linear output signal over a smaller range of concentrations.

A strip-chart recorder can be attached to the OVA to record concentrations over the sample-time interval. The chart recorder can make the OVA semipermanent, because once the strip chart and OVA are calibrated, they can be left to continually monitor the organic vapor concentration in parts per million (ppm).

Monitoring of the organic vapors in the airstream is performed for several reasons: to demonstrate compliance, to establish organic vapor removal efficiencies, and to determine optimum time intervals for adsorption and regeneration cycles.

Solvent concentration monitoring must be accompanied by adequate airflow determination to calculate solvent losses and to perform the required material balance on the system. Airflow may be determined using a standard pitot tube and inclined water-filled manometer; in cool, dry gas streams a thermal anemometer may be used.

Wastewater discharge and recovered solvent monitoring are also important to maintain the necessary effluent quality. Discharges to municipal sewers or to local streams should be minimized to reduce solvent losses. The integrity of the solvent to be reused is dependent upon individual cleanliness requirements; however, samples of the solvent should be tested periodically for moisture content and acidity as well as for cleanliness. Often, visual inspection will provide needed information pertaining to either solvent or wastewater quality. Discolorations in solvents that have been returned to the process may signal a need for further investigation. When water and immiscible solvents are present together, a clear boundary line may be visible.

Sight glasses on holding tanks can be misleading if two immiscible solvents are present in the tank, such as water and perchloroethylene. Because of the density difference, the PCE will assume the bottom position in the tank and, in conventional sight-glass installations, will fill from the bottom of the tank. The presence of water may not be reflected at all in sight-glass readings. Because of this density difference, the sight-glass levels may also be misleading, not reflecting the true level of liquid within the tank.

Certainly, a viable monitoring program is mandatory in "tracking" the performance of any air pollution control equipment. The costs to implement such a program will be returned many times over through improvements in source operations, recovery of organic solvents, and protection of the air resource through reduced emissions.

10.8 Recent Developments

The basic design of air pollution control equipment has remained relatively unchanged since first used in the early part of the twentieth century. Some modest equipment changes and new types of devices have appeared in the last few decades, but all have essentially employed the same capture mechanisms used in the past. One area that has recently received some attention is hybrid systems – equipment that in some cases operate at higher efficiency and more economically than conventional devices. Tighter regulations and a greater concern for environmental control by society have placed increased emphasis on the development and application of these systems.

Hybrid systems are defined as those types of control devices that involve combinations of control mechanisms – for example, fabric filtration combined with electrostatic precipitation. There are four major hybrid systems. These include wet electrostatic precipitators, ionizing wet scrubbers, and electrostatically augmented fabric filtration.

10.9 Conclusions

A well-designed and maintained carbon adsorption system will normally capture in excess of 95% of organic input to the bed. EPA studies indicate that carbon adsorption applied to solvent-cleaning processes will actually achieve a reduction in total solvent emissions to 40–60%. One reason for the difference between the theoretical and actual losses is found in the inadequacy of most ventilation systems and their inability to capture all of the solvent vapors and deliver them to the carbon bed. Many solvent losses occurring in manufacturing processes are in areas of drag out on parts, leaks, spills, and disposal of waste solvents, none of which are greatly affected by the ventilation system. Improved ventilation design can increase an adsorber's overall emission-control efficiency. A higher ventilation rate alone, however, will not necessarily be advantageous because increased turbulence could disrupt the air-vapor interface, causing an increase in emissions, all of which would not be captured by the collection systems. The effectiveness of the ventilation system can also be improved by the use of drying tunnels and other devices that decrease losses due to drag out.

Poor operation has been found to decrease the control efficiency of carbon adsorption systems. Examples are dampers that do not open and close properly, use of carbon that does not meet specifications, poor timing of the desorption cycles, and excessive inlet flow rates. Desorption cycles must be frequent enough to prevent breakthrough of the carbon beds, but not so frequent as to cause excessive energy consumption. The positive aspects of carbon adsorption are well known. There are, however, a few negative aspects that must be considered. These are briefly discussed below:

1) *Air impact*. No significant adverse air impact should result through the correct use of the carbon adsorber, although negligence with maintenance and operation of control devices could increase emissions in individual cases. Carbon adsorption systems operating with spent or saturated adsorbent, coupled with excessive ventilation rates, will lead to excessive solvent losses to the atmosphere. Excessive steam-strip cycles in high condensing water temperatures will also allow solvent losses to the atmosphere to increase. Consideration must also be given to the emissions from steam-generating boilers as the demand for steam increases during the desorption cycle.
2) *Water impact*. The largest impact on water quality is a result of the steam condensate discharged from the carbon adsorber. Steam used for desorption is in direct and intimate contact with the solvent vapors, and as the steam condenses into water, it will take on a small percentage of the solvent as well as the water-soluble stabilizers. When the contaminated water is discharged into the waste stream, the stabilizers and dissolved solvent are carried along as pollutants.
3) *Solid waste impact*. There appears to be no significant solid waste impact resulting from routine carbon adsorption operation; however, periodic replacement of the activated charcoal is required and may present a disposal problem. Most major activated-carbon manufacturers are equipped to reactivate spent carbon and thus eliminate any solid waste impact from normal applications.
4) *Energy impact*. A carbon adsorber consumes a large amount of energy because of the steam required for desorption; however, this energy expenditure is far less than the required energy required to manufacture replacement solvent.
5) *Noise impact*. Consideration must be given to increased noise levels due to the carbon adsorber when choosing an in-plant location for the system. Additional noise-suppression technology may be required for some installations.

Generally speaking, good operation practices and a proper maintenance program will eliminate many of the problems mentioned as environmental impacts, and allow the carbon adsorption equipment to operate effectively as intended: to reduce the discharge of pollutants into the atmosphere and to return recovered solvents to the manufacturing process.

References

1 **A.J. Buonicore, and L. Theodore**, *Industrial Control Equipment for Gaseous Pollutants*, Vol. I, CRC Press, Inc., Boca Raton, FL, 1975.

2 **OAQPS**, *Air Pollution Engineering Manual*, 2nd Ed., EPA, 1973.

3 **MSA Research Corp**, *Package Sorption Device System Study*, EPA, 1973.

4 **L. Theodore**, *Engineering Calculations: Adsorber Sizing Made Easy*, pp. 19–20, CEP, New York, NY, 2005.

5 **L. Theodore**, *Air Pollution Control Equipment Calculations*, John Wiley & Sons, Hoboken, NJ, 2008.

6 **L. Theodore**, *Chemical Engineering: The Essential Reference*, McGraw Hill, New York, NY, 2014.

7 **M. Theodore, and L. Theodore**, *Introduction to Environmental Management*, 2nd Ed., CTC Press Taylor & Francis Group, Boca Raton, FL, 2021.

8 **A.M. Flynn, T. Akashige, and L. Theodore**, *Kern's Process Heat Transfer*, 2nd Ed., Scrivener-Wiley, Beverly, MA, 2019.

9 **L. Theodore, and F. Ricci**, *Mass Transfer Operations for the Practicing Engineer*, John Wiley & Sons, Hoboken, NJ, 2010.

11

Incinerators

Incinerators can be used to control the emissions of gaseous or small-particulate air pollutants which are either combustible or thermally decompose at high temperatures. Three rapid oxidation methods are used to destroy combustible contaminants:

1) Thermal incineration
2) Catalytic incineration
3) Flares (direct combustion)

The thermal and flare methods are characterized by the presence of a flame during combustion. Catalytic incinerators serve the same purpose but utilize a metallic catalyst to promote the required rapid reaction; this unit operates at lower operating temperatures and with higher reaction rates.

Combustion is a chemical process occurring from the rapid combination of oxygen with various elements or compounds resulting in the release of heat. The process of combustion is also referred to as oxidation or incineration. Most fuels used for combustion, along with the waste, are composed essentially of carbon and hydrogen, but can also include other elements such as sulfur, nitrogen, and chlorine. Although combustion seems to be a very simple process that is well understood, in reality, it is not. The exact manner in which a fuel or waste is oxidized occurs in a series of complex, free-radical chain reactions. For example, methane undergoes the most simple combustion reaction of a hydrocarbon ($CH_4 + 2O_2 \rightarrow CO_2 + 2H_2O$) and contains over 300 intermediate free-radical reactions.[1] The precise set of reactions by which combustion occurs is termed the *mechanism of combustion*. By analyzing the mechanism of combustion, the rate at which the reaction proceeds and the variables affecting the rate can be estimated. For most combustion devices, the rate of reaction proceeds extremely fast compared to the mechanical operation of the device. Maintaining efficient and complete combustion is somewhat of an art rather than a science.

11.1 Description of Control Devices

Combustion systems are often relatively simple devices capable of achieving very high destruction efficiencies. They consist of burners, which ignite the fuel, and a chamber that provides appropriate residence (or detention) time for the oxidation process. Because of the high cost and decreasing supply of fuels, combustion systems may be designed to include some type of heat recovery. Combustion is also used for more serious emissions problems that require high destruction efficiencies, such as the emissions of toxic or hazardous gases. There are, however, some problems that may occur when using combustion. Incomplete combustion of many organic compounds results in the formation of aldehydes and organic acids, which may create an additional pollution problem. Oxidizing organic compounds containing sulfur or halogens produce unwanted pollutants such as sulfur dioxide, hydrochloric acid, hydrofluoric acid, or phosgene.

Several basic types of combustion systems are in use, and although these devices are physically similar, the conditions under which they operate may be different. Choosing the proper device depends on many factors, including the type of hazardous contaminants in the waste stream, concentration of combustibles in the stream, process flow rate, control requirements, and an economic evaluation. Three of these systems are described below.

Optimizing Air Pollution Control Equipment Performance: Operation & Maintenance, First Edition. Jay Richardson and Louis Theodore.

11.1.1 Thermal Incinerators

Thermal incinerators are the most widely used method to control the release of hydrocarbon fumes, especially smokes and solvents from coating processes. Typically, thermal incinerators are constructed of a steel outer shell lined with refractory material. The purpose of the refractory lining, which is typically 4–8 in. thick, is to protect the steel shell from direct exposure to the effects of high temperatures and corrosive materials, and also to improve the thermal efficiency of the unit by limiting heat losses. The refractory, which will be discussed in detail later in this section, serves as a thermal insulator. It lowers the temperature from greater than 2,000°F on the inside of the combustion chamber to a temperature of 180–400°F at the outer surface of the steel shell. In this temperature range, the steel shell will continue to retain its strength while also remaining hot enough to prevent the condensation of water vapor on the surface, thereby reducing corrosive attack.

These incinerators are equipped with a burner at one end that fires a fuel, typically natural gas. There is also a fume inlet located near the burner where the gas stream to be oxidized enters the incinerator. The burner may utilize the air in the process waste stream as the combustion air for the fuel, or it may use a separate source of outside air for this purpose.

11.1.2 Catalytic Incinerators

A catalytic incinerator is an alternative to a thermal incinerator as a means for oxidizing gaseous or oxygenated hydrocarbons to carbon dioxide and water. Contact of the waste stream with a catalyst bed allows oxidation reactions to occur rapidly in the temperature range of 700–900°F in contrast to the 1,300–1,600°F typically required of a thermal incinerator. The oxidation reaction that occurs at the surface of the catalyst produces the same products (carbon dioxide and water) and liberates the same heat of combustion as a thermal incinerator would.

The heat required to bring the waste stream up to the required oxidation temperature is usually supplied by a natural gas fuel burner. This type of incinerator is again constructed of a steel outer shell lined with a refractory material. The wall thickness of the refractory for a catalytic incinerator can be less than that of the thermal incinerator simply because the operating temperatures are not as great.

A catalyst bed is located at a distance downstream from the burner sufficient to allow for a uniformly preheated and distributed mixture of combustion products and waste gas stream. The catalyst bed in commercial units is typically a metal mesh mat, ceramic honeycomb, or other ceramic matrix structure with a surface coating of finely divided platinum or other platinum-family metals, such as chromium, vanadium, nickel, or cobalt. This metal coating is the actual catalyst for the oxidation reactions. The matrix serves to support the catalyst on a high geometric surface area and promotes good contact between the waste stream and the catalyst, as seen in Figure 11.1.

11.1.3 Flares

The flare system is used primarily as a safe method for disposing of excess waste gases. All process plants that handle hydrocarbons, hydrogen, ammonia, hydrogen cyanide, or other toxic or dangerous gases are subject to emergency conditions that require the immediate release of large volumes of such gases for the protection of plant and personnel. Flares are used for this purpose. In operation, the gas containing the organics is continually fed to and discharged from a stack, with the combustion occurring near the top of the stack and characterized by a flame at the end of the stack.

In general, three types of flares are in use today: elevated, ground, and forced draft. The elevated flare is made up of several components. The ground or "bilevel" flare is employed where noise, luminosity, or smoke are objectionable. They are, for all practical purposes, enclosed in combustion chambers that produce smokeless burning and conceal the flare flame. However, this type of flare has been outlawed in most states. The forced-draft flare uses air provided by a blower to supply

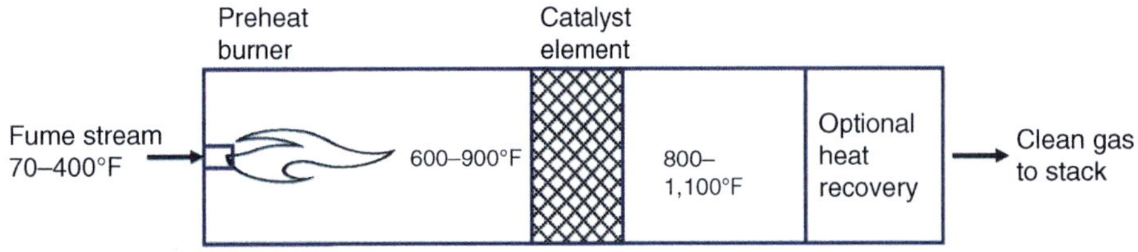

Figure 11.1 Schematic of a catalytic incinerator.

primary air and the turbulence necessary to provide smokeless burning of exhaust gases without the use of a steam injection system as used in the steam-assisted flare.[2]

The greatest variation among different afterburner designs is in how well they achieve the goal of raising all the fumes to the required temperature for the required combustion residence time. Most cases of poor performance are due to nonuniform temperatures and flows, which allow some of the pollutants to escape without adequate treatment.

11.2 Design Considerations

The design of all types of fume incinerators must take into consideration the "three T's" of combustion: time, temperature, and turbulence. These three variables will govern the speed and completeness of the combustion reaction. For complete combustion to occur, the oxygen must come into intimate contact with the combustible molecule at a sufficient temperature, and for a sufficient length of time, in order that the reaction be completed. Incomplete reactions may result in the generation of aldehydes, organic acids, carbon, and carbon monoxide.

Time and temperature affect combustion in much the same manner as temperature and pressure affect the volume of a gas. When one variable is increased, the other may be decreased with the same end result. With higher temperatures, a shorter residence time can achieve the same degree of oxidation. The reverse is also true: a higher residence time allows the use of a lower temperature. In describing incinerator operation, these two terms are always mentioned together.

For combustion processes, ignition is accomplished by adding heat to speed up the oxidation process. Heat is needed to combust any mixture of air and fuel until the ignition temperature of the mixture is reached and the combustion becomes self-propagating. By gradually heating a mixture of fuel and air, the rate of reaction and energy released will gradually increase until the reaction no longer depends on the outside heat source. In effect, more heat is being generated than is lost to the surroundings. The ignition temperature must be reached or exceeded to ensure complete combustion. To maintain the combustion of a waste, the amount of energy released by the combusted waste must be sufficient to heat up the incoming waste (and air) up to its ignition temperature; otherwise, a fuel must be added. The ignition temperature of various fuels and compounds can be found in the literature.[3–5]

11.3 Installation Procedures

Installation of fume incinerators involves mounting the unit on a concrete slab, bolting the sections together if required, and connecting the incinerator's utility lines (electric, oil, gas, and air) to plant supply lines. Small or specially designed lightweight units may be roof-mounted. If necessary, additional structural support may be provided. As previously stated, the incinerator's fume inlet is typically connected to the discharge side of the fan, which exhausts the fume-generating equipment. The incinerator's combustion chamber outlet is connected to a stack or to heat-recovery equipment preceding the stack. The concrete support pad should extend several feet beyond the incinerator on all sides, should be at least 8 in. thick, and should be properly bedded on a gravel drainage bed laid on firm, compact earth.

The incinerator discharge duct must be 100% airtight. Discharge gases may be about 1400°F and duct leakage could lead to fires. The fan and ducting to the incinerator should also be insulated to prevent loss of heat by the fumes. The fan must be firmly secured.

In areas where below-freezing temperatures are encountered, steam or electric tracer lines are required for air and fuel lines to prevent freezing of entrained water. Control valves must also be protected against freezing.

Installation is generally performed by a contractor engaged by the incinerator owner, with supervision being provided by a representative of the manufacturer.[6]

The following are excerpts from the safety startup instruction by one firm to its field personnel:

Before energizing the electrical control cabinet, make sure the lockout tags have been removed by the person who attached them. Before charging the system with any process fluids, perform a complete set of loop checks and trip checks to ensure the safety system is functioning as desired. After the control cabinet is energized, but before pushing any start buttons, check to see if all persons are standing clear of the fan. Push the start button but do not allow the motor to gain full speed. Push the stop button and check for the direction of travel. If the direction is correct, push the start button and allow the motor to reach full speed. Make a visual check of the fan for rubbing or other malfunctions. Do not attempt to run the motor with the belt guard removed.

The local gas company should be notified before opening the hand valve allowing gas into the gas valve train. Most gas companies reserve the option to visit the plant at the time of startup and supervise the opening of the valve and inspection of the gas valve train. Prior to attaching the gas to the gas train, it is possible to check the train for leaks with compressed air or nitrogen. High and low gas pressure switches and pressure reducing stations have diaphragms that are easily damaged by high pressures. These devices should be removed before air pressure is applied for test purposes. All leaks must be repaired, regardless of size, before attempting startup. Generally, one and one half times the operating pressure is accepted pressure for testing; however, never test below 3 psi.

Installation of catalytic incinerators differs from installation of thermal incinerators only in that the catalytic bed must be inserted in its holder, and temperature and/or pressure instructions relative to operating conditions preceding and following the catalyst be connected. The installation procedures presented for thermal incinerators and stacks apply, with little or no modification, to flares.

11.4 Operation

There are three distinct categories of incinerator operation:

1) Automatic
2) Semiautomatic
3) Manual

The categories refer specifically to the method of startup and then flame supervision, and have been adopted by various insurance agencies (such as Factory Mutual) and by National Fire Protection Association (NFPA). Most incinerator burners and fuel systems are designed in accordance with one of these insurance agency's guidelines, whether or not insurance is actually obtained from them. Often, valves and instruments can be purchased with preapproval by Factory Mutual.

Typically, the following procedure must be followed to start up any thermal incinerator. Each function, or group of functions, is interlocked electrically so that each must be accomplished before the next step can occur:

1) Prepurge
   a) Power on to control system
   b) Fan motor on
   c) Minimum airflow verified
   d) Fuel valves closed
2) Purge
   a) Airflow to greater than 70% of maximum rate
   b) Purge timer starts. The length of time is set to allow the furnace to be completely purged of any combustible fumes. Eight complete volume changes are considered sufficient.
3) Ignitionr sequence
   a) Airflow to light-off rate
   b) Auxiliary fuel pressure verified as sufficient; fuel temperature verified if required
   c) Ignition electric spark mechanism starts
   d) Ignition timer starts
   e) Ignition fuel valve opens
   f) Fuel is ignited in furnace by sparking mechanism
   g) Ignition flame sensed by flame scanner or flame rod
   h) Ignition timer ends time cycle, shuts off sparking mechanism. If no flame is sensed by the scanner or flame rod, ignitor fuel valve immediately closes, and post-purge cycle starts.
4) Main burner sequence
   a) Main burner fuel-control valve verified in light-off position
   b) For liquid fuel only: atomizing fluid (air or steam) valve opens
   c) Burner timer starts cycle
   d) Main burner fuel shut-off valves open allowing fuel to main burner

   e) Main flame ignites from ignitor flame
   f) Burner timer ends time cycle, ignitor fuel valve closes. If no flame is sensed by the flame scanner, ignitor fuel valve immediately closes, and post-purge cycle starts.
   g) Interlocks verifying light-off positions on fan and control valves are now bypassed, allowing fuel and air rates to be adjusted
   h) Main burner is now in service as released to modulate but subject to the following conditions:
      i) Fan on
      ii) Airflow proven
      iii) Flame detected
      iv) Fuel pressure adequate
      v) Furnace temperature not high
5) Normal operation
   a) Fuel and airflow are now adjusted to provide optimum furnace temperature for incinerator. This may be performed manually or automatically. On refractory-lined units, warm-up must be performed gradually to prevent thermal shock to the refractory.
   b) Once the incinerator reaches its operating temperature, fumes can be introduced into the furnace.
   c) Proper incineration operation is maintained by controlling incinerator outlet temperature and fume, auxiliary fuel, and airflow rates.
6) Normal shutdown
   a) Normal shutdown is accomplished by closing the fume flow valves, the auxiliary fuel valves. The systems then proceed to the post-purge cycle.
7) Post-purge
   a) Fan adjusted to maximum flow rate
   b) Post-purge timer start time cycle
   c) Keep air flowing until five complete air changes have occurred in the furnace
   d) Post-purge timer ends cycle, fan turned off, and incinerator is shut down

An automatic system requires only that an operator throw a switch to start the operation. The control mechanism then starts the incinerator, including the fans and other related equipment, in a pre-sequenced order: brings it up to the temperature; and, maintains proper incinerator operation for as long as required. Shutdown is usually initiated by another switch; the control system will then carry out all the operations necessary to shut down the incinerator.

A distinguishing feature of the "automatic" systems is the ability to "recycle." This includes recycling to reattempt to light-off if the previous attempt failed, and shutting down or starting up again as fume flow dictates. This is not a feature of the semiautomatic systems.

Semiautomatic operation is the most common method. It involves considerably more human input than does the completely automatic operation. In this type of operation, the operator must perform certain operations when called for by the control system. This input is especially required during the startup sequence. Semiautomatic control is a broadly applied term and may refer to a system that requires the operator to manually operate valves or merely to push certain buttons. However, in all cases, proper sequencing is enforced by the control systems. Emergency shutdown, however, is completely automatic.

The length of time required to bring the incinerator from "cold" to its normal operating temperature varies with the type and thickness of the lining and with the actual operating temperature. Typically, the warm-up period may require several hours. For a thick lining (6–12 in. of refractory plus any insulating block), the manufacturer's recommendation is usually to start initially with an outlet temperature of 200°F and hold for 1 hour; thereafter, increase the outlet temperature at a rate of 100°F/hr until the outlet temperature is 600°F. The outlet temperature may then be increased at a rate of 200–400°F/hr until the final operating temperature is reached. This process may take an entire shift and wastes considerable amounts of fuel; it is one of the principal disadvantages of using a refractory-lined furnace.

In actual practice, however, the manufacturer's recommended warm-up rate schedule is seldom followed, especially for relatively thin linings and for incinerators that operate only on an 8–16-hour shift. Warm-up periods of an hour or less are certainly not uncommon, regardless of lining thickness. The penalty for this is increased inspection and maintenance of patching lining. Certainly, reduced refractory life can also be expected from such a procedure, but this must be balanced against the wasted fuel usage required for longer warm-up periods. Using short warm-up periods, one can expect to replace the lining approximately once a year compared with a 10–15-year interval using the recommended procedure. A shortened warm-up period should never be used for the initial curing period or if the refractory has been wetted or soaked with water. A quick warm-up period under these circumstances will result in the immediate loss of the lining.

No special cool-down procedure is usually followed. Usually, the incinerator and fans are simply turned off and allowed to cool naturally. Under these conditions, it will take between 8 and 24 hours for the refractory to reach ambient temperature; however, the lining will usually be cool enough for inspection after a 4–12-hour period. Leaving the fan on during the cool-down period usually does not cause a serious problem but should be avoided, as it does reduce refractory life and increase operating cost. Water should *NEVER* be sprayed on the refractory to cool it down. Although this may seem obvious it nevertheless happens quite frequently if the maintenance staff is not properly trained.

Normal operation of a fume incinerator should be quite simple. A controller should be incorporated into the design to maintain the outlet temperature at a fixed value by varying the auxiliary fuel input. Combustion air (assuming limited air in the fume stream) is usually controlled by the amount required for combustion of the auxiliary fuel and the balance to control the outlet temperature of the furnace.

11.5 Maintenance

Most fume incinerators are custom-designed within certain basic parameters. Therefore, they are likely to be accompanied by a very complete instruction manual which should include the manufacturer's basic maintenance instructions from all the sub-suppliers. That is, the incinerator manufacturer will have bought equipment such as the refractories, valves, and controls from other suppliers. The operating and maintenance instructions for this equipment will be quite extensive and complete because it has been written by the original manufacturer, who is concerned only with particular items. The instruction manual is therefore a very useful document from a maintenance viewpoint and should be followed explicitly. The instruction manual may not be so useful from a system operation viewpoint because fume incinerator systems are usually custom-designed and tuned during commissioning to account for site specifics and field modifications.

There are, however, some general maintenance guidelines that can be discussed. The refractory should be inspected on a regular basis. Cracks that may develop, especially in brick joints, and thermal shock damage (spalling) should be repaired with a suitable material of the same thermal properties. The burner should be inspected at regular intervals for signs of warpage, pluggage, or corrosion. Moving parts should be lubricated with graphite or a similar high-temperature lubricant. Lubricants that carbonize under any circumstances should not be used. Also clear any dirt, mortar, carbon, or other foreign matter from the burner area. Inspect the pressure seals around any parts projecting through the burner or incinerator shell. Usually, these are asbestos rope packing glands and should be fairly tight after the adjusting/retaining screws have been loosened. Lubricate these seals only with flake or powder graphite. These should never be mixed with oil, as the oil will carbonize. On burners using gas as an auxiliary fuel, the gas nozzles should be free of corrosion and should be cleared of any deposits.

The outer shell of the incinerator should be inspected, especially when new or when a new lining has been installed for signs of thermal shock. That is, welds, especially at the outlet, should be checked for hairline cracks, which are the first signs of poor thermal design.

The auxiliary fuel piping train should be inspected in accordance with the manufacturer's instructions. Electrically operated valves and interlock switches should be inspected frequently for conditions that might cause "shorting" (e.g., dirty contacts, moisture leaks, deteriorating insulation). Air-supply lines and filters (to air-operated valves) should be inspected for dirt, blockages, or water condensation. The valves themselves are usually provided with air-supply pressure gauges, and these should be checked occasionally for accuracy. If there are shut-off dampers in the ductwork to or from the incinerator, their seals should be checked frequently.

Maintenance procedures for catalytic incinerators should include catalyst cleaning every three months to a year, and catalyst activity testing every three years or if the catalyst shows signs of reduced efficiency. Cleaning is usually accomplished by blowing clean compressed air through the catalyst element, by vacuuming, or by washing with water or a mild detergent not containing phosphates. Iron oxide deposits can be removed by soaking with a mild organic acid followed by a water rinse.

11.6 Improving Operation and Performance

Improving the performance of a thermal incinerator basically involved the optimization of fume combustion. Ideally, no more combustion air should be used than is required for the complete combustion of the fumes and the auxiliary fuel. The auxiliary fuel should be used only in amounts required to maintain the design furnace temperature.

An incinerator operating efficiently should have only 1% or 2% O_2 and 0–1% combustibles in the outlet gases. Monitors are available which can indicate these parameters and provide automatic control of the incinerator when required. Often times excess oxygen will need to be higher (on the order of 6–15% O_2) in order to maintain low outlet temperatures.[7,8]

11.7 Recent Developments

The basic design of incinerators equipment has remained relatively unchanged since first used in the early part of the twentieth century. All have essentially employed the same capture mechanisms used in the past. The main areas of advancement in incinerators are in the ancillary equipment.

With the exponential advances made to computing power, automated control systems are advancing in sophistication at a similar rate. Control systems can monitor and modulate systems at unprecedented levels of sophistication. These systems maintain a cross-limit on the fuel, air, and fumes to ensure there is sufficient air for combustion and complete burnout. They are also able to automatically trim fuel or air as required to maintain desired setpoints.

Due to the rising cost and limited availability of fuels, heat-recovery systems are becoming an integral and more advanced part of most incinerators. Heat recovered from hot flue gas can be used in multiple ways to reduce energy consumption. One way is to use an economizer heat exchanger to preheat a cooler process stream. The second way is to use the recovered heat in another process such as a drying oven. The most common way is to use the waste heat to produce steam in a heat-recovery steam generator boiler.[7,8]

Another key advancement in incineration technology is Low and Ultra-Low NO_x auxiliary burners (see Chapter 18). In order to destroy waste pollutants, fossil fuels are combusted, and this process inherently produces its own set of pollutants. One of the primary concerns of fossil fuel combustion is NO_x emissions. NO_x is formed by the oxidation of both atmospheric and fuel-bound nitrogen during the combustion process. NO_x is of environmental concern as it negatively affects the atmosphere by producing acid rain, smog, and upper atmosphere ozone depletion. NO_x formation from a burner is determined by the interaction of chemical and physical processes occurring within the incinerator furnace. There are three principal chemical processes for NO_x formation: thermal, prompt, and fuel NO_x. Thermal NO_x is the primary contributor of overall NO_x emissions and is the one considered in this discussion. Thermal NO_x results from the oxidation of atmospheric nitrogen from the combustion air in the high-temperature post-flame region of the boiler. The major parameters that influence thermal NO_x formation are: (i) temperature; (ii) oxygen concentrations; (iii) nitrogen concentrations; and (iv) residence time. If temperature, oxygen concentrations, or nitrogen concentrations can be reduced quickly after combustion, thermal NO_x formation is suppressed. This is a major issue for incinerators because it is desired to maintain elevated temperature and residence time to ensure complete destruction of the fumes. Advancements in burner design have led to a significantly reduced thermal NO_x emissions by two orders of magnitude over the last 30 years.[9]

11.8 Conclusions

Thermal incinerators, catalytic incinerators, and flares may be used in appropriate situations to control smoke, odor, and combustible fumes. They are not particularly complicated devices, but their efficiency depends, to a large degree, upon proper operation and maintenance. Improper operation and maintenance can lead to unsafe combustion conditions and premature failure of the insulation or catalyst, which can significantly increase the system's lifetime costs. It is absolutely essential, therefore, that operation and maintenance personnel be skilled in their tasks and thoroughly familiar with detailed instruction manual. The procedures provided in this chapter should be followed closely.

References

1 **S. Turns**, *An Introduction to Combustion*, 3rd Ed., McGraw Hill, New York, NY, 2012.
2 **M. G. Klett and J. B. Galeski**, *Flare Systems Study*. EPA-600/2-76-079, Research Triangle Park, NC, 1976.
3 **A.J. Buonicore, and L. Theodore**, *Industrial Control Equipment of Gaseous Pollutants*, Vol. 2, CRC Press, Inc., Boca Raton, FL, 1975.
4 **L. Theodore**, *Air Pollution Control Equipment Calculations*, John Wiley & Sons, Hoboken, NJ, 2008.
5 **L. Theodore**, *Ask the Experts: Designing Thermal Afterburners*, CEP, 2005.
6 **F.L. Cross, and H.E. Hesketh**, *Handbook for the Operation and Maintenance of Air Pollution Control Equipment*, Chapter 5, Technomic Publishing Co., Inc., Westport, CT, 1975.
7 **J. Santoleri, J. Reynolds, and L. Theodore**, *Introduction to Hazardous Waste Incineration*, 2nd Ed., John Wiley & Sons, Hoboken, NJ, 2000.
8 **L. Theodore**, *Chemical Reaction Analysis and Applications for the Practicing Engineer*, John Wiley & Sons, Hoboken, NJ, 2012.
9 **J. Richardson**, *Personal Notes*, 2023.

12

Condensers

Condensation is the process of converting a gas or vapor to a liquid. Any gas can be condensed to a liquid by sufficiency lowering its temperature and/or increasing its pressure. The most common approach is to reduce the temperature of the gas stream, since increasing the pressure of a gas can be expensive.

Condensers are simple, relatively inexpensive devices that normally use water or air to cool and condense a vapor stream. Since these devices are usually not required or capable of reaching low temperatures (below 100°F), high removal efficiencies of most gaseous pollutants are not obtained unless the vapors condense at high temperatures. Condensers are typically used as pretreatment devices. They are used ahead of adsorbers, absorbers, and incinerators (see Chapters 9–11), to reduce the total gas volume to be treated by more expensive control equipment. Used in this manner, this device can help reduce the overall cost of the control system.

When a hot vapor stream contacts a cooler medium, heat is transferred from the hot gases to the cooler medium. As the temperature of the vapor stream is cooled, the average kinetic energy of the gas is reduced. Ultimately the gas molecules are slowed down and crowded so closely together that the attractive forces (van der Waals' forces) between the molecules cause them to condense into a liquid. The two conditions that aid condensation are low temperatures so that the kinetic energy of the gas molecules is low; and high pressures so that the molecules are brought close together. The actual conditions at which a particular gas molecule will condense depends on its physical and chemical properties. Condensation occurs when the partial pressure of the pollutant in the gas stream equals its vapor pressure as a pure substance at the operating temperature.

As described above, condensation of a gas can occur in three ways[1,2]:

1) At a given temperature, the system pressure is increased (compressing the gas volume) until the partial pressure of the gas equals its vapor pressure;
2) At a fixed pressure, the gas is cooled until the partial pressure equals its vapor pressure;
3) By using a combination of compression and cooling of the gas until its partial pressure equals its vapor pressure.

Condensation of pollutant vapors has long been favored as one of the most straightforward, measurable, controllable, and effective containment technologies available. It is the simplest of the nondestructive methods and can often render pollutants available for recovery, recycling, or both. With today's high levels of mandated removal efficiencies, however, the application of simple condensation technology is becoming increasingly limited. For example, in the area of volatile organic compounds (VOCs), their implication in tropospheric ozone formation (and frequently other toxic or hazardous characteristics) assures increased government pressure for further emissions regulations. Because of the volatility of these compounds, required removal efficiencies are difficult to achieve with simple condensation. There are available, nevertheless, a number of enhanced methods of condensation that can potentially meet these requirements.

Although condensation can be achieved by increasing the pressure, reducing the temperature, or both, in practice, air pollution control condensers almost always operate by removing enough heat from the vapor to cause condensation. As a hot vapor stream contacts a cooler surface, the temperature of the gas stream lowers to a point where a pollutant's vapor pressure is at or below its entering partial pressure in the gas stream. On a fundamental level, this means the average kinetic energy of the gas molecules is reduced as the vapor stream is cooled. Eventually, the gas molecules are brought closer together, having slowed down via this cooling. Ultimately, the attractive van der Waals' forces between the molecules force them to condense into a liquid. It should be noted that typically, condensation is used in process streams from reactors, dryers, distillation columns, absorbers, etc. In these applications, the condenser is not really viewed as an emissions

Optimizing Air Pollution Control Equipment Performance: Operation & Maintenance, First Edition. Jay Richardson and Louis Theodore.

control device. However, these applications often use secondary condensers that operate at still lower temperatures and function primarily as control devices. Finally, it should be noted that condensation does have significant uses for partial contaminant removal in closed circulation systems.

12.1 Description of Control Device

As indicated in the previous section, condensation can be accomplished by increasing pressure or decreasing temperature (removing heat). In practice, air pollution control condensers operate through the extraction of heat. Condensers differ in the means of removing heat and the type of device used. The two different means of condensing are direct contact, where the cooling medium with vapors and the condensate are intimately mixed and combined, and indirect (or surface contact), where the cooling medium and vapor/condensate are separated by a surface area of some type.

Contact condensers are simpler, less expensive to install, and require less auxiliary equipment and maintenance. The condensate/coolant from a contact condenser has a volume 10–20 times that of a surface condenser. This condensate often cannot be reused and may pose a waste disposal problem unless the dilution of the pollutant is sufficient to meet regulatory requirements. Some typical contact condensers are shown in Figure 12.1.

Surface condensers form the bulk of the condensers used for air pollution control. Some of the applicable types of surface condensers are shell and tube, double pipe, spiral plate, flat plate, air-cooled, and various extended surface tubular units.[2] This chapter focuses on shell-and-tube condensers because they are so widely used in industry and have been standardized by the Tubular Exchanger Manufacturers Association (TEMA).

Condensing can be accomplished in either the shell or tubes. The economics, maintenance, and operational ramifications of the allocation of fluids are extremely important, especially if extended surface tubing is being considered. The designer should be given as much latitude as possible in specifying the condenser. A diagram of a generic shell-and-tube heat exchanger is shown in Figure 12.2.[1]

Air-cooled condensers consist of a rectangular bank of high-finned tubes, a fan, a plenum for even distribution of air to the rectangular face of the tube bank, a header for vapor inlet and condensate outlet, and a steel support structure. Fins are usually aluminum 0.5–0.625 in. in height that are applied to a bare tube by tension winding, soldering, or sold-extruded from the tube itself. Bimetallic tubes can be used to provide process corrosion resistance on the inside and aluminum

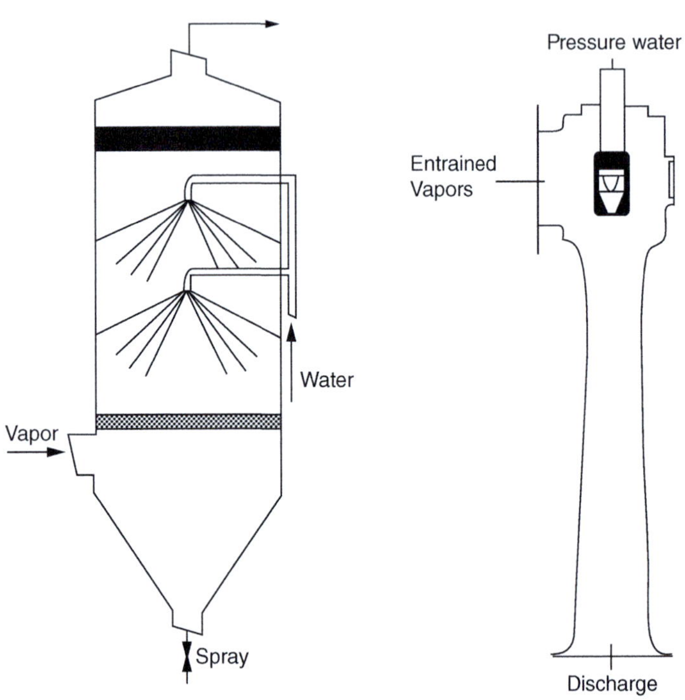

Figure 12.1 Contact condensers: spray (left) and jet (right).

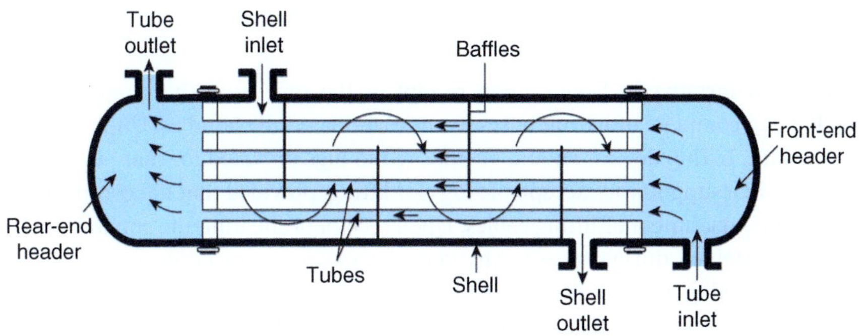

Figure 12.2 Generic diagram of a shell-and-tube heat exchanger.

extruded fins on the outside. Headers for the tube side frequently contain removable gasketed plugs corresponding to each tube end and are used for access to the individual tube-to-tube sheet joint for maintenance.

Nearly all tubular exchangers employ roller expanding of the tube ends into tube holes drilled into tube sheets as a means of providing a leakproof seal. TEMA requires that tube holes be of close tolerance and contain two concentric grooves. These grooves give the tube joint greater strength but do not increase sealing ability and may actually decrease it. Research is currently being done to clarify what is the best procedure to follow for a given set of materials, pressures, temperatures, and loadings. Welding or soldering of tubes to tube sheets may be performed for additional leak tightness or in lieu of roller expanding.

12.2 Design Considerations

A detailed description of heat exchanger equipment design is not warranted in a book of this nature. Extensive information on energy relationships, the log mean temperature difference (LMTD) driving force, overall heat transfer coefficients, and the classic heat transfer equation is available in the literature.[2–8]

The flow of heat from a hot fluid to a cooler fluid through a solid wall is a situation often encountered in engineering equipment. The heat absorbed by the cool fluid or given up by the hot fluid may be sensible heat, causing a temperature change in the fluid, or it may be latent heat, causing a phase change such as vaporization or condensation. For example, in a typical waste heat boiler, the hot flue gas gives up heat to water flowing through thin metal tube walls separating the two fluids. As the flue gas loses heat, its temperature drops. As the water gains heat, its temperature quickly reaches the boiling point where it continues to absorb heat with no further temperature rise as it changes into steam. The rate of heat transfer between the two streams, assuming no heat loss due to the surroundings, may be calculated in order to design heat transfer equipment. To calculate the required energy, one must know more than just the heat transfer rate calculated by the enthalpy (energy) balances. The rate at which heat can travel from the hot fluid through the tube walls, into the cold fluid must also be considered in the calculation of certain design variables (e.g., the contact area). The slower this rate is, for a given hot and cold fluid flow rate, the more contact area is required. The use of the overall heat transfer coefficient (U) is a simple, yet powerful concept.[2,4–6] In most applications, it combines both conduction and convection effects, although heat transfer by radiation can also be included. In actual practice, it is not uncommon for vendors to provide a numerical value for U. For example, a typical value for U for estimating heat losses from a heat exchanger is approximately 0.1 Btu/hr-ft^2-$^\circ$F.

During the heat exchange operation with liquids and/or gases, a "dirt" film gradually builds up on the exchanger surface(s). This deposit may be rust, boiler scale, silt, coke, or any number of things. Its effect, which is referred to as *fouling*, is to increase the thermal resistance, R, which results in decreased performance. The nature of the rate of deposit is generally difficult to predict a priori. Therefore, only the performance of clean exchangers is usually guaranteed. The fouling resistance is often obtained from field, pilot, or lab data, or from experience.[3] This unknown factor enters into every design. The scale of fouling is dependent on the fluids, their temperature, velocity, and to a certain extent, the nature of the heat transfer surface and its chemical composition. Due to the unknown nature of some of the assumptions, these fouling factors can markedly affect the design of the heat transfer equipment. As noted above, details on these heat exchangers are available in the literature.

12.3 Installation Procedures

Preparation of a condenser or heat exchanger for installation begins upon receipt of the unit from the manufacturer. Condensers are shipped domestically using skids for complete units and boxes or crates for bare tube bundles. Units are normally removed from trucks using a crane or forklift. Lifting devices should be attached to lugs provided for that purpose or used with slings wrapped around the main shell. Shell supports are acceptable lugs for lifting, provided that the complete set of supports are used together; never use nozzles for attachment of lifting cables. Upon receipt of the unit, the general condition should be noted to determine any damage sustained during transit. Any dents, cracks, or connections out-of-square should be reported to the manufacturer prior to attempting to install the unit. Flanged connections are blanked with plywood, masonite, or equivalent covers, and threaded connections are blanked with suitable pipe plugs. These closures are to avoid entry of debris into the unit during shipping and handling, and should remain in place until actual piping connections are made.

12.3.1 Clearance Provision

Sufficient clearance is required for at least inspection of the unit or in-place maintenance. Inspection of condensers requires minimal clearances for the following: access to inspection parts if provided, removal of channel or bonnet covers, and inspection of tube sheets and tube-to-tube sheet joints. If removal of tubes or tube bundles in place is anticipated, provision should be provided in equipment layout. Actual clearance requirements can be determined from the condenser setting plan.

12.3.2 Foundations

Condensers must be supported on structures of sufficient rigidity to avoid imposing excessive strains on nozzles due to settling. Horizontal units with saddle-type shell supports are normally supplied with slotted holes in one support to allow for expansion. Foundation bolts in these supports should be loose enough to allow movement.

12.3.3 Leveling

Condensers should be carefully leveled and squared to ensure proper drainage, venting, and alignment with piping. On occasion, condensers are purposely angled to facilitate venting and drainage, and alignment with piping becomes the prime concern.

12.3.4 Piping Considerations

The following guidelines for piping are necessary to avoid excessive strains, mechanical vibration, and access for regular inspection.

1) Sufficient support devices are required to prevent the weight of piping and fittings from being imposed on the condenser.
2) Piping should have sufficient expansion joints or bends to minimize expansion stresses.
3) Forcing alignment of piping should be avoided so that residual strains will not be imposed on nozzles.
4) If external forces and moments are unavoidable, their magnitude should be determined and made known to the manufacturer so that any necessary stress analyses can be performed.
5) Surge drums or sufficient length of piping to the condenser should be provided to minimize pulsations and mechanical vibrations.
6) Valves and bypasses should be provided to permit inspection or maintenance in order to isolate the condenser during periods other than complete system shutdown.
7) Plugged drains and vents are typically provided at low and high points of shell-tube sides not otherwise drained or vented by nozzles. These connections are functional during startup, operation, and shutdown, and should be piped up for either continuous or periodic use and never left plugged.
8) Instrument connections are provided either on the condenser nozzles or in the piping close to the condenser. Pressure and temperature indicators should be installed to validate the initial performance of the condenser as well as to demonstrate the need to inspection or maintenance.

12.4 Operation

12.4.1 Safe Working Conditions

The maximum allowable working pressures and temperatures are indicated on the condenser's nameplate. These values must not be exceeded. Special precautions should be taken if any individual part of the condenser is designed for a maximum temperature lower than the condenser as a whole. The most common example is some copper-alloy tubing with a maximum allowable temperature lower than the actual inlet gas temperature. This is done to compensate for the low strength levels of some brasses or other copper alloys at elevated temperatures. An adequate flow of the cooling medium must be maintained at all times.

Condensers are designed for a particular fluid throughput. Generally, a reasonable overload can be tolerated without causing damage. If operated at excessive flow rates, erosion or destructive vibration could result. Erosion could also occur at normally acceptable flow rates if other conditions, such as entrained liquids or particulates in a gas stream or abrasive solids in a liquid stream, are present. Evidence of erosion should be investigated to determine the cause. Vibration can be propagated by other than flow overloads (e.g., improper design, fluid maldistribution, or corrosion/erosion of internal flow-directing devices such as baffles). Considerable study and research have been applied in earlier years to develop a reliable vibration analysis procedure to predict or correct damaging vibration. At this point, the developed correlations are considered "state of the art," yet most manufacturers have the capability of applying some type of vibration check when designing a condenser. Vibration can produce severe mechanical damage, and operation should not be continued when an audible vibration disturbance is evident.

12.4.2 Startup

Condensers should be warmed up slowly and uniformly; the higher the temperature ranges, the slower the warm-up should be. This is generally accomplished by introducing the coolant and bringing the flow rate to the design level and gradually adding the vapor. For fixed-tube-sheet units with different shell-and-tube materials, consideration should be given to differential expansion of shell and tubes. As fluids are added, the respective areas should be vented to ensure complete distribution. A procedure other than this could cause large differences in temperature between adjacent parts of the condenser and result in leaks or other damage. It is recommended that gasketed joints be inspected after continuous full-flow operation has been established. Handling, temperature fluctuations, and yielding of gaskets or bolting may necessitate retightening of the bolting.

12.4.3 Shut Down

Cooling down is generally accomplished by shutting off the vapor stream first and then the cooling stream. Again, fixed-tube-sheet condensers require consideration of differential expansion of the shell and tubes. Condensers containing flammable, corrosive, or high-freezing point fluids should be thoroughly drained for prolonged outages.

12.5 Maintenance

12.5.1 Inspection

Recommended maintenance of condensers requires regular inspection to ensure mechanical soundness of the unit and a level of performance consistent with the original design criteria. A brief general inspection should be performed on a regular basis while the unit is operating. Vibratory disturbances, leaking gasketed joints, excessive pressure drop, decreased efficiency indicated by higher gas outlet temperature or lower condensate rates, and intermixing of fluids are all signs that thorough inspection and maintenance procedures are required.

Complete inspection requires shutdown of the condenser for access to internals and pressure testing and cleaning. Scheduling can only be determined from experience and general inspections. Tube internals and exteriors, where accessible, should be visually inspected for fouling, corrosion, or damage. The nature of any metal deterioration should be investigated to determine properly the anticipated life of the equipment or possible corrective action. Possible causes of deterioration include general corrosion, intergranular corrosion, stress cracking, galvanic corrosion, impingement, or erosion attack.

12.5.2 Cleaning

Fouling of condensers is the deposition of foreign material on either the interior or exterior of tubes. Evidence of fouling in operation is increased pressure drop and general decreasing performance. Fouling can be so severe that tubes are completely plugged, resulting in thermal stresses and subsequent mechanical damage to equipment.

The nature of the deposited foul determines the method of cleaning. Soft deposits can be removed by steam, hot water, various chemical solvents, or brushing. Plant experience can determine which method to use. Chemical cleaning should be performed by contractors specialized in the field who will consider the deposit to be removed and the materials on construction. If the cleaning method involves elevated temperatures, considerations should be given to thermal stresses induced in the tubes; using steam to clean individual tubes can loosen the tube-to-tube sheet joints.

Mechanical methods of cleaning are useful for soft and hard deposits. There are numerous tools for cleaning tube interiors: brushes, scrapers, and various rotating cutter-type devices. The condenser manufacturer or suppliers of tube tools can be consulted in the selection of the correct tool for the particular deposit. When cutting or scraping deposits, care should be exercised to avoid damaging tubes.

Cleaning of tube exteriors is generally performed using chemicals, steam, or other suitable fluids. Mechanical cleaning is preferred but requires that the tubes be exposed, as in a typical air-cooled condenser, or capable of being exposed, as in a removable bundle shell-and-tube condenser. The layout pattern of the tubes must provide sufficient intersecting empty lanes between the tubes, as in square pitch. Mechanical cleaning of tube bundles, if necessary, requires the utmost care to avoid damaging tubes or fins.

12.5.3 Testing

Proper maintenance requires testing of a condenser to check the integrity of the following: tubes, tube-to-tube sheet joints, welds, and gasketed joints. The normal procedure consists of pressuring of the shell with water or air at the nameplate-specified test pressure and viewing the shell welds and the face of the tube sheet for leaks in the tube sheet joints or tubes. Water should be at ambient temperature to avoid false indications due to condensation. Pneumatic testing requires extra care because of the destructive nature of a rupture or explosion or fire hazards when residual flammable materials are present. Condensers of the straight-tube floating heat construction require a test gland to perform the test. Tube bundles without shells are tested by pressuring the tubes and viewing the length of the tubes and the back face of the tube sheets.

Corrective action for leaking tube-to-tube sheet joints requires expanding the tube end with a suitable roller-type expander. Good practice calls for an approximate 8% reduction in wall thickness after metal-to-metal contact between tube and tube hole. Tube expanding should not extend beyond 1/8 in. of the inner tube-sheet face to avoid cutting the tube. Care should be exercised to avoid over-rolling the tube which can cause work hardening of the material, an insecure seal, and/or stress-corrosion cracking of the tube.

Defective tubes can either be replaced or plugged. Replacing tubes requires special tools and equipment; the user should contact the manufacturer or a contractor qualified in repair. Plugging of tubes, although a temporary solution, is acceptable provided that the percentage of the total number of tubes per tube pass to be plugged is not excessive. The type of plug to be used is a tapered one-piece or two-piece metal plug suitable for the tube material and inside diameter. Care should be exercised in seating plugs to avoid damaging the tube sheets. If a significant number of tube or tube joint failures are clustered in a given area of the tube layout, their location should be noted and reported to the manufacturer. A concentration of failures is usually caused by other than corrosion (e.g., impingement, erosion, or vibration).

12.6 Improving Operation and Performance

Within the constraints of the existing system, improving operation and performance refers to maintaining operation and original or consistent performance. There are several factors previously mentioned that are critical to the design and performance of a condenser: operating pressure, amount of noncondensable gases in the vapor stream, coolant temperature and flow rate, and mechanical soundness. Any pressure drop in the vapor line upstream of the condenser should be minimized. Deaerators or similar devices should be operational where necessary to remove gases in solution with liquids. Proper and regular venting of equipment and leakproof gasketed joints in vacuum systems are all necessary to prevent gas binding and alteration of condensing equilibrium. Coolant flow rates and temperatures should be checked regularly to ensure that they are in accordance with the original design criteria. The importance of this can be illustrated merely by comparing the winter and summer performance due to fouling will generally be exhibited by a gradual decrease in efficiency and should be corrected as soon as possible. Mechanical malfunctions can also be gradual, but will eventually be evidenced by near total lack of performance.

Fouling and mechanical soundness can only be controlled by regular and complete maintenance. In some cases, fouling is much worse than predicted and requires frequent cleaning regardless of the precautions taken in the original design. These cases require special designs to alleviate the problems associated with fouling. A leading PVC manufacturer found that the carryover of polymer reduced the efficiency of its monomer condenser and caused frequent downtime. The solution was providing polished internals and high condensate loading in a vertical downflow shell-and-tube condenser. In another case, a major pharmaceutical intermediate manufacturer had catalyst carryover to a vertical downflow shell-and-tube condenser which accumulated on the tube internals. The solution was to recirculate condensate to the top of the unit and spray it over the tube-sheet face to create a film descending down the tubes to rinse the tubes clean.

Most condenser manufacturers will provide designs for alternate conditions as a guide to estimating the cost of improving efficiency via alternate coolant flow rates and temperatures as well as alternate configurations (e.g., vertical, horizontal, shell side or tube side).

12.7 Recent Developments

The basic design of condensers has remained relatively unchanged since first used in the early part of the twentieth century. Some modest equipment changes and new types of devices have appeared in the last 20 years, but all have essentially employed the same capture mechanism used in the past. A review of the literature over the past quarter of a century produced little in terms of improvements regarding the state of the art for condensers. The following can be offered at this point in time.

The role of the heat pipes in condenser operations has been reviewed.[2–7] A heat pipe is a relatively new heat-transfer device that combines the principles of both thermal conductivity and phase transition to efficiently manage the transfer of heat between two interferences. When heat is applied to a heat pipe, the liquid in the wick (chosen for compatibility with both the heat pipe material of construction and the temperature of the hot and cold surfaces) heats and evaporates. As the evaporating fluid fills the heat pipe's hollow center, it diffuses axially along its length. Condensation of the downstream vapor occurs wherever the temperature is even slightly below that of the evaporation area. As it condenses, the vapor gives up the heat it acquired during evaporation and establishes a reverse vapor flow pattern in the pipe. The condensed liquid flows through capillary action within the wick, back to the heat source, completing the heat transfer. Heat pipes are thus capable of transferring process heat to the surroundings and have been used to recover thermal energy from process gas streams.

Defective tubes can either be replaced or plugged. Replacing tubes requires special tools and equipment. As noted earlier, the user should contact the manufacturer or a contractor qualified in repair. Plugging of tubes, although a temporary solution, is acceptable provided that the percentage of the total number of tubes per tube pass to be plugged is not excessive.

12.8 Conclusions

Condensation, which has long been a valuable operation for the chemical, petrochemical, and related industries, is now, with the advent of strict environmental concern, an important air pollution control device. Its contemporary application consists of increasing the efficiency of replacement equipment via more rigorous computerized design and consideration of vent streams, operation of condensers for streams previously vented to the atmosphere, and use of condensers as a precleaner for other air pollution control equipment.

More often than not, a vapor condenser is a secondary control device since it cannot achieve the collection efficiency required by industry using ordinarily available and economical cooling mediums. One advantage of condensation is that it can reduce the load, improve the performance and extend the operating life of more expensive devices surface condensers, and frequently return product or an intermediate to be recycled into the process. Contact condensers can return a condensate of sufficient dilution to be directly disposed of.

Virtually all heat exchangers that for decades have been used for process condensing – shell and tube, double pipe, air-cooled, flat plate, spiral pate, barometric, jet, and spray – are applicable for air pollution control.

References

1 **Y.A. Çengel, and A.J. Ghajar**, *Heat and Mass Transfer: Fundamentals & Applications*, McGraw-Hill, New York, NY, 2011.
2 **L. Farag, and J. Reynold**, *Heat Transfer, A Theodore Tutorial*, Theodore Tutorials, East Williston, NY, 1996. (originallypublished by USEPA/APTI, Research Triangle Park, NC).
3 **J.B. Fourier**, *Theori Analytique de la Chaleur*, Gauthier-Villars, Paris, 1822. German translation by Weinstein, Springer, Berlin. 1884; Ann. Chim. Et Phys, 37(2), 291, 1828; Pogg. Ann, 13, 327, 1828.
4 **D. Kern**, *Process Heat Transfer*, McGraw-Hill, New York, NY, 1950.
5 **L. Theodore**, *Heat Transfer Applications for the Practicing Engineer*, John Wiley & Sons, Hoboken, NJ, 2011.
6 **A.M. Flynn, T. Akashige, and L. Theodore**, *Kern's Process Heat Transfer*, 2nd Ed., Scrivener-Wiley, Beverly, MA, 2019.
7 **L. Theodore, and A.J. Buonicore**, *Vapor Control by Condensation: Performance Equations and Design Procedures*, paper#75-23.2. Air Pollution Control Association Meeting, Boston, MA 1975.
8 **W. Connery, N. Kafes, A.J. Bounicore, and L. Theodore**, *Evaluating Surface Condensers for Air Pollution Control Applications*, Proceedings Third An. APCA-AICHE Conference on Energy and the Environment, Pittsburgh, PA, 276–282 1975.

13

Mechanical Collectors

A mechanical collector is a device that separates suspended particles from a gas by causing the gas stream to change direction while the particles, because of their inertia, tend to continue in their original direction and become separated from the gas.

The collection efficiency of mechanical collectors depends upon the motion of the gas stream and the physical properties of the suspended particulate. Separation forces are a function of the gas velocity, rate of change in direction, and of particle size and specific gravity. The forces acting on a particle can be as high as 2,500 times gravity.

Operation of mechanical collectors depends on separating forces produced by the gas stream motion. The forces developed are primarily a function of the gas stream's speed and rate of curvature, and of particle size and specific gravity. These forces acting on a particle to carry it into a collecting zone can be made as high as 2,500 times gravity.[1]

The two designs of mechanical collectors of interest to the practicing air pollution control equipment are the gravity settler and cyclone. Each one is described below.

The gravity settler was one of the first devices used to control particulate emissions. It is an expansion chamber in which the gas velocity is reduced, thus allowing the particle to settle out under the action of gravity. One primary feature of this device is that the external force causing particle separation from the gas stream is provided free by nature. This chamber's use in industry, however, is generally limited to the removal of larger-sized particles (40–60 μm in diameter).

Settling chambers have also been used to study the flow of particles in a gas stream. The data generated from these studies can be useful in the design of other higher-efficiency particulate emission control devices. Today's demand for cleaner air and stricter emission standards has relegated the settling chamber to use in research for testing or as a pre-cleaner/post-cleaner for other particulate control devices, for example: cyclones, electrostatic precipitators, venturi scrubbers, and fabric filters. The latter three devices will be discussed in the next three chapters.

Cyclones or inertial collectors, on the other hand, depend on another effect in addition to gravity, to lead to a successful separation process. This other mechanism is an inertial or momentum effect. It arises by changing the direction of the velocity of the gas and imparting a downward motion to the particle. From a calculation point of view, this induced particle motion is superimposed on the motion arising as a result of gravity. These cyclones provide a relatively low-cost method of removing particulate matter from exhaust gas streams. Cyclones are somewhat more complicated in design than simple gravity settling systems, and their removal efficiency is accordingly much better than that of settling chambers. However, cyclones are not as efficient as electrostatic precipitators, baghouses, and venturi scrubbers but are often installed as pre-cleaners before these more effective devices.

Although mechanical collectors have been used for many years to collect suspended materials, a satisfactory theory relating to the many operating variables has never been universally accepted. The performance of collectors is in some cases almost entirely based on experience and empirical data, probably because of the complexity of the aerodynamics involved in the collector flow. However, the relatively low capital and operating costs for mechanical collectors make them an attractive choice for particulate removal where the collection efficiency requirements are not high. These collectors can be gravity chambers, dynamic precipitators, or dry inertia separators.

Optimizing Air Pollution Control Equipment Performance: Operation & Maintenance, First Edition. Jay Richardson and Louis Theodore.

13.1 Description of Control Device

13.1.1 Gravity Settling Chambers

The gravity settler depends solely on the gravitational pull on the particles to collect the dust. The separator consists of a housing and hopper in which the velocity of the gas stream is made to drop rapidly below the transport velocity of the dust particle. The velocity reduction is accomplished by the sudden expansion into the enlarged housing. The simplest settling chamber consists of a single compartment in which the dust settles.

There are basically two types of dry gravity settlers: the simple expansion chamber and the multiple-tray settling chamber, as presented in Figure 13.1. The unit is constructed in the form of a long horizontal box with an inlet, an outlet, and dust collection hoppers. As noted earlier, these units primarily depend on gravity for the collection of the particles. The particle-laden gas stream enters the unit at the gas inlet. The gas stream then enters the expansion section of the duct. Expansion of the gas steam causes the gas velocity to be reduced. All particles in the gas stream are subject to the force of gravity. However, at reduced gas velocities (in the range of 1–10 ft/s) the larger particles are acted on preferentially by gravity and fall into the dust hopper(s). Theoretically, a settling chamber of infinite length could collect even the very small particles (<10 μm), but they are limited in practice to due to size constraints.

The physical description of a gravity settler includes length, width, height, numbers of shelves (if applicable), and ancillary equipment such as inlet and outlet ducts, cleaning mechanisms, and hoppers. Regarding the latter ancillary equipment, the collection hoppers (located at the bottom of the settle) are usually designed with positive-seal valves and must be emptied as dust buildup occurs. Dust buildup will vary depending on the concentration levels of particulate matter in the gas streams, especially in the case of heavy concentrations of particles greater than 60 μm in diameter.

A variation of the gravity settling chamber is a baffle chamber, sometimes referred to as an *inertial separator*. These units have baffles within the chamber to enhance particle separation and collection. This arises by changing the direction of the gas velocity and imparting a downward motion to the particle. This induced motion is superimposed on the motion due to gravity. Thus, particle collection is accomplished by gravity and an inertial or momentum effect. Particles as small as 20–40 μm can be collected.

The multiple-tray settling chamber also called the *Howard settling chamber* (see Figure 13.2) has several horizontal collection plates N are introduced to shorten the settling path of the particle and to improve the collection efficiency of small particles (as small as 15 μm in diameter). Each shelf or tray in the unit can collect dust that settles out by gravitational forces. Since the vertical distance that a particle must fall to be captured is less than the distance in a standard horizontal settling unit, the overall collection efficiency of the Howard settling chamber can be greater than that of the horizontal chamber. However, the gas must be uniformly distributed as it passes over each tray throughout the chamber. Uniform distribution is usually achieved by the use of gradual transitions, guide vanes, distributor screens, and perforated plates. The particles settle on the individual trays, which must be cleaned periodically. The vertical distance between trays may be as little as 1 in., making cleaning much more difficult than with the horizontal settling chamber. Other disadvantages include the tendency of trays to warp during high-temperature operations or temperature excursions and the inability of the unit to handle dust concentrations exceeding approximately 1.0 gr/ft^3 (2.29 g/m^3). For these reasons, the Howard settling chamber is rarely used for particulate emission control.

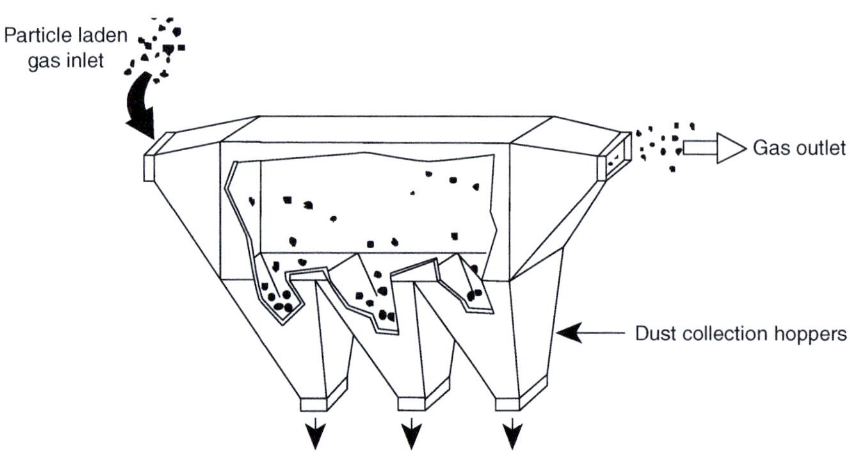

Figure 13.1 Horizontal flow settling chamber.

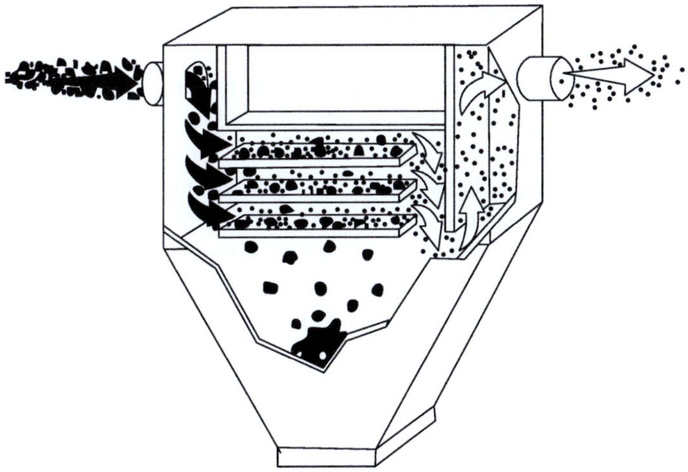

Figure 13.2 Howard settling chamber (multiple tray).

Inertial collectors, on the other hand, depend on another effect, in addition to gravity, to lead to a successful separation process. This other mechanism is an inertial or momentum effect. From a calculational point of view, this induced particle motion is superimposed on the motion arising as a result of gravity.

13.1.2 Cyclones

The most prevalent type of all mechanical collectors, especially the dry-inertia type of collection, is the cyclone. Although this type of inertial collector will rarely meet air pollution control codes when used as a primary cyclone collector, the dust collector more than justifies its popularity if applied correctly. It can increase the reliability of a secondary collector by removing excessive dust loadings from the inlet of scrubbers or electrostatic precipitators. It will act as a spark arrestor when installed upstream from a fabric filter. The cyclone can be designed in numerous arrangements and can be fabricated out of a variety of materials.

Cyclones provide a relatively low-cost method of removing particulate matter from exhaust gas streams. Cyclones have a higher removal efficiency than settling chambers. However, cyclones are not as efficient as electrostatic precipitators, baghouses, and venturi scrubbers but are often installed as pre-cleaners before these more effective devices. The low-cost cyclone device is used to reduce the cost and footprint of the larger, more efficient, and expensive devices.

Cyclones come in many sizes and shapes and have no moving parts. From the small 1 and 2 cm diameter source sampling cyclones used for particle size analysis to the large 5 m diameter cyclone separators used after wet scrubbers, the basic separation principle remains the same. Particles enter the device with the flowing gas and the gas stream is forced to turn, but the larger particles have more momentum and cannot turn with the gas. These larger particles impact and fall down the cyclone wall and are collected in a hopper. The gas stream actually turns a number of times in a helical pattern, much like the funnel of a tornado. The repeated turnings provide many opportunities for particles to pass through the streamlines, thus hitting the cyclone wall. As one might suppose, the range of particle sizes collected in a cyclone is dependent upon the overall diameter and relative dimensions of the device.

There are four major parts to a cyclone as seen in Figure 13.3. The inlet, the cyclone body, the dust discharge system, and the outlet. Gas is directed into the cyclone by the inlet, which is instrumental in the formation of the vortex. In the cyclone body, the particulate matter is forced to the wall. The gas continues down the cyclone body to the cone, which gives the gas enough rotational velocity to keep the particles against the wall. At the bottom of the cone, the gas changes direction from downward to upward. The ascending vortex enters a tube extension that is sometimes called a vortex finder and exits the cyclone. Meanwhile, the collected particulate matter drops into a hopper, where it is periodically or continuously removed. Additional details on each of these parts of a cyclone are available in the literature.[2]

The removal efficiency of a cyclone for a given size particle is very dependent on the cyclone dimensions. The efficiency at a given volumetric flow rate is most affected by the diameter. The overall length determines the number of turns of the vortex. The greater the number of turns, the greater the efficiency. The length and width of the inlet are also important, since the smaller the inlet, the greater the inlet velocity becomes. A greater inlet velocity gives greater efficiency but also increases the pressure drop.

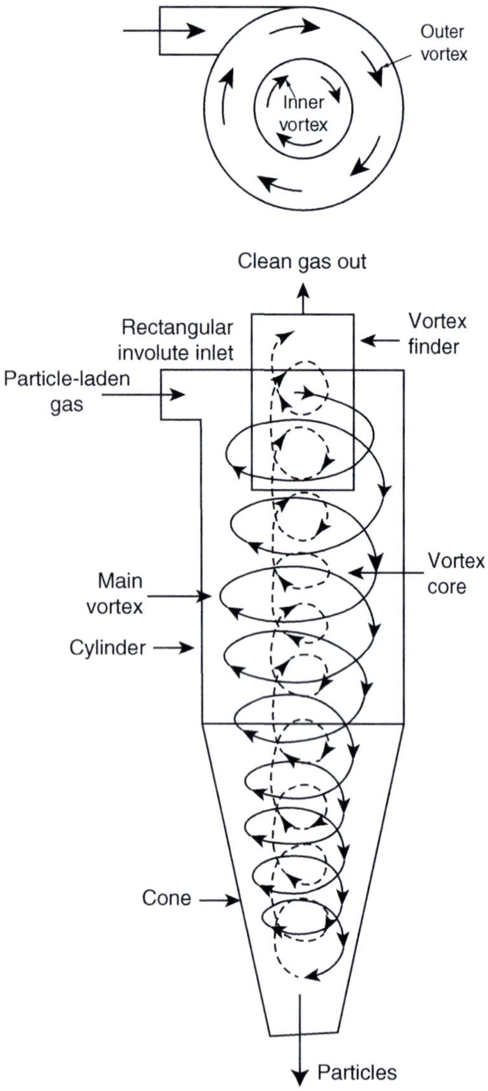

Figure 13.3 Cyclone dust collector diagram.

13.2 Design Considerations

The fundamentals governing design considerations for a gravity settler are quite simple since it is usually assumed to contain no trays. However, the analysis can easily be extended to a multiple-tray unit since the capture height between the trays can be treated as an individual collector. In addition to operating conditions and physical characteristics, the design includes the effect of the Reynolds number on particle drag, particle size distribution, and re-entrainment effects.

The analysis presented often begins by examining the behavior of a single spherical particle in a settler in which the bulk flow air velocity profile is plugged (e.g., the gas flow over the whole chamber is uniform). The particle is assumed to be located at the top of the unit – the most difficult inlet (or initial) position for particle capture. At the inlet condition, it is assumed that the vertical velocity component of the particle is at its terminal settling velocity, for example: the particle requires a negligible period of time (and consequently travels a negligible distance) to achieve a value approaching this constant terminal velocity. For capture to occur, the particle must reach a collection surface.

Considerations of cyclone geometry and vortex formation have led to the development of numerous cyclone design equations. All of these equations attempt to characterize the behavior of cyclones. Some work well, and some do not. None adequately describe performance under all operating conditions such as at high pressure and high temperatures. Three important parameters can be used to characterize cyclone design performance, including cut diameter (size of particles collected with 50% efficiency), pressure drop, and overall collection efficiency. Equations involving each of these parameters are available in the literature.[3–7] The equations should be used with caution, however, since there are strict limitations on their applicability.

13.3 Installation Procedures

The prerequisite for good collection performance is to follow proper installation procedures. To eliminate reentraiment, the collector must be airtight. Thus, when installing the equipment, access doors, inlet and outlet plenums, and dust disposal areas must be completely sealed. The same criteria as above hold true for multitube collectors, with emphasis on additional critical areas. Header sheets must be welded tight to the walls of the shell. Gasketing of tubes must seal any opening between the tube and header sheets. Outlet tubes must be tightly sealed to the top of header sheets by welds or gaskets. It should be noted that any leakage in the collector can cause a loss of over 25% in collection efficiency. One bit of caution is advised in the installation of multitube collection elements fabricated of cast-iron material. Generally, to attain the hardness required today, the tubes are cast of gray or white cast iron with a Brinell hardness of 350. This causes the tube to be very brittle. Any knock or even over-torquing when attaching the tubes to the underside of a header sheet could cause breakage around the edges of the tube. This permits leakage of dust fines to escape through the collector.

13.4 Operation

The operating variables in cyclone performance are the gas temperature, pressure, composition, and dust characteristics, including size and distribution, shape, density, and concentration.

Increasing the gas temperature decreases its density and increases its viscosity. The direct effect on efficiency by changes in gas density can be neglected since the gas density is minute compared to the dust density. If the collector pressure is held constant, the tube capacity will increase because of the lower gas density. Higher temperature increases the inlet velocity, and this increases the particle velocity toward the wall. However, because of the temperature rise, the increasing gas viscosity tends to decrease the particle velocity toward the wall. The net effect of these factors caused by the temperature rise is essentially negligible in the normal operating range of 40–700°F. Thus, at constant pressure drop, efficiency is essentially constant with temperature. As the temperature extends above 1,000°F, the viscosity effects become more predominant and there is a gradual decline in efficiency. Gas composition can also affect both gas viscosity and density.

Typical performance of conventional and high-efficiency cyclones is shown in Tables 13.1 and 13.2.

Some procedures can be instituted to increase efficiency. However, tests show that any attempt to increase efficiency by means of blades, spiral ducts, or any appliance for the purpose of directing the rotating gases is useless, as it only results in the checking of the free rotation of gases in the body of the cyclone. This rotation, which causes the separation of the dust, is continued in the surge hopper or some sort of dust bunker. Even in these, situated outside the cyclone body, the checking of the rotating gases has an unfavorable influence on the efficiency.

Table 13.1 Performance of cyclones.

Efficiency Versus Particle Size	% of Particles Below 10 μ	Efficiency Range
Fly Ash (Power)		
Spreader stoker-fired boilers	20	80–85
PC-fired boilers	42	75–90
Cyclone-fired boilers	65	55–65
Nonmetallic Minerals (When Collector is Part of Process and Collector Catch is Reusable)		
Cement (kilns and process)	40	70–85
Asphalt plant	10	80–95
Lightweight aggregate (kiln)	30–40	80–95
Refractory clays (kiln)	40–50	70–80
Lime (kiln)	40–50	75–80
Fertilizer plant (process equipment)	40	80–85
Steel (Ore Beneficiation)		
Pelletizing (vertical shaft and rotary kiln)	10–40	80–95
Foundry (general)	10–40	80–95
Chemical process (drying, calcining)	10–40	80–95
Incinerators (municipal)	20–40	65–75
Coal processing (thermal drying)	10	90–97
Petroleum (catalyst cracking process)	0.6	99+
General industrial application (in plant)	10–60	65–95

Table 13.2 Cyclone efficiency changes.

Variable	Efficiency
Dust size, increases	Increases
Dust concentration, increases	Increases
Dust density, increase	Increases
Gas molecular weight, increases	Decreases

A cyclone with a larger dust bunker or hopper will have a higher efficiency than one whose dust outlet is closed by means of a rotating valve. Similarly, the placing of vanes in the gas exhaust pipe in order to decrease the pressure drop in the cyclone will influence it unfavorably. It is most desirable, therefore, to provide the cyclone with a dust hopper and an outlet vortex tube in which the gases can continue their rotation.

The influence of gravity on the dust separation in a cycle is slight, so that efficiency is almost independent of the position of the cyclone. Experiments have shown that separation remains satisfactory not only in a horizontal position but also with the cyclone upside down with the exhaust gases directed vertically downward and the dust carried off upward from the cyclone. A good cyclone separates the dust satisfactorily in any position. Difficulties may arise, however, with coarse particles of dust that keep rotating in the conical part of the cyclone and cannot reach the dust outlet. This difficulty may arise with a cyclone in the normal vertical position, but to a lesser extent.

To understand the operation of a cyclone, it is necessary to trace the velocities and pressures within the unit and thereby understand their affect on efficiency. The gas flow is three-dimensional so that at any point within a cyclone, the velocity can be resolved into three components: tangential, radial, and vertical components of velocity.

In the turbulent and disturbed center of the cyclone, the total velocity deviates considerably from tangential. The downward flow of gas along the walls ensures the carry-off of the dust outlet and bunker of the particles that have been thrown out to the circumference. The radial velocity V_r is directed toward the center throughout most of the cyclone. This velocity tends to carry the particles to the center against the action of centrifugal force. In the center, the radial velocity is directed outward.

From the variations of velocities, it appears that the area in the center of the cyclone does not lend itself to favorable particle separation. A particle in this area will be carried by the vertical gas currents to the gas exhaust and will not drop to the dust outlet. To attain a high level of efficiency, it is desirable to keep the particles outside the center of the cyclone and to increase the tangential velocity V_t as much as possible.

Collection efficiency in a cyclone is largely dependent on particle size distribution. Efficiency can only be guaranteed if the settling rate, or alternatively, the particle size distribution is known. Generally, guarantees are made when the dust composition is constant or contains only a small fraction of particles less than 10 μm in diameter.

13.5 Maintenance

An adequate scheduled maintenance program for cyclones will repay itself many times over by reducing production time losses and costly major repairs. Effective maintenance is easily accomplished if the following basic rules are followed:

1) At initial installation, provide easily accessible access for cleaning and removal of any part of the system. Provision for convenient maintenance begins on the drawing board. The designer should allow for the eventual removal of parts that may wear out. Special material should be designated for those parts normally exposed to the greatest wear or possible abuse.
2) Establish an information file for the history of all collection equipment. The file should include actual operating data, such as gas volumes, operating temperatures, inlet and outlet velocities, pressure differentials, dust loadings, and any and all other data affecting operation.
3) Maintain data as designed on nameplate.
4) Maintain copies of arrangement or construction drawings, including parts lists and a spare parts reference drawing. These contain commercial descriptions of such items as gaskets, bolts, nuts, and washers.

At least one copy of the information file should always be available to maintenance personnel, and a master copy should always be kept in the plant engineering department.

Erosion and pluggage corrosion cause the most problems for cyclones. Under erosion conditions, metal generally fails by continuous fatigue. Maintenance features for erosion control should include the use of heavier-gauge metal for cyclones, use of abrasion-resistant removable wear plates at the impingement zone, avoidance of welds and joints that cannot be ground smooth, and the use of larger cyclones for prevention of excess gas velocities.

Two types of cyclone fouling usually occur: dust outlet pluggage and wall buildup. Outlet pluggage is usually induced by large chunks of material lodging in the outlet and forming an obstruction about which smaller particles can build up. This buildup of dust on the walls and in the hopper leads to re-entrainment. Once a hopper has filled sufficiently to pug dust outlets and has been emptied, it is not uncommon to find dust plugs remaining in the outlets after the hopper has been emptied. In large-diameter cyclones, one maintenance procedure to minimize pluggage is to provide an axial poke hole with a bolted cover plate at the top of the outlet pipe or scroll in which a rod can be inserted to break up an outlet dust plug.

Wall buildup is primarily a function of the dust. As a general rule, the finer the dust, the greater the tendency for wall buildup. Another reason for the buildup of particles below 3 μm is their inherently greater cohesive and adhesive forces. Also, it is possible for smaller particles to remain in the stagnant boundary layer. A second major cause of wall buildup is moisture, picked by hygroscopic particles at the walls of the cyclone. Buildup of these particles can be minimized by continuously flushing the dust to the outlet, but if this wet operation is used, corrosion-resistant liners or special alloys may be needed.

Sticky-material problems may be minimized by keeping the tangential velocity above the transport velocity. Also, the smoothness of the wall will greatly minimize the tenacity with which the particles stick.

Corrosion is usually caused by chlorides and sulfides in contact with the metal walls. Corrosion can be minimized by using the proper material – alloys, or in some applications, plastics. Corrosion control can be provided only in the initial stage design.

In most processes, the exhaust gas temperature is usually maintained as low as possible to obtain the best thermal efficiencies, as well as for operating economics. Precautions should be taken to avoid reaching the dewpoint. Some methods of avoiding this include: preheating the gases after shutdown until all the dust has been evacuated, insulating the collector to retain heat, or providing artificial heating units. These steps will prevent condensation of moisture and minimize the danger of corrosion.

For proper maintenance, two general rules should be followed: a regular schedule for removing the collected material from the dust collection hoppers, and inspection of all parts of the dust collector at definite intervals and cleaning or repairing of the dust collector when inspection indicates. To operate efficiently, all cyclone dust collectors should be kept airtight. Gaskets must close the gap between flanges. In axial-entry collectors, header sheets must be sealed without any breaks in the welds. Doors, ports, and poke holes must be sealed to prevent re-entrainment. A certain number of fine dust particles are always suspended in the air, and any type of leakage – no matter how slight – causes re-entrainment and escaping of dust particles from the collector.

Detection of leaks in a cyclone is not very complicated. One method for checking leakage in a collector is the use of bright floodlights. By shining a light up from the bottom of the cyclone, erosion holes, gasket leakage, and weld breaks can be detected when viewed from the outside. When directing the light upward from the hopper, for axial-entry collectors, breaks in header sheets, decomposed gaskets, and cracks in material can be detected.

Much emphasis is placed on preventing leaks. The main reason for this concern is that a cyclone operates on an inertial principle and leakage disrupts the flow pattern of the gas stream, thereby forcing the fine particles to re-entrain through the outlet vortex. It is most important to seal the cyclones, as it has been estimated that over 25% collection efficiency can be lost due to dust re-entrainment.

13.6 Improving Operation and Performance

The pressure drop within the cyclone determines the collection efficiency. The pressure drop can be increased or decreased by carrying the diameter of the cyclone body – the smaller the diameter, the higher the efficiency, or by increasing the volumetric flow rate per tube, thereby causing increased velocity and increased dropout of dust particles. These features must be designed into the system. Should it be known that the system will be turned down with less volume quite frequently, the collector should incorporate dampers so that gas velocity going to the cyclones will be increased.

Fine-tuning a high-efficiency tangential cyclone is very difficult, as there are no moving parts. With axial-entry cyclones, some manufacturers use a spiral cane that permits some control of volume by moving a vane in and out of a constricted opening of the collecting element. When fully in, maximum rotation is induced, thereby creating greater centrifugal action. When fully out, only a partial amount of gas is rotated; the rest bypasses the vanes.

Turn-down of gas flow can be controlled by the use of dampers at the inlet. This can be done if the collector is sectionalized, thus allowing the designed pressure drop to be maintained. Should the reduction in gas volume be permanent, the collector can be modified by capping off the excess tubes. This again permits the remaining tubes to handle the reduced gas volume at the designed pressure drop.

The best control of cyclone operation and performance is to initially design the collector to meet the specified criteria. It is very difficult, and in most cases impossible, to modify a cyclone once it has been improperly designed or if the operating conditions have changed appreciably.

13.7 Recent Advances

13.7.1 Gravity Settlers

As one might suspect, and understandably, the basic design of gravity settlers has remained relatively unchanged since first used in the nineteenth century. Some modest equipment changes and new types of devices have appeared in the last 10 years, but all have essentially employed the same capture mechanisms used in the past. Limited work has appeared in the literature on the effect on performance due to fouling as a function of time, and attempting to minimize this effect. Mechanical malfunctions can also be gradual but will eventually be evidenced by a decrease or lack of performance. Efforts have been made to improve the efficiency of gravity settling chambers by the use of baffles and various other methods The Howard settling chamber is an example of these efforts but problems shall remain in removing the settled dust from the horizontal trays.

13.7.2 Cyclones

The basic design of cyclones has remained relatively unchanged since first used in the nineteenth century. Some modest equipment changes and new types of devices have appeared in the last 20 years, but all have essentially employed the same capture mechanisms used in the past. The prevention of leaks is very important. Because a cyclone operates on an inertial principle and leakage disrupts the flow pattern, fine particles can be re-entrained through the outlet vortex. Collection efficiency will decrease due to dust re-entrainment, so it is very important to seal the cyclones. Batteries of cyclones in series also continue to be studied in an attempt to achieve performance levels similar to that achieved with high-efficiency equipment. The role of multicyclone arrangements and uneven flow distribution has also received some attention.

13.8 Conclusions

Because of their simplicity of construction and lack of moving parts, cyclones can be fabricated from a variety of materials, including ceramic, or lined with various materials, including ceramics. They are applicable over a broad range of temperatures from below ambient to above 2,000°F. Cyclones can also be used at high gas pressures. Generally, the only criterion for good performance is the requirement for a proportionate increase in power input to maintain separate velocities. The collector is insensitive to inlet concentration, and in fact, the efficiency can increase with increasing particle concentration because of particle interactions. Cyclones find most use when operating under the following situations: where dust is coarse, where dust lading is above $3\,g/ft^3$, where high efficiency is not a critical factor, and where stack opacity is not a criterion.

Cyclones are the most economical of the four generic classes of particulate collectors. They are also the most versatile. As unsophisticated as this type of collection device is, it does the job, and that is the only requirement for success.

References

1 **A. J. Buonicore, and L. Theodore**, *Study of Industrial Air Quality Control Methods for Particulates*, Proceedings and Am. Environ. Eng. and Sci. Conference, Louisville, KY 1972.

2 **C. Lapple**, Processes use many collection types, *Chem. Eng.*, 58, 145–151, 1951.

3 **L. Theodore**, *Air Pollution Control Equipment Calculations*, John Wiley & Sons, Hoboken, NJ, 2008.

4 **L. Theodore, and R. Allen**, *Air Pollution Control Equipment, A Theodore Tutorial*, Theodore Tutorials, East Williston, NY, 1993.

5 **A.J. Buonicore, and L. Theodore**, *Air Pollution Control Equipment, Particulates*, CRC Press, Boca Raton, FL, 1988.

6 **A. J. ter Linden**, *Investigations into Cyclone Dust Collectors*, Proceedings of the Institute of Mechanical Collectors, Amsterdam, Holland 1949.

7 **B. G. van der Heffe Zigneu**, *Investigations into Cyclone Dust Collectors*, Proceedings of the Institute of Mechanical Collectors, Amsterdam, Holland 1949.

14

Wet Scrubbers

Wet scrubbers are devices that use a liquid for removing particles or polluted gases from an exhaust gas stream. Water sprays can be injected into the gas stream; gas can be forced to pass through sheets or films of liquid; or, the gas can move through beds of spheres covered with liquid. Each of these techniques can effectively remove particulate matter from process exhaust gases. They can also effectively remove gases such as HCl or SO_2, but removal conditions must be right. In addition, gas–liquid contact can bring about gas conditioning, and to a lesser extent, liquid conditioning.

The particulate collection mechanisms involved in the wet scrubbing operation may include some or all of the following[1–6]:

1) Inertial impaction
2) Direct interception
3) Diffusion (Brownian movement)
4) Gravitational forces
5) Electrostatic forces
6) Condensation
7) Thermal gradients

Inertial impaction occurs when an object (the droplet), placed in the path of a particulate-laden gas steam, causes the gas to diverge and flow around it. Larger particles, however, tend to continue in a straight path because of their inertia; they may impinge on the obstacle and be collected (as in Figure 14.1a).[1] Direct interception also depends on inertia and is merely a secondary form of impaction. A collision occurs due to direct interception if the dust particle's center misses the target object by some dimension less than the particle's radius (see Figure 14.1b). Direct interception is, therefore, not a separate principle, but only an extension of inertial impaction. Diffusion is another extension of the impaction principle. Very small particles (submicron) suspended in a gas stream have an individual oscillatory motion known as Brownian movement (see Figure 14.1c). In this case, the particle and target collide as a result of relative motion within the limited space. As in all diffusional processes, the rate of diffusion is favored by large areas for diffusion, thereby necessitating small liquid droplets with high surface-to-volume ratios for high collection efficiencies. While collision or impaction may be the result of either inertia or Brownian movement, the results are the same. Gravitation forces can also cause a particle, as it passes an obstacle, to fall from the streamline and settle on the surface of the obstacle. Such forces come into play, however, only when dealing with larger particles (usually greater than 40 μm). Electrostatic forces result when particles and liquid droplets become electrically charged (see Figure 14.1d). In addition, when only the particle or obstacle is charged, a charge may be induced on the uncharged component, resulting in a polarization force that can also affect particle removal. The electrostatic charge may be acquired, for example, by liquid droplets during their formation. An electrical charge may also be induced by flame ionization, friction, or the presence of charged matter. The effect of the electrostatic mechanism may be significant when the charge on the particle or obstacle is high and when gas velocity is low. Condensation effects may also come into play. Condensation occurs if the gas or air is rapidly cooled below its dew point. When moisture is condensed out of the gas stream, fogging occurs, and the dust particles can serve as condensation nuclei. The dust particles can become larger as a result of the condensed liquid, and the probability of removal by impaction is increased. Particle collection may also be affected by thermal gradients. Such forces can drive particles from hotter to colder regions. The motion

Optimizing Air Pollution Control Equipment Performance: Operation & Maintenance, First Edition. Jay Richardson and Louis Theodore.

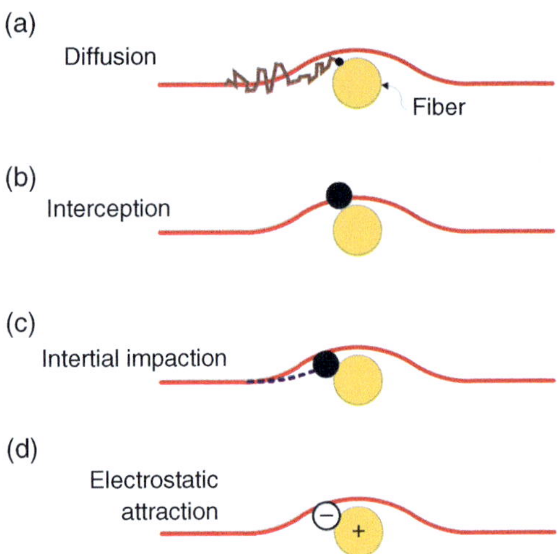

Figure 14.1 Filtration collection mechanisms. (a) diffusion, (b) interception, (c) inertial impaction, and (d) electrostatic attraction.

can be caused by unequal gas molecular collision energy on the surfaces of the hot and cold sides of the particle; it is directly proportional to the temperature gradient.

It should be noted that the two principal collection mechanisms are inertial impaction and diffusion, both of which are again described below:

1) *Impaction*. When a gas stream flows around a small liquid droplet, the inertia of the particles causes them to continue to move toward the object, and particles will be collected by the liquid. Inertial impaction customarily describes the effects of small-scale changes in flow direction.
2) *Diffusion*. When particles are small enough ($<1\,\mu m$), they are "thrashed" about by gas molecules, such that they act similarly to gas molecules. They "diffuse" randomly through the gas. As these small fines are diffusing randomly through the gas, the probability of contact with a liquid droplet increases. This phenomenon explains why the collection efficiency rises for extremely small particle diameters.

A variety of wet collectors are commercially available. The evaluation and selection of a scrubber system is somewhat simplified by the observation that collection efficiency is a function of the amount of energy required to operate the scrubber. This simply means that independent of design, the more power put into the system, the greater the collection efficiency. Therefore, for systems of nearly equivalent power inputs, the collection efficiencies should be nearly equivalent. The selection or evaluation procedure between two systems could then concentrate on ease of operation, potential maintenance problems, and comparative costs.

14.1 Description of Control Devices

The description of several scrubbing devices is briefly reviewed below. The two most commonly used scrubbers in the industry are the spray tower and venturi scrubber.

14.1.1 Spray Tower

The most common low-energy scrubbers are gravity spray towers in which liquid droplets are made to fall through rising exhaust gases and are drained at the bottom of the chamber. The droplets are usually formed by liquid atomized in spray nozzles. The spray is directed into a chamber shaped to conduct the gas through the finely dispersed liquid. The relative velocity between the droplets and the gas in a vertical tower is eventually the terminal settling velocity of the droplets. However, to avoid spray droplet reentrainment, the terminal settling velocity[6] of the droplets must be greater than the velocity of the rising gas stream. In practice, the vertical gas velocity typically ranges from 2 ft/s to 5 ft/s. For higher velocities, a mist eliminator must be used at the top of the tower.

14.1.2 Venturi Scrubbers

To achieve high collection efficiency for particles by impaction, a small droplet diameter and high relative velocity between the particle and droplet are required. This is accomplished in a venturi scrubber by introducing the scrubbing liquid at an angle to a high-velocity gas flow in the venturi throat (vena contracta). Very small water droplets are formed, and high relative velocities are maintained until the droplets are accelerated to their terminal velocity. Gas velocities through the venturi throat typically range from 12,000 ft/min to 24,000 ft/min and the velocity of the gases alone causes the atomization of the liquid. The energy expended in the scrubber (except for the small amount used in the sprays) is accounted for by the gas stream pressure drop through the scrubber. Another factor important to the effectiveness of the venturi scrubber is the conditioning of the particulates by condensation. If the gas in the reduced-pressure region in the throat is fully saturated, or supersaturated (preferably), some condensation will occur on the particulates in the throat due to the Joule-Thompson effect. Because of the cooling effect of the scrubbing liquid, condensation will be more pronounced if the gas is hot.

14.1.3 Centrifugal Collectors

As noted, particulate collection efficiency is improved by increasing the relative velocity between the droplets and the gas stream. The particulates are assumed to travel at the same velocity as the gas stream. An increase in this relative velocity may be achieved by utilizing the centrifugal force of a spinning gas. The spin may come from introducing gases to the scrubber tangentially or by directing the gas stream against stationary swirl vanes. Although normally designed for specific installations, cyclonic spray scrubbers can accommodate gas entrance velocities up to 100+ ft/s.

14.1.4 Dynamic Collectors

Dynamic collectors serve primarily as air movers (centrifugal blowers). Spray nozzles at the inlet offer an opportunity for impaction. Liquid droplets from the nozzles and uncaptured dust must pass through the many specifically shaped blades of the rotating fan wheel. Centrifugal forces and impingement on the blades are utilized for further collection and water separation. Problems have arisen, however, due to instability attributed to sludge buildup on the fan wheel. Slurry is removed and discharged separately from the cleaned air.

14.1.5 Packed-bed Scrubbers

Packed-bed scrubbers find application in the removal of pollutant gases and vapors. However, such systems have also been considered for the removal of particulate matter. There are four general ways in which these beds may be operated to effect particulate removal. Cross-flow-packed scrubbers will efficiently remove the relatively large particulates; they are also used for the removal of liquid particulates. The co-current flow-packed scrubber is somewhat more efficient than the cross-flow and is effective in removing both solid and liquid particulates. For the collection of particulates as small as 3–5 μm, countercurrent packed scrubbers have been found to be most efficient; these scrubbers operate generally at much higher liquid rates than either the cross-flow or the co-current-flow types. Bed depth in countercurrent packed scrubbers is typically 2–6 ft; gas velocities typically range between 3 and 6 ft/s. Cross-flow beds are usually less than 2 ft in depth.

The difficulty of removal of particulate material increases rapidly with a decrease in particulate size. A 1 μm particle is only 10 times larger in diameter than a 0.1 μm particle, but it will have 100 times more surface area and 1,000 times more weight. As particle size decreases, the surface area/mass ratio increases so rapidly that eventually, the surface properties can dominate over the mass properties. When this happens, higher velocities and more acute changes in direction are required to permit the particle to be separated from the gas stream by impingement. Cross-flow, co-current-flow, and countercurrent-flow-packed scrubbers are not capable of achieving a high enough gas velocity to remove effectively particulates smaller than 3–5 μm in diameter. To remove particles in this small size range, much more energy must be used to attain the velocities needed, and to avoid flooding, the liquid should be introduced in the same direction as the gas (i.e., the bed is operated in a co-current flow mode). Liquid requirements for cross-flow operation typically range from 1 gal/1000 ft^3 to 4 gal/1000 ft^3 of gas, with pressure drop from 0.1″ WC/ft to 0.25″ WC/ft of bed depth. Liquid requirements for countercurrent packed scrubbers usually range from 10 gal/1000 ft^3 to 20 gal/1000 ft^3 of gas, with pressure drops typically 0.2–1.5″ WC/ft of bed depth. However, some adaptations, which make use of a highly efficient turbulent layer above the bed, require considerably less scrubbing liquid, typically 2–2.5 gal/1,000 ft^3 of gas at a pressure drop ranging from 4″ WC to 6″ WC. In co-current flow, a liquid requirement of 7–15 gal/1,000 ft^3 of gas is fairly common. Coarsely packed beds are used for removing coarse dusts and mists 10 μm and larger, with velocities through the bed on the order of 400 ft/min. Finely packed beds may be used for removing contaminants in the smaller size range, but because of pressure drop considerations, the velocity through the bed must be kept relatively low, preferably below 50 ft/min. Finely packed beds have a greater tendency to plug, and their applications are generally limited to gas streams with relatively low particulate loadings.

14.1.6 Moving-bed Scrubbers

Moving-bed (fluid-bed) scrubbers incorporate a zone of movable packing where gas and liquid can mix intimately. The system uses packing consisting of low-density polyethylene or polypropylene spheres about 1½ inches in diameter; these are kept in continuous motion between the upper and lower retaining grids. Such action keeps the spheres continually cleaned and considerably reduces any tendency for the bed to plug. Pressure drops typically range from 3″ WC to 5″ WC (per stage), and collection efficiencies can be in excess of 99% for particles down to 2 μm. Particle collection may be enhanced by using several moving-bed stages in series.

14.1.7 Atomizing-inertial Scrubbers

The atomizing-inertial category of wet scrubbers probably represents the largest grouping of commercial dust collectors. Available designs vary more than in any other category. For the sake of convenience, this category has been further divided into those that utilize the kinetic energy of the gas stream to fragment or atomize the liquid on which the particle impacts and those that utilize a perforated plate with an impingement baffle over each perforation.

14.1.8 Orifice-type Wet Scrubbers

In orifice-type wet scrubbers (sometimes referred to as self-induced spray scrubbers), the gas stream comes into contact with a pool of liquid at the entrance to a constriction (submerged orifice). Liquid is entrained and carried into the restriction where greater liquid-particulate interaction occurs, resulting in a high frequency of particulate impaction on the droplets. Upon leaving the restriction, most of the water droplets (those large enough) are separated by gravity since the gas velocity is reduced from what was in the restriction. Smaller droplets are subsequently removed by centrifugal force and impingement on baffles located in the upper part of the unit.

14.1.9 Impingement Plate Scrubbers

Impingement plate scrubbers utilize perforated plates with an impingement baffle over each perforation. The intention here is to expand the impingement area of the liquid through the use of the gas stream's kinetic energy. Gas flowing up is divided into thousands of jets by the orifices. Gas velocities of 15–20 ft/s through the orifices are common. Each jet aspirates liquid from the blanket and creates a wetted surface on the baffle, located at the point of maximum jet velocity. The directed impingement on the wetted target dynamically precipitates particles and entraps them in the scrubbing liquid. On impingement, each jet forms minute gas bubbles which rise through and create turbulence in the liquid blanket. This provides extremely close gas–liquid contact for maximum cleaning. Continuous violent agitation of the blanket by the bubbles prevents the settling of entrapped particles and flushes them away in the scrubbing liquid.

14.1.10 Atomizing Mechanical Scrubbers

In mechanically induced scrubbers, high-velocity sprays are generated at right angles to the direction of gas flow by a partially submerged rotor. The dirty gas stream passes through the area of the collector that contains mechanically produced droplets. Scrubbing is achieved by impaction because of both high radial-droplet velocity and vertical gas velocity. Liquid atomization occurs at both the rotor and the outlet wall.

14.2 Design Considerations

Wet scrubbers are constructed with such a multiplicity of designs that no single type can be considered representative of the category as a whole. Some units simply consist of an existing dry-type collector modified by the introduction of a liquid phase to assist in particulate removal and to prevent particulate reentrainment; other units are specifically designed to operate as wet collectors.

It is difficult to generalize about relationships among operating parameters such as pressure drop and liquid flow rate. It is also difficult to classify wet scrubbers according to particulate collection mechanisms since the usual case finds many of the collection mechanisms working simultaneously in each type of scrubber. However, in a very general sense, wet scrubbers may be loosely categorized by pressure drop (or energy consumption). Low-energy scrubbers are those with typical pressure drops less than 5″ WC; medium-energy scrubbers are those with typical pressure drops from 5″ WC to 15″ WC; and high-energy scrubbers are those with typical pressure drops greater than 15″ WC. Spray chambers and spray towers, for example, provided the lowest pressure drop and, correspondingly, the lowest collection efficiencies; they would be classified as low-energy scrubbers. The medium-pressure-drop group could include centrifugal fan wet scrubbers, atomizing impingement collectors, and certain packed-bed scrubbers. The most familiar of the high-pressure-drop (high-energy) group is the venturi-type collector. Many scrubbers can conceivably be listed in more than one of the categories listed above, and each type of collector with its eventual application, determines whether it will be classified as a low-, medium-, or high-energy scrubber.

Ultimately, the design of wet scrubbers usually focuses on those parameters affecting collection efficiency and pressure drop. In most cases, the scrubber must be designed to guarantee a specified collection efficiency, which in turn, is strongly dependent upon pressure drop (among other parameters). The system pressure drop also dictates the power requirement and the size of auxiliary equipment such as the fans.

Collection efficiency equations as a function of particle diameter typically are of the form originally suggested by Johnstone et al.[2, 3] However, a number of other performance and design equations have been developed from basic particle movement principles (theory) to explain the action of wet scrubbing systems.[4] Many of these start from firm scientific concepts, but give only qualitative results when predicting collection efficiencies or pressure drops. The interaction of particulate matter having a given particle size distribution with water droplets having another size distribution is not easy to express in quantitative terms. As a result of this complexity, experimentally determined parameters are usually required in order to perform engineering calculations.

To achieve a high collection efficiency of particulates by the aforementioned capture mechanisms, a small droplet diameter and high relative velocity between the particle and droplet are normally required. This is often accomplished in a venturi scrubber by introducing the scrubbing liquid at right angles to a high-velocity gas flow in the venturi throat (vena contracta). Very small water droplets are formed, and high velocities are present. Additional details are available in the literature.[5–7]

14.3 Installation Procedures

Prior to installing the scrubber system, a careful inspection of the equipment should be made for shipping damage. The packing list should also be checked to ensure that all parts are accounted for.

In the event that the scrubber is not installed immediately, attention must be directed to proper methods of storage. Table 14.1 lists shelf-life requirements, under specific conditions, for typical vendor-supplied equipment. Indoor storage is always preferable. Storage practices are to a large extent common sense. For example, components and equipment should be stored in an upright position, fan belts should be removed, fans should be rotated once a month, and so on.

Table 14.1 Storage shelf life.

	Storage Time (months)		
Equipment	**0–6**	**6–18**	**18–36**
Electric components, control equipment	3	4	4
Gates, mechanical assembly, machine casting	2	3	4
Closed crates and boxes	1	2	3
Vessels and tanks	1	2	3
Rough castings, structural steel	1	2	3

Note 1: Unprotected outdoor storage; Note 2: Protected outdoor storage (elevated and covered); Note 3: Unheated indoor storage; Note 4: Heated indoor storage.

Installation procedures should follow the vendor specifications precisely. Deviations from vendor instructions may cause later problems and invalidate a warranty.

After the scrubber system is installed, it is important that all of the major items of equipment, connecting pipe, and auxiliaries be inspected, cleaned, tested, and calibrated (if necessary) before startup.

General pre-operation practices include checks to ensure that piping is free of debris, oil levels are correct, fans and pumps rotate in the proper direction, and alignments appear proper. Operating instructions should be reviewed with all appropriate plant personnel. A typical checkout might include:

1) Utilities
   a) Power supply
   b) Instrument air
   c) Process air
2) Pumps
   a) Belt tensions, pump rotation, pump alignment, lubrication, seal-water operation, and electrical interlocks
   b) Recycle pumps – suction and discharge valves
   c) Flush-water pumps
   d) Pneumatic pumps for scrubber variable volume and liquid level
   e) Spare pump availability and operation
3) Valves/dampers (stack, isolation, bypass)
   a) Bypass
   b) Density control
   c) Water purge control
   d) pH elements for flush water
   e) Freshwater makeup
4) ID/FD fan
   a) Electrical controls
   b) Fan-bearing coolant
   c) Lubrication
   d) Vibration sensors
   e) Bearing temperature sensors
5) Process water
   a) Level detector calibration
   b) Mist-eliminator sump-level alarm calibration
   c) Mist-eliminator sump agitator
   d) Recycle pump
6) System controls/feedback controls
   a) Stack gas flow
   b) Makeup-water control
   c) Slurry pH
   d) Slurry density
   e) Sludge drain off and disposal
7) Safety system
   a) Interlocks
   b) Alarms for various system components in accordance with design

14.3.1 Routine Startup

1) Allow scrubber vessel to fill with liquid to a predetermined level through normal level controls, if practical.
2) Start control liquid to all pump glands and fan sprays.
3) Start recycle pumps with liquid bleed closed.
4) Adjust liquid rates to manufacturer's specifications.
5) Check system isolation dampers and place scrubber in series with primary operation.

6) Energize fan and check vibration; if the fan has an inlet control damper, it should normally be almost closed until the fan reaches speed.
7) Check the most important operating variables (e.g., gas saturation temperature, liquid flows, liquid levels, fan static pressure differentials, duct pressure drops, and scrubber pressure drop).
8) Slowly open the bleed to drain the system so that the slurry concentration is allowed to build up slowly; check the final concentration as a cross-check on bleed rate calculations.
9) Adjust the makeup liquid flow rate to maintain liquid reservice requirements.

14.3.2 Routine Shutdown

1) Deenergize the fan and the fan spray water and isolate the scrubbing system from operation.
2) Allow the liquid system to operate for as long as practical; this will cool the scrubber and will reduce scrubbing liquid slurry concentrations.
3) Shut off the makeup water to the system, allowing the system to bleed normally.
4) When levels reach minimum settings (indicated by pump cavitation noise), the pump should be shut down immediately; the pump water can then be shut off.
5) Open system manholes, bleeds, and other drains, and the system may have to be purged first with inert gas if explosive conditions could be created; in any event, purge the system with fresh air before entering to inspect it.

14.4 Operation

Although wet scrubbers are relatively simple devices, they do require proper care to ensure long service life and trouble-free operation. Properly designed and installed units can do an excellent job in helping plants comply with air pollution control regulations. However, as with most processing equipment, periodic reviews and inspections must be made to ensure that scrubbers are operating most efficiently. Improper performance of a unit can usually be traced to any of the following:

1) Inadequate scrubber system design and/or operation
2) Process conditions have changed so that the scrubber system cannot operate efficiently
3) Scrubber system's mechanical condition has deteriorated to a condition where performance is below design specifications

The first step in solving a scrubber-performance problem is to assess and evaluate the problem. The more thoroughly the problem is defined, the easier it is to find the proper solution. Problem evaluation can be greatly helped by plant personnel. Operating experience and performance records can provide valuable clues. Data that must be studied include production process variables (flow rates, temperatures, and pressures), stack opacity, and the scrubber's operating conditions.

A number of possible solutions usually suggest themselves for any problem. As noted, the choice can be narrowed only if the situation is thoroughly investigated and sufficient and accurate data are available.

Often, the solution to the problem can be relatively simple. For example, minor modifications in the production process may sufficiently upgrade a scrubber's performance and quickly eliminate the problem. However, the overall effect on the process must be carefully studied, because each process variable can have a critical influence on an individual operation. Process engineers should always be consulted when such evaluations are made.

If process flow rates have increased significantly – or the pollutants have changed in character, size, and quantity – then scrubber upgrading or new equipment may be necessary.

If equipment modification is justified, efforts must be made to ensure that the desired results are obtained at a minimum cost and operating expense. The job must also be accomplished with the least disruption to plant layout and operation. The achievement of these goals requires that detailed tests be conducted. Extensive evaluation of materials and data is needed before meaningful specifications can be prepared. If erosion or corrosion is a problem, specimens of different materials and coatings should be exposed to the exhaust stream and studied.

Rebuilding rather than replacing scrubbers is an approach rapidly gaining in popularity. A properly planned rebuilding program can add years of service to scrubbers and save a plant considerable time and money. Scrubbers operating in parallel may be a wise choice if flow rates are high and maximum operating flexibility is needed.

Assuming that the proper unit was selected for the given application, most scrubber problems usually involve spray nozzle plugging, liquid circuit restrictions, and/or entrainment of droplets from the vessel. Such problems may include:

1) *Wet/dry zone buildup*. Scrubber design may improperly allow dry, dust-laden gas to contact the juncture of the scrubbing liquid and the vessel, causing dust buildup. Proper design prevents this contact by extending ductwork sections sufficiently into the scrubber and thoroughly wetting all scrubber surfaces through reliable means (usually gravity flush and sometimes sprays).
2) *Nozzle plugging*. Nozzles plug through improper selection; too small orifices through which too dense scrubbing liquid must pass; improper header design; drawing off a sump that also settles (and concentrates) solids; erratic pump operation; chemical scaling sealing; and mechanical failure.
3) *Flow imbalance*. The headers external to the scrubber must send the required flow to the proper location at the proper rate. Many problems can be solved through the simple adjustment of the flow using existing valves or dampers.
4) *Buildup (scaling)*. Scaling is the plating out of deposits on a scrubber surface and is most often related to the chemical composition, solubility, temperature, and pH of the scrubbing liquid. Scaling is usually not operationally significant unless the surface in question is a functional one. Proper control begins with the scrubber design and process control.
5) *Localized corrosion*. Corrosion is a major factor in reducing the operating life of a scrubber, whether properly designed or not. Wells or pockets of liquid should be avoided and points of particulate buildup should be adequately flushed.
6) *Instrumentation fitting blockage*. One problem that can cause major trouble is instrumentation blockage or pluggage. Often, a standard fitting is not adequate in a scrubber. Specifically designed fittings and connections are often recommended.
7) *Sump swirling*. As one might suspect, sump swirl problems are most often associated with cyclonic devices. The swirling of the scrubbing liquid can cause severe wear and draining problems unless arrested by anti-swirl plates in the scrubber or rapid continuous draining.
8) *Entrainment*. Entrainment occurs when the droplet separator is not functioning properly. Nearly all scrubbers produce entrainment; only the properly designed systems reduce it to acceptable levels prior to discharge.
9) *Reentrainment*. Reentrainment occurs beyond the droplet removal device through improper draining or erratic flow patterns. It can also occur in stacks of very high velocity or where fittings protrude into a high-velocity area.
10) *Liquid–gas maldistribution*. The gas and liquid must be properly distributed for the given application. Each affects the other and is aggravated by the influence of baffles (designed or accidental), buildup, mechanical failure, wear, scaling in headers, or improper design.
11) *Thermal shock*. Where hot gases meet cold scrubber liquor, thermal shock may occur. Proper design permits gradual cooling rather than abrupt changes. Thermal shock is a relatively simple problem to prevent (e.g., through the use of multiple cooling zones).
12) *Loss of seal*. The juncture of the scrubber liquid circuit with its surroundings is often a liquid seal. This seal may be at the top of a quencher or from an overflow connection. These lines must have seals able to prevent gas movement to or from the ambient surroundings. Loss of seal can cause entrainment or plugging, and instrumentation malfunction.
13) *Wear*. Wear can be tolerated unless it is localized. Unfortunately, a scrubber's functioning parts are also the wear parts. One should expect to replace high wear points at frequent intervals. Excessive wear is often a result of excessively high solids concentration.
14) *Vibration*. This is most common in pumps and fans associated with wet scrubber systems. Vibration problems are controlled by monitoring and scheduled preventative maintenance.

Since packed- and moving-bed scrubbers have additional operational problems associated with them by virtue of their packings, these are considered separately.

14.4.1 Packed-bed Scrubbers

One of the most common operational difficulties noted with packed-bed scrubbers is high-pressure drop on the gas side. As the liquid rate increases, the pressure drop will increase until, at the flood point, the liquid will actually begin to form a layer above the packing. Typical packed bed scrubbers for air pollution control are designed to operate at values well below the flood point. Many are equipped with spray nozzles that make it extremely difficult to overload the scrubber with liquid.

The first step in avoiding flooding rates is to measure either the inlet or discharge liquid flows. If a spray nozzle is used, its liquid pressure should be checked as well as its condition. If a trough- or weir-type distributor is used, there should be little restriction on liquid flow through the unit.

A quick check for high liquid rate can be obtained by turning off the liquid temporarily. If the pressure drop immediately falls to the dry value, the liquid rate is in excess of the design. If the pressure drop remains high, other problems, such as fouling or plugging, are present.

Assuming that the liquid rate is correct, a check should be made of the gas rate through the unit. As this rate increases at any fixed liquid rate, the pressure drop through the column will increase.

A pitot tube or similar device can be used to obtain stack velocity readings needed for calculating the total volume leaving the column. Exit gas volume can differ significantly from that of the inlet. Whenever possible, inlet measurements should be taken to ensure that the gas volume is equal to or less than the design volume. High inlet temperatures, or changes in water content, can alter the volume flow rate considerably.

Changes in gas composition from the design values can also alter the pressure drop through the column. As the gas density increases, the pressure drop will tend to increase.

Next to incorrect gas or liquid rates, the presence of the particles themselves becomes the greatest concern. Solids or vapors that precipitate within the scrubber can cause plugging in the mist eliminator, packed bed, or support plates. These items can be visually inspected. Typically, plugging results from inlet material accumulating on the bottom of the packing, or from liquor (containing suspended solids) recirculated to the top of the bed.

Since many packed-bed scrubbers use injection-molded plastics, warnings should be posted against steam cleaning. Many problems in scrubber operation can be traced to a maintenance program involving extremely hot solutions used to clean the internals. Although the packing may not actually melt if such solutions are sprayed on the bed, they can cause deformation because of the weight of the packing resting on the bottom layer.

In scrubbers having fiberglass-reinforced plastic (FRP) walls, any incipient corrosion normally attacks the inner resin layer. Exposed fiberglass may then begin to shred. If such material is picked up in the recycle loop, it can cause plugging at the top of the packed bed. Bedsides raising the pressure drop, the existence of such material is a sign that more serious structural damage can be occurring in the scrubber. The cause should be determined immediately. If a particular area is affected, care should be taken during repairs to eliminate the contributing conditions. If these are due to chemicals not previously suspected, additional washdown sprays may prove to be a temporary answer. Serious damage may call for alternative materials of construction.

Previous operation history helps but is not always available. Repeated upsets can shift the arrangement of the internals. When packing is installed randomly, it is normally loose. Gas velocities through such an arrangement are so low that they do not cause problems under most conditions. During startup or upset, the gas flow rate can suddenly rise and fluidize part of the packing, resulting in a nonlevel bed. With a spray- or weir-type distributor, the liquid will then be unevenly distributed. The result is channeling of the gas and liquid, with low particulate collection or gas-adsorption efficiencies. A quick visual check of the packing will determine whether leveling is necessary.

Mist carryover usually follows high gas or liquid rates, or a defective mist eliminator. Normally, a high-pressure drop will also ensue. In some cases, a temporary upset may force the mist eliminator out of its normal position within the column. Visual inspection to ensure that no gaps appear between the various sections of the separator is advised.

Sometimes, mist carryover can even be experienced at low gas velocities. Mist eliminator systems are normally designed for minimum velocities of 3 ft/s. If the spray system in the column is not properly designed, it may create a certain amount of fine mist and induce carryover through the mist eliminator at gas velocities below the initial design point. For this reason, a low nozzle pressure is strongly recommended. If an existing scrubber has a high nozzle pressure (over 25 psig), changing the nozzle may eliminate the carryover problems.

If high gas velocity is the culprit, it may be possible to adjust the inlet gas rate. If not, substitute an alternative scrubber packing to reduce the overall pressure drop. For example, a more open design 2 in. packing may replace the existing 2 in. saddles. Although the surface areas may be similar and the overall efficiency the same, the pressure drop can be reduced to the original design conditions. Care should be taken when considering such substitutions so that efficiency is not sacrificed.

Packed-bed scrubbers can operate over a wide range of gas rates. The maximum rate is normally the design condition. As the rate decreases, the effective liquid/gas ratio will increase. This means that the efficiency will remain constant or actually increase slightly. At some point, the efficiency will begin to decrease again; this occurs when pressure drop through the column becomes so low that the gas is not forced to distribute itself across the entire packed bed. Such a condition often results in values less than 30% of the design gas rate.

If wide variations in the gas rate are anticipated, the manufacturer of the column should be consulted to assure that it can handle these variations. A spray-type distributor is strongly recommended for such a service. In trough- or weir-type distributors, the gas velocity and accompanying turbulence normally ensure a good distribution of liquid. But, as the gas velocity decreases, the distribution of the liquid can be disturbed. In existing columns, a spray-type distributor and a mist-eliminator section can be added where variations in the gas rate are causing problems with efficiencies.

14.4.2 Moving-bed Scrubbers

Moving bed scrubbers are often used where packed-bed scrubbers would be expensive or impossible to maintain due to plugging problems. The rolling action of the spheres in the moving-bed scrubbers can prevent scaling and plugging in many instances. In those cases where scaling does occur, it frequently takes place in the bed section. Material accumulates in the grid interstices and between balls. Generally, as the scaling becomes more abundant in one area, the velocity through the open area of the scrubber increases, causing more rapid ball movement in that area. At some point, the ball activity becomes so accelerated that further scaling is prevented. Nozzle plugging may also result in scaling of that portion of the spheres covered by the nozzle.

Sphere movement is an important maintenance characteristic. If the distribution of air entering the bed is uneven, spheres in the high-velocity zone will tend to gravitate to the low-velocity area. This, in turn, creates a greater resistance in the low-velocity area than in the high-velocity area. Since the pressure must equalize, the airflow increases in the lower depth sphere zone and causes liquid carryover. This problem is difficult to correct with baffles underneath the bed.

Nozzles are another potentially high-maintenance area in moving-bed scrubbers. This problem is less severe in the plastic ball scrubber because of the more open nozzle design and lower pressure requirement.

Wear and deterioration of the balls also pose a potential problem. When exposed to abnormally high temperatures, the balls may deform. Most applications are designed with safety precautions to prevent loss of water and resultant high-temperature conditions. Nevertheless, occasional malfunctions can occur.

Support grids may also be a source of maintenance. The motion of the spheres tends to wear and vibrate the grids. These problems, however, have been quite minor in most applications.

Entrainment elimination is quite necessary for moving-bed scrubbers because of the quantity of fine droplets created. The horizontal-flow chevron[6] mist eliminator has proven very effective in eliminating droplets with a minimum of maintenance. These eliminators can be sprayed with water, if necessary, to keep them clean.

14.5 Maintenance

Many of the items on the pre-operation checklist should be checked during routine maintenance. Access doors should be operational and scrubber internals visually inspected. The maintenance performed generally includes unplugging lines, nozzles, pumps, and so on; replacing worn pump parts, erosion/corrosion prevention liners, and instruments (level indicators, pH indicators, etc.); and, repairing damaged components when this is practical from the standpoint of labor and materials.

The following checklist is based on problems encountered in typical scrubber operation. These should be checked routinely and corrected in accordance with the manufacturer's recommended procedures.

1) Check for wear (abrasion/erosion); heavy wear usually occurs in areas downstream of gas and liquid acceleration.
2) Check for corrosion on all scrubber internal surfaces.
3) Check for excessive buildup, in particular in the wet/dry zone.
4) Check for excessive scaling; this may be caused by process changes such as changes in temperature, pH, chemical composition of the dust, or chemical composition of the makeup water; reduced liquor recycle rate; increase in the inlet loading; or failure of solids removal system.
5) Check the nozzles for buildup and/or damage; repair or replacement may be necessary.

6) Check for solids buildup in blowdown lines; cleaning may be effected without system shutdown, and a flush connection may be installed to prevent this condition in the future.
7) Check for corrosion and leaks in lines and vessel, and in particular, where protective liners may have deteriorated.
8) Check operation of the mist eliminator; formation of droplets can be caused by excessive gas flow rate, plugged drains from the droplet eliminator, or condensation in the outlet duct; check structural supports for structural integrity and smooth operation.
9) Check pumps for wear, seal water, packing, and smooth operation.
10) Check dampers for damper linkages for proper positioning, wear, and/or buildup.
11) Check fan for lubrication, fan-bearing coolant, belt wear and belt tension, and impeller erosion/corrosion.
12) Inspect all interior surfaces and the condition of holding tanks during major outages.
13) Inspect the exterior for leaks in all process and control lines, ductwork, and expansion joints.
14) Inspect the condition of all instruments (e.g., level probes, pH elements) with regard to solids buildup; it is impractical and usually impossible to remove solids buildup from the probes and the probes must be replaced.
15) Inspect for weather-related problems; freezing weather can cause dampers to lock up and can freeze water and slurry lines (process lines); all lines (process and control) should be properly insulated and heat-traced, if necessary.

An adequate supply of spare parts should also be a part of the prevention maintenance program. The minimum inventory for a venturi scrubber system is indicated in Table 14.2.

Table 14.3 indicates the types of personnel generally required to perform maintenance on various parts of the venturi scrubber system.

Table 14.2 Spare parts inventory for venturi scrubber.

	Scrubber	Separator	Fan	Pumps	Mist Eliminator
Motors			x	x	
Mist eliminator modules					x
Seals			x	x	
Bearings			x	x	
Impeller			x	x	
Reamers	x				
Packing material	x				
Adjustable throat damper		x			

Table 14.3 Type of maintenance required – venturi scrubber systems.

	Type of Maintenance Worker				
System Section	Laborer	Electrical	Plumber	Wastewater Operator	Mechanical
Scrubber	X				
Separator	X				
Fan		X			X
Pump		X	X		x
Piping/valves	X		X		
Water treatment	X	X	X	X	X

14.6 Improving Operation and Performance

To troubleshoot air pollution control equipment performance problems adequately, a multiphase program should be implemented. As a guideline, the following procedures are recommended:

Phase 1. Problem identification. This phase should incorporate a detailed inspection of the system and culminate with a report listing all observations (positive and negative), provide interpretations of these observations (why things were the way they were), and finally, include recommendations of methods and items to improve performance. Troubleshooting guidelines for venturi scrubbers, dry scrubbers, and pack-bed scrubbers are provided in Tables 14.4–14.6.

Phase 2. Implementation. After thorough analysis and discussions, the recommendations provided in phase 1 should be implemented. These recommendations may be in the form of design modifications, component and accessory part replacement, and fabrication of new equipment. Following this aspect of the program, repair and replacement of the procured and fabricated components must be instituted. Finally, the entire system must be started up and debugged.

Phase 3. Testing and sampling. A performance test must be executed to evaluate the work done. This may include a stack sampling program and/or measurement of the system in continuous operation.

Phase 4. Preventive maintenance. When the modified and overhauled system has been shown to demonstrate satisfactory performance, the preventative maintenance program should be implemented to help maintain continuous and reliable operation.

14.6.1 Troubleshooting Problems Associated with Scrubber Ancillary Equipment

A great deal of the operating problems associated with a particulate scrubbing system is due to the scrubber's auxiliary equipment. The auxiliary equipment includes everything it takes to make a scrubber system operate, such as any pumps, fans, clarifiers, vacuum filters, motors, agitators, ducts, dampers, piping, valves, recirculation tank, instruments, and motor starters.

Table 14.4 Troubleshooting venturi Scrubbers.

Problem	Possible Solutions
Particulate buildup in scrubber	1) Clean liquid distribution system 2) Clean drain 3) Check gas flow and composition against design specifications 4) Increase scrubbing liquor flow rates 5) Improve flushing system
Scrubber differential pressure drop too low	1) Repair plugged, broken, or leaking static tap lines 2) Check gas flow against design specifications 3) Adjust fan drive belts 4) Adjust fan speed 5) Clean inlet duct 6) Check inlet static pressure against design specifications
Scrubber differential pressure drop too high	1) Eliminate leakage 2) Check gas flow against design specifications 3) Adjust fan drive belts 4) Adjust fan speed 5) Clean scrubber
Entrainment	1) Improve mist elimination 2) Reduce gas flow 3) Reduce liquid flow 4) Clean (or enlarge) drains 5) Clean mist eliminator 6) Reweld any loose vanes 7) Improve flushing

Table 14.5 Troubleshooting tray scrubbers.

Problem	Possible Solutions
Weeping	1) Select open area proper to eliminate weeping 2) Bleed in air if possible 3) Blank off excess area
Plugging	1) Use spray wash heater; check operation 2) Reduce solids content in scrubbing liquid 3) Clean periodically 4) Raise flow rate
Poor distribution	1) Check, inspect trays; lack of uniformity of condition indicates poor distribution; baffle as required to offset poor gas flow 2) Use end weirs on the downcomers to retain extra water on the tray 3) Clean weir boxes
Weir sizing	1) Size for full flow, maximum conditions; if too small, enlarge or install supplemental external weirs 2) Keep weirs clean
Mechanical	1) Make certain trays are securely in place 2) Stiffen warped trays 3) Tighten all fittings 4) Reweld baffle strips, caps, keeper plates, etc., or replace trays if broken

Table 14.6 Troubleshooting packed-bed scrubbers.

Problem	Possible Solutions
Poor gas distribution	1) Use injection-type support grid 2) Allow extra vertical height-inlet to grid 3) Use less than 8 ft/s vertical velocity
Poor liquid distribution	1) Install redistributors every 4–6 ft of packing 2) Rearrange headers and liquid entry 3) Use a reflux distributor grid
Packing sized improperly	1) Check with packing manufacturer, not scrubber designer; many companies ignore recommendations
Too high velocity	1) Put another tower in parallel with existing unit 2) Cut back in flow rate if possible
Too low velocity	1) Restrict bottom part of grid 2) Use small packing 3) Use redistributors 4) Raise water rate

As in any air pollution control system, a scrubber system is subjected to a harsh environment. The abrasiveness of some types of particulate causes equipment such as fan wheels and valves to wear away. Stack gases may also contain a variety of components that, when wetted, produce compounds that corrode equipment; this is particularly true of any gas stream from a boiler that fires sulfur-bearing fuel and of gases from many chemical processes. All scrubber components must be designed and maintained with these considerations in mind.

14.6.2 Rotating Machinery

Rotating machinery consists of all items such as fans, pumps, clarifiers, rotary vacuum filters, motors, and agitators. Because these items are in constant motion, a great deal of wear can take place. These components require some amount of regular attention to ensure the proper operation of the system. Key areas of maintenance on rotating machinery are the bearings and any components rotating in the fluid stream.

There are also many other areas of maintenance unique to each type of equipment. Wherever practical, spares should be included. This is especially true to high-maintenance items such as pumps. Wear resulting from any abrasive solids in the liquid stream often dictates frequent inspection and repair. When mounted before the scrubber, fan wheels may also exhibit a great deal of abrasion from the particulate in the gas stream. Unfortunately, however, it is rather impractical to utilize spare fans in the scrubber system due to the difficulties in the duct arrangements and the high capital cost of fans.

A stock of the more failure-prone components should be maintained. These include such items as bearings, seals, bushings, pump housing liners (if pumps with replaceable liners are used), pump impellers, fan wheel gaskets, makeup-water control valves, pH control valves, pH probes, recirculation tank level control sensors, and vacuum filter cloths.

14.6.2.1 Fans

As in all rotating machinery, various components of the fan must be periodically lubricated. These components include all bearings on the wheel shaft, couplings, and dampers. Lubrication schedules and types of lubricants should follow the equipment manufacturer's guidelines. Generally, the lubrication schedule for these items ranges from quarterly to annually. The frequency of inspection of wear-prone components of the fan will vary with the severity of the application. On a fan handling relatively clean gas, the wheel and housing may operate for up to one year without any need for inspection. Fans on dirty applications may require monthly or even weekly inspection. The fan wheel and housing should be inspected for signs of corrosion, abrasion, and particulate buildup. Corrosion is generally not a problem if the fan is located upstream of the scrubber and the gas temperature is above 250°F. However, if the fan is downstream of the scrubber, moving the fan upstream of the scrubber may eliminate the problem of corrosion if the gas temperature is high enough. Should rearranging the fan not be feasible, the fan will have to incorporate more corrosion-resistant materials of construction. This requires going to a higher alloy on the fan wheel and possibly installing a corrosion-resistant liner on the housing. Attempting to reduce corrosion by collecting the corrosive gas in the scrubber system generally will not work. This is due to the fact that any condensation on the fan wheel will tend to pick up any residual corrosive gas that gets through the scrubber system. This corrosive component, when collected on condensation, will eventually react with the fan materials no matter how small the concentration.

Abrasion on the fan may be lessened by the installation of replaceable wear-liners on the wheel. Replacing the existing fan with a larger fan capable of the same flow rate and static pressure but at a lower speed may help to reduce the abrasion problem. If the fan is located before the scrubber, the installation of a high-efficiency mechanical collector before the fan may help to reduce any abrasion problems. Should a mechanical collector already be installed prior to the fan, the mechanical collector's tubes should be checked for pluggage or excessive wear. Should the existing mechanical collector be found to be working properly, consideration might be given to replacing it with a higher-efficiency mechanical collector or going to a two-stage mechanical collector.

Material buildup on the fan wheel may require a spray wash system to be installed in the fan. If this fails to improve the situation enough, moving the fan upstream of the scrubber may alleviate the problem.

Fan bearings should be inspected once per shift, if possible, for oil level, oil color, oil temperature, and vibration. Bearings found to be lubricated inadequately may require a greater frequency of lubrication or the installation of a forced lubrication system. The bearing housing should be inspected weekly for leaks, cracks, and loose fittings. Excessive bearing wear may be caused by a higher fan operating temperature than was originally specified or by excessive vibration due to material buildup on the fan wheel. Misalignment of the bearing mountings could also account for this problem. A major annual inspection should be made to check all bearing clearances and to detect any sign of wear, pitting, or scoring on the bearings.

Dampers should be inspected semi-annually for wear, corrosion, cracking, or looseness. All stuffing boxes should also be inspected. Annual damper inspection should include the tightening of all fastenings. The travel of the actuator and linkage versus damper leaf and linkage travel should also be checked at this time. Shaft seals on the fan should be inspected visually for wear monthly.

14.6.2.2 Pumps

All bearings on the pumps and pump couplings should be lubricated periodically in compliance with the manufacturer's instructions. Many pumps include oilers, which should be maintained at a specified oil level. Some pumps are equipped with grease-lubricated bearings. These should also be lubricated according to the manufacturer's specifications.

As with fans, the frequency of inspection will vary with the severity of the application. Normally, pump bearings should be inspected at least as frequently as fan bearings with minor inspections once per shift, if possible, for bearing oil levels and for seal leaks and vibration. Semi-annual inspection should be made for leaks, cracks, and loose fittings. A major inspection for wear, pitting, scoring, and clearance should be made at least once per year.

The pump housing and impeller, along with the seals, are components that are subject to abrasion and corrosion. A properly operating mechanical collector preceding the scrubber will aid in reducing the amount of particulate to be collected by the scrubber. This, in turn, will reduce the amount of abrasive solids required to be pumped through the scrubber recirculation system and should reduce any abrasion in the pump.

Increasing the bleed stream from the scrubber system will also reduce the concentration of suspended solids handled by the pump. Should neither alternative be available, a pump with a replaceable rubber liner may be of benefit. This should result in minimizing pump replacement costs. The installation of a water flush in the seals may help to reduce wear on the seals.

Corrosion may be combated by installing a pump constructed of a high-grade alloy, or a rubber-lined pump, depending on the chemistry of the recirculation system. Increasing the amount of chemicals added to a scrubber system can help neutralize the recirculation stream and help preserve the equipment. However, this may require a significant expenditure for purchasing chemicals. For example, the addition of sodium hydroxide will increase the scrubber's efficiency to collect a corrosive gas such as sulfur dioxide. As more sulfur dioxide is collected the pH of the solution drops and increased amounts of sodium hydroxide are needed for neutralization until an equilibrium is reached between the scrubbing efficiency and the desired pH level. At a recirculating solution pH of about 7, the scrubber's sulfur dioxide removal efficacy using caustic is about 90% as compared to a 60% efficiency when scrubbing with water at a pH of 4. Increasing the scrubber's bleed rate may help to lessen corrosion problems by decreasing the concentration of the corrosive compounds in the recirculation system.

14.6.2.3 Clarifiers and Drag Chain Tanks

Clarifiers and drag chain tanks fall into a general category of gravity separation devices. A clarifier or drag chain tank may be used to remove the particulate matter from the scrubber bleed stream when the particulate is sufficiently large and has a short settling time.

Clarifiers rake the settled particulate to the center of the unit, where it is removed from the bottom in a concentrated slurry. The slurry from the clarifier is pumped to a final discharge or to an additional dewatering device if a drier final discharge is required. The pipeline from the clarifier must be periodically inspected to make sure that it is not plugged.

The rake mechanism in a drag chain tank moves the solids up an incline. As the rake breaks the surface of the liquid, water falls off the rake and a relatively dry solid is moved up the incline. The solids continue to be pushed up the incline and over the edge of the tank, where they all onto a pile along the side of the tank. The pile of solids must be regularly removed from the side of the drag chain tank at a rate dependent on the application. A front-end loader may be used for this chore; however, it is usually better housekeeping to have the solids discharged into a dumpster and then have the dumpster hauled away when full. The rakes of this equipment require little maintenance since the speed of rotation is very slow. However, sludge buildup may put excessive torque on the rake mechanism. If this should occur, the clarifier or dray chain tank must be shut down, drained, and hosed out.

Bearings and gear reducers must be lubricated and inspected according to the vendor's schedule. This is generally done semi-annually. In addition to the inspection of the bearings, the drive mechanism must be inspected for temperature rise and excursions, sprocket alignment, sprocket wear, chain tension, and oil level. These inspections should generally be done after the first 100 hours of operation and every 500 hours thereafter.

14.6.2.4 Rotary Vacuum Filters

Rotary vacuum filters are becoming a more common part of wet scrubber particulate control systems, most often used on the bottom discharge from a clarifier to obtain a more concentrated sludge discharge before ultimate disposal. In a particulate control system, the fine particulate matter may pass through an ordinary rotary vacuum filter. A precoat on the rotary vacuum filter may be required for such an application. In a precoat filter, a layer of filter aid is formed on the filter drum. Once the filter has been coated with the filter aid, the liquid to be clarified is fed into the unit. The knife edge then shaves off the particulate from the filtered mass together with a small portion of the precoat. After the precoat has been removed down to a depth where it can no longer effectively filter, the filter must be shut down and recoated. Precoating the filter may take as long as 1 hour and the filtering cycle may last from 16 hours to 1 week, depending upon how fast the precoat is removed from the filter. The depth of the precoat must be visually checked periodically. A spare filter or multiple filters should be installed to ensure continuous operation of the particulate removal system.

Periodic lubrication of all knife-edge bearings, filter bearings, motor bearings, gear-speed reducers, and vacuum pump bearings must be done in accordance with the vendor's schedule. The filter cloth, knife-edge alignment, and drive belt must be inspected occasionally and adjusted or replaced as required. The frequency of inspection will again vary with the severity of the application, ranging from once per week to once per year.

14.6.2.5 Motors – Mechanical

Motors are basically quite simple and reliable from a mechanical standpoint. However, as with all rotating machinery, the bearings require some degree of care and attention. Motor bearings should be lubricated in accordance with the manufacturer's specifications. The schedule for this may vary from 750 hours to 8,000 hours of operation depending on the motor speed and the severity of the conditions to which the motor is exposed. Extreme care must be exercised to avoid excessive greasing of the motor bearings. Excessive greasing can cause the bearings to overheat and can contaminate the motor's windings. When greasing motor bearings, the bearing drain or flush holes must be unplugged to help guard against over-greasing. Some manufacturers supply bearings that are pre-lubricated and sealed for life. This is advantageous in that over- or under-lubrication is perhaps the greatest cause of motor-bearing failure. Bearings should be inspected in the same manner as described for fan and pump bearings.

14.6.2.6 Agitators

Agitators may be required on systems where sedimentation of the particulate in the recirculation tank is a problem, or if additional mixing is required to mix any additives into the scrubbing liquor. The bearings and gear reducers should be inspected and lubricated according to the manufacturer's specifications. The agitator's shaft and propeller should be inspected occasionally for wear and corrosion. Should the shaft or propeller exhibit excessive wear or corrosion, it should be replaced with a higher alloy. On side-entering agitators, the packing and seals should be inspected for signs of leakage and wear.

14.6.3 Static Equipment

14.6.3.1 Ductwork

Ductwork should be inspected occasionally for leakage and excessive flexing. The frequency of these checks will vary with the severity of the application, but should be done no less than semi-annually. Holes in the ductwork can be the result of a number of causes, but are most often the result of corrosion or abrasion. For ductwork located after the scrubber, corrosion will be a primary cause of failure. The deterioration of the duct due to corrosion may be alleviated by lining the duct with a corrosion-resistant liner. This, however, may not be possible if the duct has corroded too greatly or if the gas temperature is too high. Should this be the case, the duct will have to be replaced with an appropriate alloy or with fiberglass if the temperature is low enough. Fiberglass has proven to be an appropriate material for construction when located downstream from the scrubber. The duct can be protected from temperature excursions due to water outages in the scrubber and supplying it with water independent of the normal scrubbing recirculation system.

Abrasion is a problem common to ductwork located upstream from the scrubber. The severity of this problem can be lessened by increasing the duct diameter, thus slowing the velocity through the duct. The addition of a mechanical collector prior to the scrubber system will lessen the particulate loading through the ductwork and therefore aid in reducing ductwork abrasion. Should a mechanical collector already be in the system and abrasion is a problem, the collector should be inspected to ensure that it is in proper operating condition. If it is found to be working properly, consideration should be given to installing a higher-efficiency mechanical collector. If excessive flexing or cracking should develop, an expansion joint should be added to the ductwork at or near that point.

14.6.3.2 Dampers

Because they are operated infrequently, dampers present special maintenance problems. Guillotine zero-leakage dampers are perhaps the most effective for air pollution control systems. Louver and butterfly dampers, however, may be appropriate for less severe or demanding service. Since the maintenance on all types of dampers is similar, concentration will be given to guillotine-type dampers.

Lubrication is an important part of any damper maintenance program. The scheduling and types of lubricants should follow the manufacturer's specifications. Lubrication is normally done annually on all components of the drive mechanism and bearings. The blower motor bearings should also be lubricated at that time if one is included with the damper.

A program of regular inspection must be instituted to help ensure that the damper will operate when required. The damper blades should be moved in and out of the duct 6–12 in. several times, at a frequency of no less than once per month. More severe services may require a greater frequency of this checking. Checks for leakage around the flanges should be made at least once per month. All portions of the damper's sealing elements should be inspected for wear semi-annually. On a yearly basis, all drive components, bearings damper blades, and blowers should be inspected for wear or leakage. These inspections should be carried out with strict adherence to the manufacturer's instructions.

Guillotine dampers should be equipped with an air purge to prevent the buildup of debris in the blade seating surface trough when the blade is in an open position. It is important, to ensure proper damper operation, and that the manufacturer's schedule for activating these purges be followed. The frequency of these purges will depend again on the severity of the application; however, it is recommended that it be done at least daily using six air bursts of 5 s duration with 1–1.5 minutes between each burst.

14.6.3.3 Piping

Piping should be inspected at regular intervals depending on the application. Piping is prone to three common problems: abrasion, corrosion, and pluggage. Abrasion is common to any scrubber system collecting a hard, insoluble particulate, such as coal-fired boiler fly ash. Abrasion problems in the recirculation piping are best alleviated by reducing the concentration of the particulate in the recirculation system. This can be done by increasing the bleed rate from the system, by installing a mechanical collector ahead of the scrubber if none already exists, or by ensuring the proper operation of an existing mechanical collector. Reducing the pipe velocity by installing a larger-diameter pipe may also help relieve an abrasion problem. It must be noted, however, that a velocity of 4–7 ft/s is a reasonable compromise between preventing abrasion and preventing pluggage.

Corrosion will be a problem in recirculation systems where the scrubber is operating on a gas stream containing corrosive gases. The methods of alleviating a corrosion problem in the recirculation system are similar to those discussed for pumps. The addition of caustic, for example, can ultimately raise the pH and neutralize the scrubber solution. Replacing the existing piping with piping constructed of a more corrosion-resistant material may be a more economical alternative in the long run. FRP pipe or plastic-lined cast-iron pipe has been successfully used where corrosion is a problem and the particulate being collected is water-soluble, such as sodium sulfate. Plastic-lined cast iron has an advantage of FRP in that it is more resistant to accidents on the outside, such as with forklift crashes. On a recirculation system, where both corrosion and abrasion are problems, stainless steel and rubber-lined case iron pipes have been shown to be effective.

Pluggage can be a problem on recirculation systems handling a slightly soluble particulate or if the pipe velocities are too low, thus promoting settling. On systems where the former is the problem, procedures such as increasing the bleed rate or raising or lowering the pH may help to alleviate the pluggage. Should the latter be the case, an increase in the pipe velocity by decreasing the pipe diameter may help to eliminate this problem. Piping on a system that is prone to pluggage should be flushed out with clean water on a yearly basis at a minimum. Consideration should also be given to rodding out the piping if this is possible.

14.6.3.4 Valves

Valves will be subject to most of the same problems that have been discussed for piping: i.e., abrasion, corrosion, and pluggage. The solution to these problems in valves will be similar to those solutions for the equivalent problems in piping. Installing larger-diameter valves, increasing the bleed rate, and installation of a mechanical collector before the scrubber will all aid in reducing abrasion. Increasing the system's pH and installing valves of corrosion-resistant material will help to eliminate corrosion problems. Pluggage can be alleviated by increasing the system bleed rate, adjusting the pH, or increasing the valve velocity in cases of settling. On a system that exhibits both corrosion and abrasion, rubber pinch valves have been found to be effective. The rubber provides resistance to corrosion, and the design of the valve proves a smooth passage for the liquid stream containing the particulate, thus minimizing abrasion.

Air-operated control valves may exhibit sticking problems. The problem may stem from one or two sources. The valve may exhibit some pluggage due to the particulate in the recirculation system, or it may have some water buildup at the air connection. The latter problem may be reduced by ensuring a supply of dry/oil-free air to the valve or by regular inspection and drying of the air connection. The frequency of this cleaning will depend on the amount of moisture in the air supply. Since a control valve can be a rather vital component in the operating of a scrubber system, it is recommended that the piping be arranged so that any system containing a control valve can be temporarily placed in manual operation while the control valve is repaired or replaced. This is particularly important in the caustic addition system, where loss of caustic may severely damage the scrubber system's components. It is also important in the scrubbing liquor makeup system, where a malfunction of the control valve may cause the scrubber to run dry, thus damaging the equipment from excessive temperature. A malfunctioning control valve in the makeup line may also cause the recirculation tank to overflow. A failure of either of these control valves and/or the inability to resort to manual operation may require a shutdown of the entire scrubbing system.

14.6.3.5 Recirculation Tank

The scrubbing liquor recirculation tank in most systems will require very little maintenance other than an occasional inspection of the general condition of the tank. Corrosion may be exhibited on a tank handling an acidic scrubbing medium. Adding sodium hydroxide, lining the tank with a corrosion-resistant liner, or replacing the existing tank with one constructed of FRP or an appropriate alloy can alleviate the problem. If the scrubber is handling a large, easily settleable particulate and if the tank has a relatively long retention time, particulates may build up on the bottom of the tank. This buildup may plug the recirculation pump nozzle and foul the instrument probes. The addition of an agitator may help to solve this problem. Lowering the retention time by lowering the liquid level may also help, provided that this can be done without causing cavitation in the pump. Certain contaminants collected in the scrubber may cause a foaming problem. This problem may be reduced through lowering the scrubbing liquor's pH by reducing the amount of sodium hydroxide added to the system or by adding a slight amount of acid if the scrubbing solution is originally basic. Should the scrubber system not be able to tolerate a lower pH, foam-breaking sprays can be installed above the tank or antifoaming agents may be added to the tank. Adding more freeboard to the recirculation tank may also be of some benefit in relieving a foaming problem.

Pump cavitation may be a problem if the liquid level in the tank is not high enough for the pump flow rate and the diameter of the recirculation pump nozzle. A baffle or bell mouth may be used to prevent such cavitation.

14.6.4 Electrical

14.6.4.1 Instruments

Instruments usually require very little maintenance. The philosophy is generally to replace them as they fail. The accuracy and calibration of instruments should be checked annually. All probes should be cleaned at a frequency dependent on the application. Level controller tanks handling highly soluble particulate may never require cleaning, whereas the pH probes on a lime slaker, for example, should lime be used as a neutralizing agent, may require daily cleaning. It is therefore important to select probes that will be appropriate for the frequency of maintenance. A flange-mounted differential pressure transmitter, used for level control and located at the bottom of a recirculation tank, may be appropriate for a scrubber system operating on a highly soluble particulate. However, if the particulate tends to settle easily, this type of level controller might require that the tank be frequently drained in order to clean the probe when a sufficient amount of solids have settled around it. A top-entering float-type level controller may be more appropriate for this type of application. All pneumatic lines should be cleaned and dried of moisture regularly at intervals depending on the quality of air supplied to them. Orifice plates should be inspected at least annually and may require more frequent inspection should the fluid contain any abrasive solids. Excessive wear on orifice plates may necessitate replacement with a harder material of construction. Ink and paper in the recorders must be checked frequently.

14.6.4.2 Motor Starters

Motor starter switches should be lubricated occasionally according to manufacturer's specifications. Cleaning of all electrical equipment is important to ensure reliable operation. Dirt in electrical equipment may decrease heat dissipation and cause poor ventilation. Overheating and insulating breakdown may result if the dust is electrically conductive. All contacts, transformer insulation, and cooling fans should be inspected, cleaned, and dried annually. Dirty environments may require more frequent inspection and cleaning.

14.6.4.3 Motors – Electrical

Motors require little attention from an electrical standpoint except for an occasional cleaning and check of temperature rise. Dust may be removed by using dry low-pressure compressed air and a dry cloth. When mixed with dirt, grease can form a gum that is difficult to remove. Oil can cause a deterioration of the insulating varnish, which in turn can lead to a burnout. Any grease or oil can be removed with a solvent such as naphtha. A fresh coat of insulating varnish should be reapplied after such a cleaning.

The measurement of the temperature rise can provide an indication of problems such as overloading, shorts, or burned insulation. The three methods available to measure the temperature rise are thermometer, resistance thermocouple, and embedded thermocouple. The thermocouple method simply consists of placing a thermocouple as close as possible to the windings and core of the motor. This may be done by removing a head bolt or eye bolt and inserting a common mercury or alcohol thermometer. The resistance method requires that a Wheatstone or Kelvin bridge be used to measure the winding resistance at room temperature and then once again after the motor has operated for 4–6 hours at full load.

The embedded thermocouple method involves connecting a bridge-type instrument to a thermocouple that has been embedded in the windings. This will usually give a reading directly in degrees Celsius. The actual temperature rise should then be compared against the manufacturer's recommended temperature rise for that type of motor. A motor can be too hot to touch and still be within the proper operating temperature. Therefore, it is important not to attempt to measure the motor temperature by hand. On motors of 300 hp and above, the thermocouple method should be used with a recorder to monitor the temperature continuously.

14.7 Recent Developments

The basic design of venturi scrubbers has remained relatively unchanged since first used in the early part of the twentieth century. Some modest equipment changes and new types of devices have appeared in the last few decades, but all have essentially employed the same capture mechanisms used in the past.

The role of an effective preventive maintenance program maximizes the time between scheduled shutdown periods, significantly minimize unscheduled breakdowns, extend equipment life, and reduce maintenance and repair costs. Adherence to a routine maintenance program will go a long way toward both eliminating many of the problems that have plagued the operation of particulate scrubber systems and maintaining/improving equipment performance.

It is generally believed that increasing the operating pressure drop of a venturi scrubber produces an increase in collection efficiency (or a decrease in penetration). This is almost always the case for pressure drops below 75″ WC. However, based on a computer model study, there appeared to be a leveling out of efficiency once the pressure drop reached the 75–100″ WC range. For pressure drops greater than 100″ WC, the efficiency was found to decrease marginally (and not increase) with increasing pressure drop. These strange phenomena may be explained in terms of the effect of gas velocity on the two principal mechanisms of particle capture – inertial impaction and molecular diffusion. As the velocity increases, capture by impaction for large particles increases, with little effect on small particles. However, the effect of capture by diffusion on small particles is just the opposite: efficiency decreases because of the shorter residence time brought about by increased velocity. Since the larger particles are normally captured with 100% efficiency, a point is reached at which increases in velocity (and accompanying increases in pressure drop) produce a decrease in efficiency. Thus, contrary to data and information that have appeared in the literature, a plot of collection efficiency versus pressure drop does not increase or decrease monotonously but rather goes through maximum.

14.8 Conclusions

Wet scrubbers are probably the most widely used and most often abused of all air pollution control devices. The proper operation and maintenance of these systems, in many cases, can enhance particulate removal control efficiency and reduce operating costs. An effective preventative maintenance program will also maximize unscheduled breakdowns, extend equipment life, and reduce maintenance and repair costs. Proper system design and adherence to a routine maintenance program will go a long way toward eliminating many of the problems that have plagued the operation of particulate scrubber systems.

References

1 **A. Jarvis**, *Current Strategies for Engineering Controls in Nanomaterial Production and Downstream Handling Processes*. U.S. National Institute for Occupational Safety and Health, pp. 14–15, Washington, DC, 2013.

2 **H.F. Johnstone, and M.H. Roberts**, *Deposition of Aerosol Particles From Moving Gas Streams*. Ind. Eng. Chem., 2417, New York, NY, 1949.

3 **H.F. Johnstone, R.B. Field, and M.C. Tassler**, *Gas Absorption and Aerosol Collection in a Venturi Atomizer*. Ind. Eng. Chem., 46/8, New York, NY, 1954.

4 **L. Theodore, and A.J. Buonicore**, *Industrial Air Pollution Control Equipment for Particulates*, Chapter 6, CRC Press, Inc., Boca Raton, FL, 1976.

5 **L. Theodore, C. LoPinto, L. Llama, and C. Murray**, *Pressure Drop in Venturi Scrubbers: Experimental Results*. Proceedings Fourth An. Env. Sci. Conf., Louisville, KY, 248–256, 1974.

6 **L. Theodore**, *Air Pollution Control Equipment Calculations*, John Wiley & Sons, Hoboken, NJ, 2008.

7 **L. Theodore, P. Byrne, S. Murtha, A. Buonicore**, *Pressure Drop in Venturi Scrubbers: Describing Equations*. Proceedings Fourth An. Env. Sci. Conf., Louisville, KY, 236–247, 1974.

15

Electrostatic Precipitators

Electrostatic precipitation is perhaps the most popular method in use for removing fine solids and liquids from gas streams. The first successful commercial-scale precipitators were installed by Frederick G. Cottrell on two California chemical plants in 1907. One of these units treated 5,000 cfm of gas and recovered precious metals and sulfuric acid. In 1910, Cottrell installed a successful precipitator on a 250,000 cfm smelter. The technology grew rapidly, and successful applications were soon found in a variety of industrial processes and in the power generation industry. Precipitators have been used to treat over 5,000,000 cfm of gas and have attained removal efficiencies approaching 99.99%.

Compared to other methods of gas cleaning, electrostatic precipitators (ESP) are as elegant as they are efficient. Instead of performing work on the entire gas stream in the cleaning process, the forces in a precipitator are applied directly to the suspended particles themselves. The result is high collection efficiency at a moderate cost of power. This chapter is by far the longest due to a combination of the complexity, unique nature, and widespread use of the ESP systems.

There are three fundamental steps in the electrostatic precipitation process:

1) Charging the particles suspended in the gas stream
2) Collecting the charged particles
3) Removal of the collected particles into an external receptacle

Charging is accomplished by applying a high DC voltage to an electrode system. The presence of a grounded electrode near the charged electrode gives rise to the formation of a corona and a unidirectional electric field. Particles traveling between the two electrodes acquire charge from the corona discharge and are driven toward the collecting electrode by the electric field. All commercial precipitators use negative polarity.

Particles arriving at the collecting electrode do not have to be captured immediately for successful operation and may actually be recollected several times before they are removed. Current flows in the collection process are large enough to describe the operation as electrical in nature, the term "electrostatic" being one of convention rather than description. Precipitators may be single stage (Cottrell type) or two stage, in which the charging and collecting fields are formed independently. Also, the gas stream may flow through tubular collecting electrodes or between parallel plates. The parallel-plate single-stage precipitator is by far the most common.

Particles remain on the collecting electrode, or plate, until they are dislodged by a mechanical blow, called rapping, and fall by gravity into an external vessel, usually some form of hopper. Rapping is not as simple as it may sound, and indeed must be as effective as the charging and collecting steps for efficient precipitation.

The widespread use of precipitators in the electric power industry has uncovered problems resulting from variations in the chemical composition of the coal and hence in the properties of the fly ash (the electrical resistivity). Most of the early control installations were made on boilers that burned relatively high-sulfur coal. As new installations were made, problems were encountered with boilers firing low-sulfur coal. This coal produces fly ash that has a high resistivity and is difficult to collect.

The physical arrangement of precipitators differs depending on the type of application. Wire and cylinder electrodes are used in some applications; however, for reasons of space economy, most commercial precipitators use plates as collection electrodes. The majority of precipitators are constructed so that the charging and collection steps take place within the

Optimizing Air Pollution Control Equipment Performance: Operation & Maintenance, First Edition. Jay Richardson and Louis Theodore.

same region. Precipitators of this type are termed single stage. For some applications, charging takes place in one section which is then followed by a section consisting of alternately charged plates. The collecting electric field is established independently of the corona field and such precipitators are termed two stage.

Electrostatic precipitation occurs in the space between the discharge electrode and the collection surface. A high-voltage, pulsating direct current is applied to an electrode system consisting of a small-diameter discharge electrode that is usually negatively charged, and a collecting plate electrode that is grounded. This produces a unidirectional, nonuniform electric field whose magnitude is highest near the discharge electrode.

Resistivity is related to the ability of a particle to take on a charge. In most industrial applications, the resistivity of the particle is such that the charge on the particle is only partially discharged on contact with the grounded collection electrode. A portion of the charge is returned and contributes to the intermolecular cohesive and adhesive forces that hold the particles to the collection surfaces. The dust layer builds up on the collection plate to a thickness of 0.5 in. (1.27 cm). If the dust layer becomes too thick, it is possible for the accumulated layer to act as an insulator, reducing the flow of the electric field lines.

The negative gas ion movement has two main charging effects on dust particles in the interelectrode region. These effects are called field charging and diffusion charging. Each type of charging is used to some extent in particle charging, but one dominates depending on the particle size. Field charging dominates for particles with a diameter >1.0 μm, while diffusion charging dominates for particles with a diameter <0.3 μm. A combination of these mechanisms occurs for particles in the range between 0.3 and 1.0 μm in diameter. It is also possible to charge particles by electron charging. In this case, free electrons that do not combine with gas ions are moving at an extremely fast rate. These electrons hit the particle and impart a charge. However, this effect is responsible for very little particle charging.

In field charging, as particles enter the electric field, they cause local dislocation of the field. Negative gas ions traveling along the electric field lines collide with the suspended particles and impart a charge on them. The ions continue to bombard the particle until the charge on the particle is sufficient to divert the electric lines away from the charged particle. This prevents new ions from colliding with the dust particle. When a particle no longer receives an ion charge, it is said to be saturated. Saturated charged particles then migrate to the collection electrode and are collected.

Diffusion charging is associated with the random Brownian[1] motion of the negative gas ions. The random motion is related to the thermal velocity of the gas ions; the higher the temperature, the more movement occurs. Negative ions collide with the particles because of the random thermal motion of the ions and impart a charge on the particle. The charged particles migrate to the collection electrode. The mechanism is important for charging particles in the submicron range. As noted above, in the intermediate size range from 0.3 μm to 1.0 μm in diameter, both field and diffusion charging are important.

In practice, the applied voltage is increased until it produces a discharge (corona), which can be seen as a luminous blue glow around the discharge electrode. The corona is a discharge phenomenon in which gaseous molecules are ionized by electron collisions in the region of a high electric field. The intense electric field close to the discharge electrode accelerates the free electrons that are present in the gas. These electrons acquire sufficient velocity to ionize gas molecules upon collision, producing a positive ion and an additional free electron (see Figure 15.1).

The additional free electrons create more positive ions and free electrons as they collide with additional gas molecules. This process is called avalanche multiplication, and occurs in the corona glow region (see Figure 15.2). Avalanche multiplication will continue until the local electric field strength decreases to the point where there is insufficient energy to perpetuate ionization. The sluggish positive ions migrate back to the negative discharge electrode and form new free electrons on impaction with the discharge wire or gas space around the wire. The electrons produced during the avalanche multiplication process follow the electric field toward the grounded collection electrode.

With avalanche multiplication, the electrons leave the corona region and enter the interelectrode region. The magnitude of the electric field is diminished and the free electrons' velocity decreases. When electrons impact gas molecules in the interelectrode region, they are captured, and negative gas ions are created. These negative ions serve as the principal mechanism for charging the dust.

Negative gas ions migrate toward the grounded collection electrode. A space charge, which is a stable concentration of negative gas ions, forms in the interelectrode region. Increases in the applied voltage will increase the field strength and ion formation until sparkover occurs. Sparkover refers to internal sparking between the discharge and collection electrodes. It is a sudden rush of localized electric current through the gas layer between the two electrodes. Sparking causes an immediate short-term collapse of the electric field. In general, it is very desirable to operate at voltages high enough to cause some sparking but not at a frequency such that the electric field constantly collapses. The average sparkover rate for

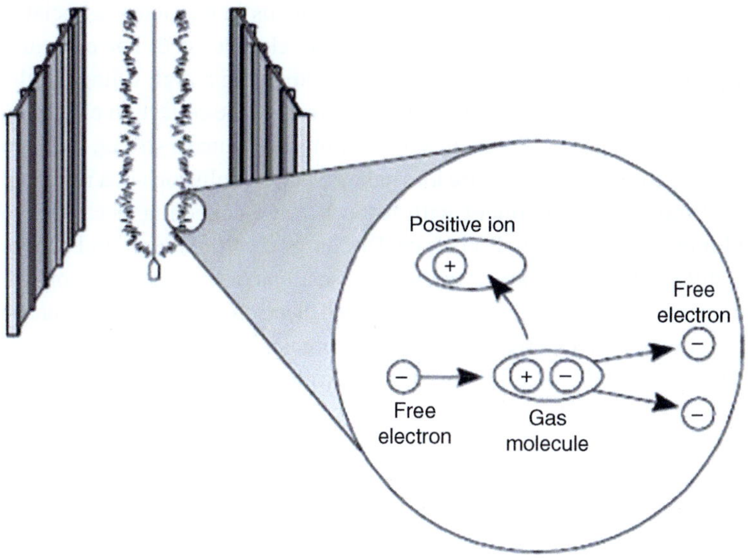

Figure 15.1 Generation of corona.

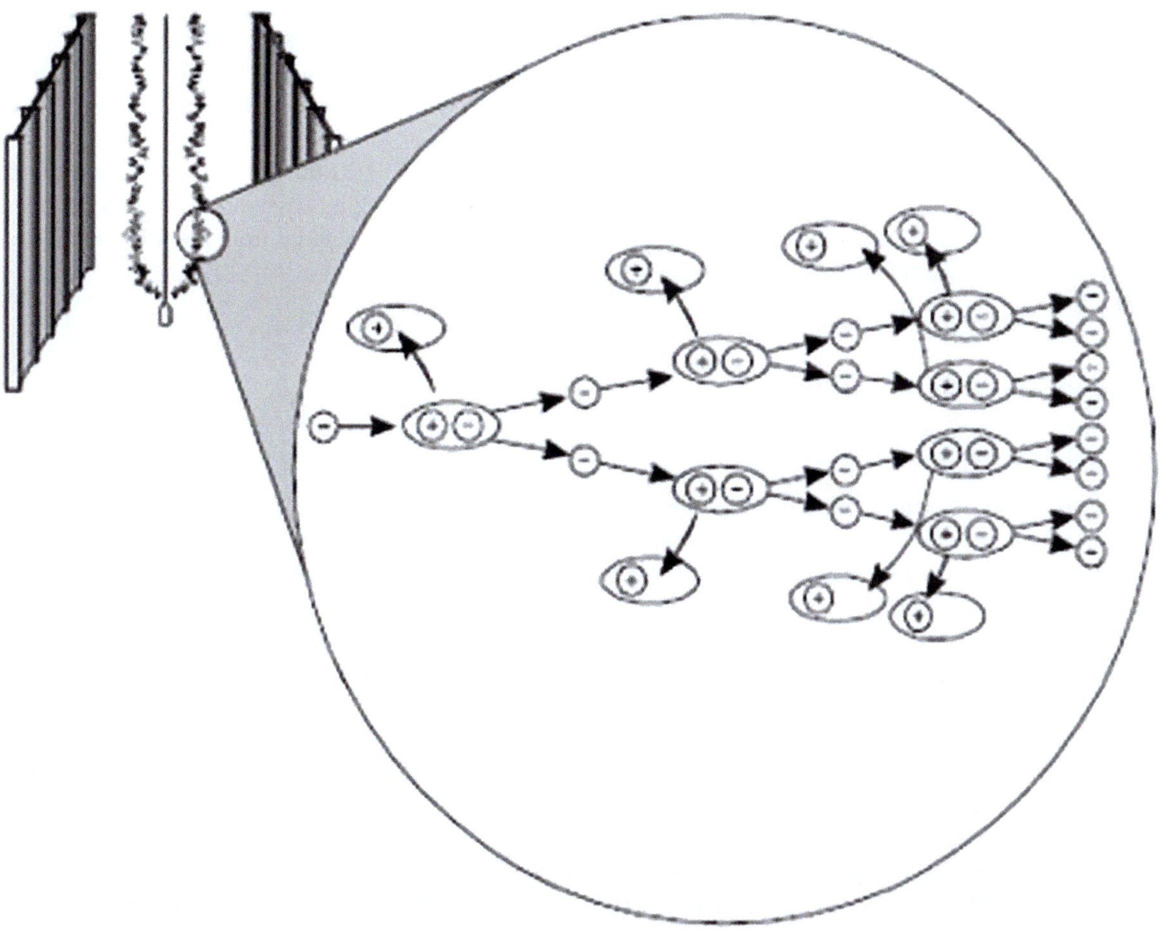

Figure 15.2 Avalanche multiplication.

optimum precipitator operation is between 50 and 100 sparks/min. At this spark rate, the gains in efficiency associated with increased voltage compensate for decreased gas ionization due to the collapse of the electric field. For optimum efficiency, the electric field strength should be as high as possible. This is accomplished by applying a high voltage to the discharge electrode with the consequent high corona current flow from the discharge electrode to the collection electrode.

The heart of the electrostatic precipitation process is the discharge electrode system. It must produce a strong, uniform corona (the corona is a discharge phenomenon in which gaseous molecules are ionized by electron collisions in a region of a high electric field) while maintaining the correct distance and alignment with respect to the collecting electrodes to prevent imbalances in the electric field and to avoid unnecessary arcing discharges. Discharge electrode sizes and shapes vary mainly by manufacturer, but variations among different applications or in different sections of the same precipitator are feasible. Round, straight wires about 0.1 in. in diameter are the most common discharge electrodes in use. They may be hung individually and free with a suspension weight at the bottom end (early electrostatic precipitator design), or they may be held in a structural framework that is rigidly attached to the precipitator structure, which is how almost all new precipitators are designed.

High-voltage rectifiers providing pulsating DC waveforms are in use almost exclusively because of the higher voltage and current attainable under sparking conditions when compared to pure direct current. With few exceptions, the discharge electrode system is subdivided into discrete sections, each being energized by a separate transformer–rectifier (TR) set. This "sectionalization" is important in matching corona currents and voltages to the TR set, and to promote reliability and stability under arcing conditions. A TR set consists of a high-voltage transformer and bridge rectifier, with typical secondary winding root mean square (rms) ratings between 53 and 66 kV and 250–2,000 mA. Most TR sets can be connected to the precipitator discharge electrode system in either full-wave or double half-wave.

An important adjunct to the discharge electrode system is the automatic regulation of the high-voltage input to the precipitator, because only in rare cases does nature permit ideal operation. The earliest precipitators had no means of voltage regulation, but state-of-the-art advancements such as the silicon-controlled rectifier (SCR), solid-state construction, and digital control circuits have enabled precipitators to perform efficiently under the most adverse conditions.

Most designs incorporate baffles to provide quiescent zones where the probability of particle collection is enhanced. These baffles are integrated into vertical stiffeners, which are required because collecting plate heights of up to 50 ft are occasionally in use.

Hoppers are best thought of as temporary holding bins to store collected particulate until permanent disposal can be scheduled. They take a variety of shapes, and are also sectionalized to facilitate handling large quantities of dust. It is a mistake to think the precipitation process stops at the hoppers. Removing collected material from the hopper is just as important as getting the material to fall into the hopper in the first place.

15.1 Description of Control Device

15.1.1 Physical Description

The heart of the electrostatic precipitation process is the discharge electrode system. It must produce a strong, uniform corona while maintaining the correct distance and alignment with respect to the collecting electrodes to prevent imbalances in the electric field and to avoid unnecessary arcing discharges. Discharge electrode sizes and shapes vary mainly by manufacturer, but variations among different applications or in different sections of the same precipitator are feasible. Round, straight wires about 0.1 in. in diameter are the most common discharge electrodes in use in the earlier ESP designs. They were hung individually and freely with a suspension weight at their bottom end, or they may be held in a structural framework that is rigidly attached to the precipitator structure. Modern discharge electrodes utilize a rigid frame design and may be mast-type, or formed elements, where rigidity, mechanical strength, and corona properties are all embodied into the same member.

High-voltage rectifiers providing pulsating wavefronts are in use almost exclusively due to the higher voltage and current attainable under sparking conditions when compared to pure direct current. With few exceptions, the discharge electrode system is subdivided into discrete sections, each being energized by a separate TR set. This "sectionalization" is important in matching corona currents and voltages to the TR set, and to promote reliability and stability under arcing conditions. TR sets are comprised of a high-voltage transformer and bridge rectifier, with typical secondary winding rms ratings between 53–66 kV and 250–2,000 mA. Most TR sets can be connected to the precipitator discharge electrode system in either full-wave or double half-wave.

An important adjunct to the discharge electrode system is the automatic regulation of the high-voltage input to the precipitator, because only in rare cases does nature permit ideal operation. The earliest precipitators had no means of voltage regulation, but state-of-the-art advancements such as the SCR, solid-state construction, and digital control circuits have enabled precipitators to perform efficiently under the most adverse conditions.

The discharge electrodes, collecting plates, and hoppers are all contained and supported by the casing, or shell. This structure must provide a gas-tight envelope in which the process takes place and must also hold the two electrode systems in proper alignment, sometimes under cyclical load conditions where the gas temperature (and hence the inner shell temperature) may vary over several hundred degrees F. In nearly all economical designs, essentially all the important auxiliary equipment are attached somewhere directly to the casing. Insulators that support the discharge electrodes system are made almost exclusively from porcelain of fused alumina, and are contained in individual or grouped insulator compartments or all the high-tension support insulators may be housed in a top housing or penthouse. In many cases, the manufacturer requires these insulators to be heated and ventilated at certain times during the operation of the precipitator.

The widest variation in design among manufacturers comes from rapping. In the simplest sense, rappers are either impulse (single blow) or vibrating (multiple blow) types. One impulse type consists of swing hammers striking the lower portions of the collecting electrodes or an intermediate portion of a rigid frame discharge electrode system. The swing hammers are actuated by a camshaft driven by an electric motor, and swing from the peak of the cam action by gravity to strike the electrode systems in a horizontal direction. Another type of impulse rapper is the drop hammer, which may be actuated by an electromagnetic solenoid located on top of the precipitators, or a motor-driven mechanical linkage. The rapping blow is gravity-actuated or may be spring-assisted, striking the tops of the systems in a vertical direction, or nearly any combination of these practices. Vibrating types are used almost exclusively on the discharge electrode systems, especially in industrial processes where weighted wire geometries are most common. Vibrators may be electric or pneumatic, the latter being extremely powerful and potentially destructive. Some precipitator designs utilize what is called "gentle rapping" where a gentle rapping of the plates at regular intervals causes the collected dust to move down the plate over a period of time before being discharged into the collecting hopper.

Other key components in a precipitator include the high-voltage bus and guard that deliver the TR output to the discharge electrodes; gas distribution diffusers at the inlet and outlet face; access systems, and a host of other systems, subsystems, and components necessary to support the operation by providing electrical power distribution, control, insulation, and so on. Most of these features are shown in Figure 15.3.

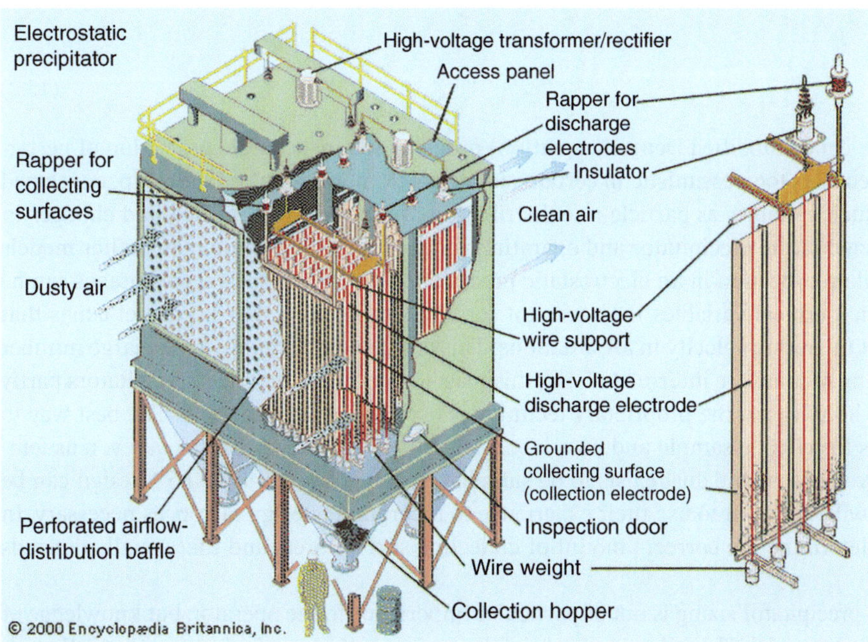

Figure 15.3 Electrostatic precipitator diagram.

15.2 Design Considerations

There are many ways to describe the design of any industrial machine, and precipitators are no exception. However, in almost all cases, precipitator design can be effectively divided into the four following aspects, each equally important over the life of the plant:

1) Performance
2) Reliability
3) Operability
4) Maintainability

Performance design begins with sizing, or selection of the correct amount of collecting surface area, its arrangement, how it is energized, and how it is rapped. A fundamental knowledge of sizing is important to the operator for understanding the basic variables that control precipitator performance. From empirical observations by Anderson[2,3] in 1919 and theoretical derivation by Deutsch[2,3] in 1922, the first expression for precipitator performance (Deutsch–Anderson equation) took the form

$$E = 1 - e^{-\left(\frac{A}{V}\right)w} \tag{15.1}$$

where E = collection efficiency
A = total collecting surface area
V = gas flow rate
w = migration velocity

Since E and V are usually known, the amount of collecting surface A can be calculated from a judicious selection of the term w. For decades designers selected w by analogy, that is, from experience with similar types of dusts and the performance that resulted. For each of these cases, w was presumed constant. Sizing by analogy and some inherent inaccuracies in the Deutsch–Anderson equation related to particle size and uniformity led to some disappointing results, especially for very high-efficiency requirements. It is virtually impossible to find a truly analogous situation, since industrial dusts are composed of many different-size particles, and w is in fact not constant but dependent upon many process variables.

An attempt to account for the sensitivity of w on process variables, especially particle-size distribution, appeared in 1957 and was later revised by Allander, Matts, and Ohnfeldt[4], who derived the expression

$$E = 1 - e^{-\left(\frac{Aw_k}{V}\right)^n} \tag{15.2}$$

The second exponent (in this, the so-called modified Deutsch equation) provides a more accurate prediction of performance at high-efficiency levels, but becomes too pessimistic in certain situations. A more useful relationship, developed by Feldman in 1975,[5] accounted for such variables as particle-size distribution at the precipitator inlet, and changes in size distribution as particles are collected in the precipitator and operating voltage levels. These and many other models have been proposed to predict particulate collection in an electrostatic precipitator, and while special advantages can be argued for each depending upon which critical variables they account for, the fact remains that no model exists that accounts for all the variables that limit migration velocity in all situations. This is attributable mainly to the large number of variables that exist and because many of them are interrelated. For this reason, manufacturers size precipitators partly by theory and partly by analogy, and in all cases, the proprietary techniques used are closely guarded. The best way to determine the resistivity of the dust is to collect a sample and, in a laboratory, develop a curve of resistivity versus temperature. Where this curve can be developed (if the dust exists to be sampled) then a fairly accurate ESP design can be developed. The ESP design company will still want to use their experience to incorporate design factors as necessary. In the final analysis, the goal is still to determine the correct amount of collecting surface area, and that usually depends upon the proper selection of w.

The method used to determine w in precipitator sizing is not of particular importance to the operator, but knowledge of the variables that influence its value can be useful. The following is a brief discussion of the principal factors that influence effective migration velocity and which manufacturers must contend with in the design stage. The operator should be able to identify which factors he has the most operational control over and which ones, although fixed by design, provide information in understanding precipitator behavior under different operating conditions.

1) *Energization*. Usually expressed as power density related to the total collecting surface area or unit of gas flow rate, the amount of useful power delivered to the precipitator is one of the chief variables that control the performance. Useful corona power is the power absorbed in collecting particulate, not the power consumed in sparking or simply the theoretical rating of the TR set. Voltage is an important parameter, but reasonable levels of corona current are necessary, since without current flow, collection would not take place.
2) *Sectionalization*. Energization levels are sensitive to differences in gas temperature, dust concentration, particle-size distribution, and resistivity. Variations in these parameters are common and even expected across the width and along the length of a precipitator. No sound theoretical relationship between sectionalization and migration velocity is in use, but empirical results have been observed that strongly support the need for proper sectionalization. Theodore reported a relationship between efficiency and the sectionalization of precipitators.[1]
Additionally, sectionalization gives an added safety factor to the precipitator design. Since each ESP section will have its' own TR set, the loss of power to that individual TR set will have an effect on the overall efficiency of the ESP. Sectionalization reduces the detrimental effect on efficiency that could result if a TR set is taken off line through maintenance or failure.
3) *Dust resistivity*. It is determined by ash chemical composition, gas temperature, and moisture levels. The resistivity of collected ash limits the rate of current flow and thereby affects the total useful corona power that can be delivered to the precipitator. Sulfur content of the coal plays a major role in determining the resistivity level.

15.3 Installation Procedures

To a large extent, long-term reliability for a precipitator can be assured by the dimensional stability of the casing and all internal components attached to it. This means that all design tolerances, especially those affecting electrode alignment, must be built into the final structure in such a way that thermal expansion, wind loads, and all other stresses the precipitator will experience during its lifetime will not cause dimensional distortion. Most precipitators are large structures assembled in the field using a variety of trade labor crafts and heavy equipment. Construction of them is a project, often requiring many months to complete, with the natural concerns for cost and time taking priority. Quality of construction and dimensional tolerance under these conditions can be assured best by people not responsible for schedules or cost overruns. This approach is generally the one taken. Of all the categories of interest in a construction project, the ones that influence dimensional stability are procedures, tolerances, and quality assurance.

15.3.1 Procedures

Construction.is a stepwise process, beginning after the foundations have been poured, cured, and checked for elevation and perimeter dimensions. A supporting open structure is then completed and the main precipitator structural members are built upon it, followed by the remainder of the shell and finally the internal components. Most of these activities overlap each other in time, with major checkpoints along the way to ensure that necessary prerequisites for tolerances are met before construction of the remainder of the unit gets too far along.

The erection sequence ordinarily requires that the load-bearing columns of the precipitator make full, flat contact with the slide plates on top of the supporting structure before hoppers or internals are attached. When the open shell is essentially complete, struts or sway bracing is generally attached and the entire structure is checked for plumbness and squareness. This check is extremely important since it will control the ease of gaining alignment of all subsequent parts. Benchmarks are established prior to installation of the electrode systems. The positioning of the electrodes is made relative to these benchmarks. Collecting and discharge electrode systems may be hoisted into the casing at the same time, but the collecting electrodes are normally completed first, and the discharge electrodes are positioned relative to them. When the electrodes have been installed, slide plate contact may again be checked, and a major check for square, plumb, and level electrodes and the alignment between electrode systems is completed. Rapping equipment and insulators are installed and aligned, the roof is set, and the remainder of the precipitator construction proceeds toward mechanical completion. Mechanical completion usually includes the items listed in Table 15.1.

Table 15.1 Mechanical completion items.

1) All collecting and discharge electrodes complete, within alignment tolerances, and welded out
2) All rapping equipment complete, aligned, and ready for test
3) Protection removed from insulators; no tracking or weld spatter
4) Pressure test completed
5) Field wiring complete and verified, megger tests done
6) Instrumentation and rotating equipment checked and ready for test
7) Key interlocks checked out

15.3.2 Tolerance

It comes as a surprise only to the uninitiated that a precipitator structure 200 ft wide, 90 ft long, and 80 ft high can be built within tolerances that require almost perfectly flat and level foundations, elevations on vertical steel members to within 1/8 in., and dimensions on squares and diagonals within 1/4 in. throughout the entire structure. These tolerances are necessary, and although care must be taken in the construction process, they are usually achieved.

The American Institute of Steel Construction allows the correction of minor misfits by reaming, chipping, cutting, and the use of drift pins. Methods other than these are not recommended, but if they appear necessary, construction may have to be interrupted until the problems are corrected. Construction methods have to be reviewed during the progress of work to prevent distortion of previously obtained tolerances. Backstep welding may be required to avoid excessive heat distortion of components essential to proper alignment. Care must be exercised when handling or erecting discharge or collecting electrode panels and pieces to prevent warping or bowing. Labor forces must also be cautioned against using alignment-sensitive precipitator components as fulcrums or jacking attachments.

15.3.3 Quality Assurance

Assuring structural integrity and dimensional stability of the precipitator takes careful planning and dedication throughout the construction project. Quality assurance interfaces with all aspects of the construction phase, from material delivery to final air-load test. Receipt of materials is an important aspect because in the interest of cost and schedule, materials may be shipped to the field site up to 12 months before they will be erected. The site logistics may require a dedicated laydown area adjacent to the crane-operating side of the precipitator and a preassembly area nearby. Erection is easier and quality better controlled when the largest practical components can be shipped to the site or assembled on the ground. Material storage areas may require up to 20,000 ft^2, including warehouse storage for electrical equipment, and protective coverings for electrode components. Most other materials can be stored in the open. Care must be exercised in unloading material, and defective or damaged shipments should be corrected at once.

Usually, dimensional tolerance checks and tests for mechanical completion are performed by the manufacturer's service department near the end of the construction period. All prior checks are usually made by the erection superintendent. It is a customary and sound practice to avoid major corrections by holding off final welding until the service representative has completed his work.

15.4 Operation

In the sense of operations, it is probably better to consider the precipitator as a process system rather than as a piece of hardware. Outwardly simple in design, the electrostatic precipitator takes part in a process that is complex. Many things can influence its behavior, but most of these influences can be overcome if the precipitator is well designed and properly operated. Operation is, in fact, equally important to design. To obtain maximum results, the operator must know how to properly start up, shut down, and operate the precipitator; learn how to distinguish between normal and abnormal behavior; and know what steps to take in case of trouble.

15.4.1 Safety

Personnel assigned to operate or maintain precipitators must be well trained in special safety procedures to avoid injury. Like many other industrial machines, precipitators have claimed lives, but in every one of these cases, tragedy could have been avoided if logical safety precautions had been observed. Safety is everyone's responsibility, but one person in the plant should be assigned the responsibility for monitoring and upgrading safety, standards and equipment, and for training others whose duties include working with or inside the precipitator system. Table 15.2 describes important special safety precautions for precipitator.

15.4.2 Pre-startup

Prior to energizing the precipitator and its auxiliaries for the first time, a preoperational inspection should be conducted to verify that all equipment is ready to operate. Basic inspection items are listed in Table 15.3. This list can be adopted into a routine checklist, as many of the inspection items are applicable following any unit shutdown of more than 24 hours, or following a shutdown where work was done inside the precipitator. When the checklist has been completed, an air load should be performed.

Table 15.2 Important safety precautions.

Wiring and Controls

1) Prior to startup, double-check that field wiring between controls and devices (TR sets, rapper prime motors, etc.) is correct, complete, and properly labeled.
2) Never touch exposed internal parts of control system. Operation of the transformer–rectifier controls involves the use of dangerous high voltage. Although all practical safety/control measures have been incorporated into this equipment, always take responsible precautions when operating it.
3) Never use fingers or metal screwdrivers to adjust uninsulated control devices.

Access

1) Use a positive method to ensure that personnel are out of the precipitator, flues, or controls prior to energization. Never violate established plant clearance practices.
2) Never bypass the safety key interlock system. Always keep lock caps in place. Use powdered graphite only to lubricate lock system parts; never use oil or grease. Never tamper with a key interlock.
3) Use grounding chains whenever entering the precipitator, TR switch enclosure, or bus ducts. The precipitator can hold a high static charge up to 15 kV after it is deenergized. The only safe ground is one that can be seen.
4) Never open a hopper door unless the dust level is positively below the door. Do not trust the level alarm. Check from the upper access in the precipitator. Hot dust can flow like water and severely burn or kill a person standing above the door. Wear protective covering.
5) Be on firm footing prior to entering the precipitator. Clear all trip hazards. Use the back of the hand to test for high metal temperatures.
6) Avoid ozone inhalation. Ozone is created any time the discharge electrodes are energized. Wear an air line mask when entering the precipitator, flues, or stack when ozone may be present. Do not use filters, cartridge, or canister respirators.
7) Never rod hoppers with an uninsulated metal bar. Keep safety and danger signs in place. Clean, bright signs are obeyed more than deteriorated signs.

Fire/Explosion

1) In case of boiler malfunction that could permit volatile gases and/or heavy carbon carryover to enter the precipitator, immediately shut down all transformer–rectifier sets. Volatile gases and carbon carryover could be ignited by sparks in the precipitator, causing fire or explosion, damaging precipitator internals.
2) If high levels of carbon are known to exist on the collecting surface or in the hoppers, do not open precipitator access doors until the precipitator has cooled below 125°F (52°C). Spontaneous combustion of the hot dust may be caused by the inrush of air.
3) If a fire is suspected in the hoppers, empty the affected hopper. If unable to empty the hopper immediately, shut down the transformer–rectifier sets above the hopper until it is empty. Use no other method to empty the hopper. Never use water or steam to control this type of fire. These agents can release hydrogen, increasing the possibility of explosion.

Table 15.3 Precipitator inspection items. Preoperational checklist.

Collecting Plates
1) Free longitudinal and horizontal bows
2) Free of burrs and sharp edges
3) Support system square and level
4) Spacer bars and corner guides free
5) Free from excessive dust buildup
6) Gas sneakage baffles in place and not binding

Discharge Electrodes
1) No breaks or slack wires
2) Wires free in guides and suspension weight free on pin
3) Rigid frames square and level
4) Rigid electrodes plumb and straight
5) Free from excessive dust buildup and grounds
6) Alignment within design specifications

Hoppers
1) Scaffolding removed
2) Discharge throat and poke holes clear
3) Level detector unobstructed
4) Baffle door and access door closed
5) Heaters, vibrators, and alarms operational

Top Housing or Insulator Compartments
1) Insulators and bushing clear and dry with no carbon tracks
2) All grounding chains in storage brackets
3) Heaters intact, seal-air system controls, alarms, dampers, and filters in place and operational
4) Seal-air fan motor rotation correct, or vent pipes free
5) All access doors closed

Rappers
1) All swing hammers or drop rods in place and free
2) Guide sleeves and bearings intact
3) Control and field wiring properly terminated
4) Indicating lights and instrumentation operational
5) All debris removed from precipitator
6) All personnel out of unit and off clearances
7) All interlocks operational and locked out
8) No broken or missing keys
9) Covers on all locks

Transformer—Rectifier Sets
1) Surge arrestor not cracked/chipped and gap set
2) Liquid level satisfactory
3) High-voltage connections properly made
4) Grounds on: precipitator, output bushings, bus ducts, conduits

Rectifier Control Units
1) Controls grounded
2) Power supply and alarm wiring properly completed
3) Interlock key in transfer block

Air load is simply the test energization of the discharge electrodes in air. The purpose of this test is to check the proper operation of the rectifier control unit, verify that all grounds, misalignments, and debris have been cleared, and to check safety and alarm conditions. It should be performed with only natural stack draft or minimum forced-air flow, and it also provides a convenient time to test and operate all auxiliary equipment and the ash conveying system.

Air-load electrical readings of the TR sets are important, and the data should be recorded and retained for future reference. Readings should be taken of the voltage and current levels in a stepwise fashion from corona onset (the voltage at which corona current flow is indicated by the panel secondary current meter or current waveform ripple if tested with an oscilloscope) to maximum power attainable without excessive sparking. Air-load readings are most useful when secondary voltage (kV) can be read, but primary voltage is an adequate substitute. The readings should be more frequent at lower values of corona current, since this is where the voltage changes most rapidly.

The air load voltage–current (V–I) curve is characteristic of a certain bus section under local ambient conditions of temperature and humidity in air. On subsequent tests under similar conditions, similar curves should result, and if the discharge and collecting electrode geometries and spacings are the same for each bus section, similar curves should be generated. A difference between curves indicates a difference inside the precipitator. On some days it is impossible to obtain an air load without sparking, and any similarity between air-load and actual gas-load curves are usually coincidental. When the air-load tests are complete, the precipitator is ready to start up.

15.4.3 Startup

Startup should be a relatively simple operation, requiring the turning of a few switches to the ON position, and letting the automatic controls take over. Under ideal conditions, startup should not be initiated until the process has reached stable, normal full-load conditions for several hours. However, environmental regulations do not usually permit the discharge of pollutants at any time when the gas process is on-stream, including startup and shutdown. Startup problems are compounded when starting up on oil when oil is not the primary fuel. It is better to start up on gas or the primary fuel.

If the precipitator must be energized prior to reaching minimum operating temperature, high gas humidity and unburned carbonaceous material prevailing under transient startup conditions can lead to fouled electrodes and tracked insulators. Clinker formation on discharge electrodes is common. To avoid these hazards, the startup should be programmed sequentially.

Bushing and hopper heaters, seal-air systems, and rapping equipment should be started up to 12 hours prior to light-off and left in operation. Rapper intensity should be higher than normal at light-off. At light-off, the ash-removal system should be started and operated continuously. The TR sets can also be energized (after purging), but only to the extent necessary to remain in emissions compliance. Startup mass emissions are much lower than the precipitator was designed for, and low corona power levels are sufficient to control them. This can generally be achieved by setting controls to operate at the spark threshold and energizing as few fields as possible.

Energization starting with the inlet field is frequently favored since subsequent operation with coarser ash will help scour off remaining startup deposits. However, this procedure will also foul downstream fields by deposition of penetrating charged particulate, and coarser ash, if it is eventually generated, will reach the outlet field as well. The main concern should be to minimize the potential for fouling buildups and corrosion, and this can only be done by starting with the outlet field.

Startup of a boiler or furnace on non-distillate fuels for prolonged periods of time presents a more difficult problem due to the higher fouling tendency of the emission, especially in a system not designed to burn oil under normal conditions. The potential for fire is increased, or explosion if a flame-out occurs, and the operation has to be monitored closely, keeping the spark rate at or near zero. Energization beginning with the outlet field helps confine the potential for fire in a known area. Hot precipitators present the worst case. If process startup is difficult, requiring prolonged use of heavy oil, shutdown, inspection, and water washing may be required. Under cyclic operations, the situation is similar, but generally less hazardous with regard to fire and explosion.

15.4.4 Normal Operation

When normal full-load operation is achieved, precipitator energization, rapping, and ash removal can be operated normally. For energization, this still means using only as many bus sections as required, mainly for economy, keeping spark rates very low (less than 10 per minute), and optimizing voltage and current on the TR sets in use. Energization can be either full- or half-wave, depending upon which mode yields the best results. Contrasting to air-load characteristics, gas-load V–I curves will likely be different by field, as particulate physical and chemical properties, will largely control the power levels that can be attained.

Rapper operation is second only to energization in controlling precipitator performance. Often, energization and rapping must be optimized together. There are four fundamental variables in rapper operation: pattern, intensity, duration, and time interval between raps. Since precipitator performance is least sensitive to pattern, this variable can be fixed by design.

The duration of impulse rappers is instantaneous. This leaves intensity and interval, or cycle time. "Theoretically" each field of a precipitator has the same efficiency. So, if the efficiency of a section is (for this example) 90%, then the first section collects 90% of the inlet dust, the second section will collect 90% of the remaining 10% (i.e., 9% of the ESP inlet dust loading) and a third section would collect 90% of the remaining 1% (i.e., 0.9% of the ESP inlet dust loading) for an overall efficiency of 99.9%. As can be seen by this example most of the collection in a normally operating precipitator occurs in the inlet field. Thus, when compared to downstream fields, the inlet rapping intensity should be higher and the cycle time should be shorter. In addition, higher-resistivity dusts, or dusts that display a relatively small particle-size distribution, will generally require higher rapping intensities and longer cycle times than will coarse, conductive particulates. Vibrating-type rappers are most often used on weighted wire discharge electrode systems. In addition to cycle time and intensity considerations, the duration of each vibration is important, since it must be long enough to establish the small oscillations and random nodal points critical to the cleaning action. A minimum necessary duration of 8 seconds was observed on one 30-ft weighted wire system. For a rigid frame system, shorter durations will be possible due to the lack of swaying of the discharge electrode systems.

Finally, evacuating ash from hoppers is an important operational variable that, unlike other parameters, has little available control. The interval between pulling cycles is the major variable. A compromise between operating costs for conveying equipment and hopper storage capacity is a logical approach to plant practice, but it is difficult to recommend storage times greater than 12 hours. The reason for this concern becomes obvious when one considers the effort that would be required to remove, for example, up to 20 hours of collected ash from a plugged hopper. Part of the operation should include frequent walking tours of the hopper deck. If a hopper level alarm is not energized and the hopper throat is cool, the hopper is almost certainly plugged.

15.4.5 Performance

Precipitator performance is dependent upon many variables, but fortunately, process by-product or fly ash and operating conditions are reasonably consistent within the plant, lessening the problem for the operator. However, changes in particulate resistivity, size, and quantity can change even in the same process or from firing coal from the same mine. Precipitators of different manufacturers may be installed at the same plant, and if their electrode geometries are appreciably different, their operating characteristics will be different. When operating conditions deviate from the norm, the effects of the abnormality will almost always be detectable from energization symptoms. Table 15.4 contains common operating problems and their probable causes that can be inferred from energization characteristics.

Table 15.4 Troubleshooting guide – precipitator energization symptoms.

Symptom	Probable Cause
No primary voltage No primary current No secondary current	TR control set failure, such as reversed wiring to or shorted silicon-controlled rectifier (SCRs), blown fuse, failed automatic voltage controller (AVC) component, misadjustment of control or internal ground, causing trip-out
No primary voltage Normal primary current Normal secondary current	Short circuit in system after TR. Check for ground in bus duct, insulator housing or precipitator; full hopper; broken discharge electrode; broken support insulator.
Low primary voltage Normal primary current Normal precip. current	High-resistance short to ground in precipitator. Check for insulator tracking; clinker or foreign material on discharge electrode; full hopper.
Low primary voltage Low primary current Low secondary current No apparent sparking	Control set misadjustment or maloperation. Spark rate indication, if installed, may be high. Check AVC and remainder of TR control.
Low primary voltage High primary current Low secondary current	Fault in TR secondary, such as rectifier diode failure, secondary winding failure, shorted surge arrestor, or ground in control circuit.

Table 15.4 (Continued)

Symptom	Probable Cause
High primary voltage No primary current No secondary current	Open TR secondary. Replace TR
Unstable voltage and current	Control circuit misadjustment or maloperation.
Normal primary voltage Normal primary current Half-normal secondary current	Shorted surge protector.
Rhythmic fluctuation in voltage and current	Broken or slack discharge electrode, unstable discharge electrode system, or dust buildup near discharge electrode
Spark bursts at regular intervals	Dust buildup near discharge electrodes or on collecting plates.
Higher-than-normal voltage, decreasing current	High-resistivity dust buildup on discharge electrodes or collecting plates.
Gradual voltage drop with steady current and sparking	Discharge electrode erosion.
Gradual voltage drop and corresponding current rise	Tracking insulator, hopper fouling.
Abnormally low or no voltage, high current	If hopper throat cool – plugged hopper. If hopper throat warm – shorted insulator or broken discharge electrode.
Higher spark rate in one chamber only	Ambient air in-leakage.

15.4.6 Shutdown

When the process gas source equipment is temporarily shut down, it is always desirable to deenergize the precipitator and its auxiliaries, mainly to prevent waste of the large amounts of power that precipitators can consume.

The shutdown process under routine operating circumstances is the reverse of startup. Fields can be deenergized beginning with the inlet field as the process gas source is being turned down, leaving only as many bus sections in service as necessary to maintain emissions compliance or by-product recovery rates. Usually, shutdown progresses more rapidly than startup, but the deenergizing sequence should be followed, especially if the precipitator is upstream of induced-draft fans. When a TR set is deenergized, the electric field that was forcing collected particulate to adhere to the collecting electrode suddenly disappears, but repulsive charges on adjacent particles will remain. In many cases, this residual charge can force large amounts of material off the collecting electrodes and back into the gas stream. If the outlet field is deenergized first, all the reentrained material will reach the ID fan and can cause overload or fan blade erosion over a period of time. Shutdown usually does not require the firing of an auxiliary fuel, so fouling is generally not a concern during shutdowns.

The last bus section can be deenergized as soon as particulate is no longer being emitted from the gas source equipment. Seal-air fans can be shut down as soon as the main process air or gas stream is terminated. Ash removal systems and rappers can be left on for several hours to enhance electrode cleanliness and assure empty hoppers in preparation for the next startup.

On cyclic processes, shutdown is rapid and time may not be available to follow the sequence recommended above. A summary of startup and shutdown recommendations is provided in Table 15.5.

15.4.7 Record Keeping

Maintaining efficient and reliable precipitator operation is dependent on the operator's knowledge and the amount, type, and use made of the data collected and retained. Analysis of properly recorded data can heighten the operator's knowledge of precipitator behavior. Fortunately for the operator, the parameters that can be measured and adjusted are not complex, and to the extent that the manufacturer has followed good design practices and provided the means for adjustment, these parameters can be effectively controlled to maintain precipitator performance at optimum levels.

Table 15.5 Startup and shutdown procedures.

Normal operation startup

Preoperational checks – at least 2 hr prior to gas load:

1) All maintenance/inspection items complete.
2) All debris removed from ESP.
3) Safety interlocks operational and all keys accounted for.
4) No personnel in ESP.
5) Lock out ESP and insert keys in transfer blocks.

Prestart (at least 1 hr prior to gas load):

6) Check hoppers
   a) Level indicating system operational.
   b) Ash-handling system operating and sequence check – leave in operational mode.
   c) Hopper heaters on.
7) Check top housing seal-air system
   a) Check operation of seal-air fan – leave running.
   b) Bushing heaters on.
8) Check rappers
   a) Energize control, run rapid sequence, ensure that all rappers operational.
   b) Set cycle time and intensity adjustments, using installed instrumentation – leave rappers operating.
9) Check T/R sets
   a) Check half-wave/full-wave operation (half-wave operation is recommended for lignite on cold-side ESP).
   b) Keys in all breakers.
   c) Test-energize all sets and check local control alarm functions.
   d) Set power levels and deenergize all T/R controls.
   e) Lamp and function-test all local and remote alarms.

Gas load:

10) After gas at temperature of 200°F has entered ESP for 2 hr:
    a) Energize T/R sets.
    b) Check for normal operation of T/R control.
    c) Check all alarm functions in local and remote.
    d) Within 2 hr, check proper operation of ash removal system.
    e) Deenergize bushing heaters after 2 hr (hopper heaters optional). Shutdown
11) When boiler load drops and total ash quantity diminishes:
    a) Deenergize ESP by field, starting with inlet field to maintain opacity limit.
    b) Deenergize outlet field when all fuel flow ceases and combustion air flow below 30% of rated flow.
    c) Leave rappers, ash removal system, seal-air system, and hopper heaters operational.
    d) Four hours after boiler shutdown, deenergize seal-air system and hopper heaters. Secure ash removal system.
    e) Eight hours after boiler shutdown, deenergize rappers.

Note: Normal shutdown is a convenient time to check operation of alarms.

Cold start and emergency shutdown

Cold start

When it is not possible to admit flue gas at 200°F for 2 hr prior to energizing controls, proceed as follows:

12) Perform steps 1–9 above. Increase rapping intensity 50%.
13) Energize T/R sets, starting with inlet field, setting Powertrac voltage to a point just below sparking.
14) Successively energize successive field as load picks up to maintain opacity, keeping voltage below normal sparking (less than 10 flashes/min or spark indicator).
15) Perform step 10d above.
16) After flue gas at 200°F has entered ESP for 2 hr, perform steps 10b–e above. Set normal rapping.

Emergency shutdown

17) Deenergize all TR sets.
18) Follow steps 11c–e above.

Source documents for precipitator internal conditions include electrical energization readings V–I curves, rapper operating data, ash-removal-system operation log, internal inspection reports, and a replacement-parts log. Rapper data should include the cycle time and intensity by field, and a chart should be kept of the current rapping pattern. Recording descriptions, failure dates, and replacement dates in a parts-replacement log will show whether components fail on a regular or random basis, and can be used to ascertain the cause of failures. There is an understandable and natural reluctance for an operator to generate what may seem to be a mountain of useless paper. If the data, once recorded, gets filed in a drawer never to be seen again, a useless pile of paper is what it will be. However, if data is collected and stored electronically, it is more easily retrieved. It can then be used to identify and correct real occurrences of precipitator malperformance.

Table 15.5 Startup and shutdown recommendations.

15.4.7.1 Startup

Preoperational checks – at least two hours prior to gas load:

1) All maintenance/inspection items are complete.
2) All debris removed from ESP.
3) Safety interlocks operational and all keys accounted for.
4) No personnel in ESP.
5) Lock out ESP and insert keys in transfer blocks.

Prestart (at least one hour prior to gas load):

6) Check hoppers
   a) Level indicating system operational.
   b) Ash-handling system operating and sequence check – leave in operational mode.
   c) Hopper heaters on.
7) Check top-housing seal-air system
   a) Check operation of seal-air fan – leave running.
   b) Bushing heaters on.
8) Check rappers
   a) Energize control, run rapid sequence, and ensure that all rappers are operational.
   b) Set cycle time and intensity adjustments, using installed instrumentation – leave rappers operating.
9) Check T/R sets
   a) Check half-wave/full-wave operation (half-wave operation is recommended for lignites on cold-side ESP).
   b) Keys in all breakers.
   c) Test-energize all T/R sets and check local control alarm functions.
   d) Set power levels and deenergize all T/R controls.
   e) Lamp and function-test all local and remote alarms.

15.4.7.2 Gas Load

10) After gas at a temperature of 200°F has entered the ESP for 2 hours:
    a) Energize T/R sets.
    b) Check for normal operation of T/R control.
    c) Check all alarm functions in local and remote.
    d) Within two hours, check proper operation of ash-removal system.
    e) Deenergize bushing heaters after two hours (hopper heaters optional).

15.4.7.3 Shutdown

11) When boiler load drops and total ash quantity diminishes:
    a) Deenergize ESP by field, starting with inlet field to maintain opacity limit.
    b) Deenergize outlet field when all fuel flow ceases and combustion air flow below 30% of rated flow.
    c) Leave rappers, ash-removal system, seal-air system, and hopper heaters operational.
    d) Four hours after boiler shutdown, deenergize seal-air system and hopper heaters. Secure ash-removal system.
    e) Eight hours after boiler shutdown, deenergize rappers.
    Note: Normal shutdown is a convenient time to check operation of alarms.

15.4.7.4 Cold Start and Emergency Shutdown

15.4.7.4.1 Cold Start When it is not possible to admit flue gas at 200°F for 2 hours prior to energizing controls, proceed as follows:

12) Perform steps 1–9 above. Increase rapping intensity 50%.
13) Energize T/R sets, starting with inlet field, setting secondary voltage to a point just below sparking.
14) Successively energize successive field as load picks up to maintain opacity, keeping voltage below normal sparking (less than 10 flashes/min or spark indicator).
15) Perform step 10d above.
16) After flue gas at 200°F has entered ESP for 2 hours, perform steps 10b, c, d, and e above. Set normal rapping.

15.4.7.4.2 Emergency Shutdown

17) Deenergize all TR sets.
18) Follow steps 11c, d, and e above.

15.5 Maintenance

While pollution control regulations provide for some upsets in process operation, they have little tolerance for malfunctions in air pollution control equipment. Forced outages to repair defective equipment are costly. For this reason, specifications demand high component reliability and design redundancy, with the expectation that sufficient performance reliability will be attained so that major maintenance can be performed during regularly scheduled outages. However, assurance of adequate performance reliability transcends design. Maintenance proficiency at the plant, reluctance or inability of manufacturers to provide experienced service advice or personnel when needed, and similar inertia on the part of operating and maintenance personnel at the plant to learn essential methods and procedures on their own all contribute to the actual reliability of the equipment. Precipitator designs vary according to manufacturer and they have all demonstrated a relatively low maintenance requirement when properly applied, designed, and maintained. In those areas where reliability has been disappointing, it is reasonable to conclude that additional operator knowledge and well-planned and executed preventive and corrective maintenance programs would result in measurable improvement.

15.5.1 Preventive Maintenance

Any action on a piece of equipment that is planned and executed to prolong availability by detecting or preventing impending failures can be defined as preventive maintenance. In addition to potential reduction in downtime cost and improved reliability, a major advantage of preventive maintenance is that since it is planned, it can be scheduled. Key elements in a successful preventive maintenance program are promoting the concept of recognizing trouble before it happens, following the preventive maintenance schedule without exception, and recording essential data at appropriate times.

Table 15.6 illustrates the rank order of the most common precipitator problem components, taken from a 1975 survey conducted by the Air Pollution Control Association (now the Air & Wastes Management Association)[6], and the consequences

Table 15.6 Rank order of precipitator component failures from 1975 APCA survey and failure consequences.

	Frequency (%)	Impending Failure	Actual Failure
Discharge electrodes	35.2	Diminished capacity	Outage
Ash-removal system	31.8	None	Outage
Collecting plates	13.6	None	None
Rappers	5.7	Diminished capacity	Diminished capacity
Insulators	1.1	Diminished capacity	Diminished capacity

resulting from impending and actual failure. It is apparent from the table that impending failure can be detected in three of these five areas, and in many cases, it is possible to detect impending failure of ash removal systems. A preventive maintenance program can readily be developed for these and more precipitator components. Table 15.7 provides an example of a precipitator-specific personnel and dates to each task. In addition to the potential for improved reliability and lower operating cost, a well-executed preventive maintenance program can be an excellent vehicle for training, for enhancing operator knowledge, streamlining workflow and communication, and it is relatively simple to implement.

Table 15.7 Preventative maintenance checklist for a typical fly-ash precipitator.

Daily
1) Take and record electrical readings and transmissometer data.
2) Check operation of hoppers and ash-removal system.
3) Examine control room ventilation system.
4) Investigate cause of abnormal arcing in TR enclosures and bus duct.
Weekly
1) Check rapper operation.
2) Check and clean air filter.
3) Inspect control set interiors.
Monthly
1) Check operation of standby top-housing pressurizing fan and thermostat.
2) Check operation of hopper heaters.
3) Check hopper level alarm operation.
Quarterly
1) Check and clean rapper and vibrator switch contacts.
2) Check transmissometer calibration.
Semiannual
1) Clean and lubricate access-door dog bolt and hinges.
2) Clean and lubricate interlock covers.
3) Clean and lubricate test connections.
4) Check exterior for visual signs of deterioration, and abnormal vibration, noise, leaks.
5) Check TR liquid and surge-arrestor spark gap.
Annual
1) Conduct internal inspection.
2) Clean top housing or insulator compartment and all electrical insulating surfaces.
3) Check and correct defective alignment.
4) Examine and clean all contactors and inspect tightness of all electrical connections.
5) Clean and inspect all gasketed connections.
6) Check and adjust operation of switchgear.
7) Check and tighten rapper insulator connections.
8) Observe and record areas of corrosion.
Situational
1) Record air-load and gas-load readings during and after each outage.
2) Clean and check interior of control sets during each outage of more than 72 hours.
3) Clean all internal bushings during outages of more than five days.
4) Inspect condition of all grounding devices during each outage over 72 hours.
5) Clean all shorts and hopper buildups during each outage.
6) Inspect and record amount and location of residual dust deposits on electrodes during each outage of 72 hours or longer.
7) Check all alarms, interlocks, and all other safety devices during each outage

15.5.2 Corrective Maintenance

As indicated by the earlier APCA survey, maintenance on precipitators will almost certainly include discharge electrodes and their suspensions, hoppers, collecting plates, rappers, and insulators. Other areas that require occasional attention or modification work are casing access doors, control cubicles, alignment, and corrosion.

1) *Discharge electrodes.* As indicated earlier, the four major causes of discharge electrode failure are mechanical fatigue, electrical arc erosion, wire instability, and corrosion. Mechanical failures usually occur near the points of attachment to the high-voltage support structures, where the wire is twisted about itself in some designs, or near the terminus of a shroud if one is used. Most fatigue failures occur as a result of cyclic movement of the electrode, causing excessive cold working of the electrode material. Cyclic movement can be caused by rapping, energization conditions, high gas velocity jetting, or a combination of these. In the majority of cases, the problem can be eliminated by obtaining a tighter fit at the points of attachment or on wire electrodes themselves. The tight fit must be accomplished without creating stress concentrations. Later designs of ESP's that have utilized rigid frame discharge electrodes have largely eliminated this failure mode although this is still a potential failure point. Failures in shrouds usually occur where the wire emerges from the shroud because design or quality flaws leave a sharp edge to work against the wire. Failures where the wire pulls out at the shroud occur occasionally. In the majority of these cases, adequate design and quality assurance testing should eliminate the problems, and users should contact the manufacturer if the problem is severe.

   The majority of discharge electrode failures occur as a result of arc erosion. Arcing can occur at localized high-field regions due to sharp edges or protrusions on collecting plates or between the wire and its support if the connection is not tight and a small air gap exists. This type of failure can be visually recognized by observing a flat and rough surface of the wire directly exposed to the arc. The erosion mechanism is most detrimental to small-diameter electrodes. When localized high-current arcing persists at one point on the discharge electrode, some electrode material is carried away in the discharge. The intense heat cannot be dissipated by the small wire mass, and the electrode flexes and creeps, aggravating more arcing. Eventually, enough material is eroded and sufficient cycling occurs to cause breakage and grounding outage of the affected bus section. This type of failure is commonly found near the bottom of the wire in the vicinity of the edge of the collecting plate, caused by misalignment or wire instability. However, a major factor contributing to arc erosion is the nature of the energization equipment.

   The magnitude of the arcing current is a function of the TR set size and the stored energy in the bus section. The energy in a system comprised of a 2,000-mA set serving 50,000 ft^2 of surface is enormous. If spark rates must be kept high to provide sufficient corona voltage for emissions compliance, or if the sectionalization and arc-quenching capability of the control are inadequate, electrode failures will be higher. The high inrush currents associated with sparking can be especially detrimental to a design that contains many small connections between electrodes and their supports. Remedial actions for these types of failures include identifying the nature of the failure and isolating the cause, which will usually be realignment or modifications to gain better stability. Use of a properly designed shroud and improving the energization design are appropriate measures also.

   Instability problems arise when the total unrestricted length of the discharge electrode and its support and steadying frames is excessive. Accumulated tolerances, thermal expansion, and structural deflections that occur during normal operation make it difficult for systems using long wires and many long couplings to maintain alignment. In a noteworthy work by Casiglia and Fletcher,[7] it was found that the fundamental mechanism of instability in weighted wire systems involves the resonant interaction of the wire and its tensioning weight. When the fundamental wire frequency and the tensioning weight pendulum-mode frequency approach the same value, large-amplitude oscillations readily occur. Separating the natural frequencies of these two components was successful in reducing instability-induced arc erosion failures in one fly-ash precipitator from several per week to less than two per month. Better stability was gained by increasing the mass and lowering the center of mass of the tensioning weight, and decreasing the linear mass density (weight per unit length) of the wire itself. It was also found that any weighted wire system could be made stable if the suspension weight were restrained from movement, but this is not a practical solution. Some manufacturers use suspension designs where stabilization of the weighted wire structure is close-coupled, such as placing a stabilizer bar between the bottom of the collecting plates and the lower steadying frame.

   In a properly erected precipitator, the maintenance and modification actions described above should minimize discharge electrode failures, even in weighted wire designs. Electrode failures of one or more per week have been reported for designs that are identical to those reporting as few as one or two per year. A well-executed maintenance program should keep the electrode failure frequency low.

2) *Hopper*. Inadequate ash removal systems can permit buildups in hoppers to the point where they overflow. Of all the causes for misalignment and subsequent discharge electrode arc erosion, overfull hoppers are by far the biggest offenders. Weighted wire systems are least sensitive to misalignment by this mechanism, while rigid frame designs appear to be most sensitive. In either case, grounded bus sections and diminished precipitator capacity result in whether there is any electrode damage or not.

   Major causes of ash removal problems include low gas temperatures, promoting condensation of tacky condensable matter or free water vapor, insufficient insulation, inadequate hopper heat, internal obstructions, air in-leakage, inoperative vibrators, shallow valley angles, faulty level detectors, and evacuation system faults. Evacuation system capacity problems are more prevalent on industrial precipitators employing mechanical conveyors than on dusting equipment using pneumatic conveyors.

   The most direct attack against the formation of condensable uses external heat and better insulation. In the worst cases, it may be necessary to apply heat to almost all of the hopper surface instead of the nominal 15% of surface found on many installations. If hang-up at the very bottom is the common fault, heating and insulating the hopper throat and transition pieces can be helpful. Additional heat should be applied in vertical zones and thermostatically controlled. Location of thermostats is important, because if housed in a normally high-temperature zone, they may never activate. Care must be taken when selecting heaters to ensure that they have sufficient design margin for any temperature excursion if the heaters are energized and the hopper does not evacuate. Internal obstructions can cause relatively minor residual accumulations of ash to remain after a pulling cycle. These deposits flatten the effective valley angle, promoting additional hang-up on successive cycles and eventual blockage. Anti-circulation and sneak baffles should be terminated as far above the throat as possible. In hopper designs where the ratio of the vertical height to the horizontal span is high (say greater than 2.5), the distance between the baffle and throat can be increased to several feet instead of the 18 in. many standard designs call for. Ladder rungs are an excellent access feature, but must be confined to vertical baffles or the steepest wall, and in no event should they be near corners. Hinged, flexible gas baffles have been used overseas with some success, and rounded corner fillets are recommended in the most severe problem cases. Air in-leakage is most common at access doors and dump valves. Door gaskets, damaged most frequently by overtightening, should be inspected regularly and changed if worn. Dump valve seats and gates should be checked for erosion and replaced if air leakage can be heard in this area during normal operation. Two vibrators operating on opposite walls at a relatively low frequency are better than one vibrator operating at a higher frequency. Level detectors should be placed on either side of a gas baffle. If only one detector is installed, a hole can be cut in the gas baffle to allow free communication of dust from one side to the other, but care has to be taken to prevent gas sneakage. This may require the installation of additional baffle plates. Poke holes are rarely useful but should be provided for rodding clinkers. Aeration stones should only be used if the air is heated and dried.

3) *Collecting plates*. Misalignment and fatigue failures are the most common problems associated with collecting plates. Fatigue failures generally result from the use of rapping forces that the plate attachments were not designed to survive. Manufacturers should be consulted for repairs if fatigue failures of this nature occur.

4) *Rappers*. Problems associated with internal swing-hammer rappers include bearing failure due to wear and abrasion by dust in the gas stream, misalignment of the hammers and anvils, and wear or breakage of the anvils themselves. Swing-hammer failures on collecting plates can be detected only by changes in precipitator operation, principally energization. On discharge electrode frames, failure of the shaft insulator will stop shaft rotation, which can be detected fairly easily. Failures of electromagnetic drop hammers are associated primarily with the packing gland or boot seal. Gas leaking by the gland can soften the rod tube and corrode mounting bolts. Better packing and seal materials and the use of clamps on the seals can eliminate the problem. For any rapping equipment, frequent inspection is a good maintenance practice. On vertical drop hammers, there is a popular notion that binding between the rod and guide tube is detrimental to delivery of the rapping blow. Actually, measurements have shown that the attenuation caused by this binding is less than 5%, and the condition, if it is noted, can probably be left unmodified. These studies show that a dust layer not much thicker than this page, allowed to accumulate between the hammer and the anvil, can attenuate the blow as much as 50%. If this condition is noted, immediate correction is required. In some installations using magnetic drop hammers, there is a tendency for the hammer to magnetize and stick in the dropped position at low energy levels. Insertion of a flat aluminum plate or a shrink-fit stainless steel cap between the hammer and rapper rod will eliminate this problem. In no event should rubber pads be used.

   For the rapper input force to perform the most work, it is essential that the collecting plate attachment to the casing be rigid and tight, and that no energy absorbers are built or allowed to creep into the system. Manufacturers of highly active and "cleanable" plates work hard at evolving designs that ring like a bell when rapped. If loose attachments are found, the manufacturer should be contacted for methods to correct the problem.

5) *Insulators*. These components can be made of tile, porcelain ceramic or alumina, and in some cases Teflon or nylon. To be successful, they all must be kept relatively dry, and kept from being exposed to conditions that can cause tracking, a condition where part of the material is carbonized by high-voltage stress.
Drying is accomplished mainly by heated air. In designs that use local resistance heaters, these devices must not be too close to the insulator, or local high temperatures will reduce the stress-carrying capability of the insulator and failure will occur. Thermostat controls must be located near the heaters to sense the heat near the insulator and prevent overheating. A 5-in. clearance between insulator and heater and a thermostat setting around 250°F, but certainly not more than 300°F, are satisfactory. Some support insulators are ventilated. If the original design calls for 100 cfm per insulator and corrosion is evident on the precipitator internal parts below the insulator, this flow rate can probably be reduced to 25 cfm with a corresponding drop in heater temperature to maintain dry insulators and minimize corrosion simultaneously.
   Some insulators are contained in small compartments while others are grouped in large penthouses or top housings. Top housings provide better maintenance access, but maintenance activities should pay attention to insulation, ventilation, and ambient dust. If insulation is installed on the floor of a top housing, metal walkways should always be used, and kept in good repair and in their proper location. Ventilation pipes should not impinge on insulator surfaces. If insulator problems occur near ventilation openings, some ducting or diffusers should be installed. In-leakage of dust from the precipitator indicates a deficiency in the seal-air system. Fan filters, vent diffusers, and operating pressures should be checked to verify that the seal-air system is operating properly and providing positive ventilation. Annuli between support insulators and hanger pipes should be checked for excessive leakage or mechanical binding. Most insulators can tolerate incredibly high compression forces even in dirty flue gas atmospheres, but they generally cannot sustain much shear force. All support and transmission insulators should be checked periodically for misalignment and pinching.
6) *Doors*. Access doors are usually gasketed to prevent air in-leakage to the precipitator. Overtightening can cut and eventually destroy the gasket material. If left uncorrected, local thermal gradients on the door can cause it to warp and the whole door will require replacement for lack of a properly maintained and inexpensive gasket. Doors with interlocks are the primary precipitator safety feature. Frequent inspection and powdered graphite lubrication are mandatory to maintain key interlocks in proper working order.
7) *Control cubicles*. The biggest enemies of control enclosures are dust and heat. For this reason, open cubicles are usually equipped with fans and filters, but these may be insufficient, especially in controlling dust. If problems arise, consideration should be given to changing or eliminating the vent fan. If the fan intake is near the floor, reversal of the airflow with an outdoor intake and additional filters can be considered. In the worst case, the cubicle can be sealed, but in TR controls, the linear reactor may have to be removed.
8) *Alignment*. One of the major maintenance requirements for a precipitator is the inspection for and correction of misalignment between electrode systems. Misalignment can be caused by mishandling of materials during construction, by improper construction techniques or quality checks, by overfull hoppers, or thermal excursions. Misalignment may result from mislocated electrode suspension systems or bending and bowing of collecting electrodes, discharge frames, or the discharge electrodes themselves. Manufacturers have design standards for alignment which are developed from the clean-air sparking-voltage characteristics of their design geometries. Maintenance alignment efforts should strive to obtain these tolerances.
9) *Corrosion*. Low-temperature fouling and corrosion of metal surfaces is a problem that affects nearly all industrial process machinery. The process of corrosion itself is a natural one. The makeup of the earth's environment permits the forces of corrosion to be continuous and powerful, and short of attempts to prevent it completely, maintenance efforts are often directed toward identifying and minimizing the rate at which corrosion or fouling takes place. Corrosion and fouling agents vary by process. Although much fieldwork and research has been performed in the general areas of corrosion and fouling, there is no concise formula that enables operators and maintenance people to predict the amount of metal loss or particulate buildup that will occur under changing process conditions. In many processes, particularly in the power industry, corrosion depends upon the presence of free water vapor in the gas stream. For example, in power boilers firing sulfur-containing fuels, practically all of the SO_3 generated in the combustion process will react with available water vapor to form gas-phase sulfuric acid. In order for this acid to attack metal surfaces, the gas temperature ordinarily must be below the dew point of the specific concentration of acid formed. However, estimation of dew points in particulate-laden gas streams is highly uncertain, and the presence of particulates complicates the corrosion process. Acids can absorb first onto particle surfaces before the dust is deposited onto a metal surface. If the particle surface contains neutralizing chemicals, the concentration of acid available will be reduced, and although the ash will be the

vehicle by which the corrosive agent contacts the metal surface, its presence can impede the rate at which acid reacts with the metal. The rate of corrosion, being uncertain, can only be roughly estimated in most precipitators.

Preventing the rate of corrosive agent attack on metal surfaces depends upon good design and thoughtful operations and maintenance. The greatest amount of corrosion in precipitator systems occurs when cold pockets are allowed to exist. The maintenance program should inspect the precipitator and fluework for potential problem areas and modify them to the best degree possible.

Other physical aspects of the precipitator design that promote corrosion are air in-leakage and inadequate insulation. Older units and designs that contain many expansion joints are especially sensitive to areas of inadequate insulation. Insulation thickness in these areas should be measured and compared to modern industry standards for the normal precipitator operating gas temperature. Air in-leakage sources are structural cracks, access doors, hopper valves, and seal-air systems. Air in-leakage through cracks can be detected audibly during routine walking tours around the precipitator, and even closer audible or visual detection is possible for leakage through warped doors or damaged gaskets. Hopper valve leakage is more difficult to detect this way, but during a shutdown, inspection from an empty hopper with the ash-removal system operating can reveal the presence of excessively worn valves. Seal-air flow through support insulators can cool gases and metal temperatures below the acid dew point in localized areas where the corrosion rate will be much more rapid than in any other part of the system. Areas affected include the upper ends of collecting and discharge electrodes, support systems for these electrodes, local structural components, and the underside of the shell roof. Air in-leakage through packing glands or boot seals will operate the same corrosion mechanism.

Obviously, maintaining gas temperatures well above the acid dew point will accomplish more of the desired results than any other means, but this is usually not a practical solution, especially in cyclic operations, where corrosion can be greatly aggravated. In the most severe cases, special coating or metal preparations may be required.

15.6 Improving Operation and Performance

The knowledge that precipitator performance is controlled by many and often interrelated variables may discourage the operator into thinking that there is little he can do to improve the performance of a precipitator is already in commercial operation. On the contrary, the sensitivity of precipitators to the quality of energization and rapping provides the operator with controls over performance that can even overcome some design deficiencies. At the very least, examination of these and a few other operating parameters permits rapid and accurate diagnosis of precipitator mal performance.

Efforts at improving operation and performance are as dependent upon approach as they are upon technique. The first steps should always be the complete correction of mechanical/component deficiencies and excessive dust buildups, since their existence will almost certainly inhibit later efforts, utilizing subtle techniques that can identify problems associated with fuels, boiler, or process operation, or those problems complicated by the ravages of age, corrosion, or neglect. In most cases, several different problem areas will exist simultaneously, and proper interpretation of available data, possible only through a systematic approach, is essential for reliable performance. Once mechanical and component deficiencies are corrected, optimization of energization, rapping, and gas flow distribution can proceed toward the goal of optimum precipitator performance within the inherent limitations of precipitator size, or toward the definition of alternative methods and contingent technologies necessary to achieve a minimum level of performance.

15.6.1 Energization

Since a large number of process variables directly affect the quality of energization, it is difficult to prescribe a best sequence for its optimization. However, a systematic approach would suggest that examination of the design and application of energization equipment is an appropriate first step since this will define the limitations of the equipment and assist further optimization efforts. The basic energization package consists of a high-voltage transformer, a group of silicon rectifiers, a current-limiting reactor (CLR), and a voltage controller, in this case, a SCR or thyristor, although a saturable core reactor could also be used as a control. Of these components, the function of the CLR seems to be the least understood and often abused. Linear reactors are used to control the precipitator current waveshape, add stability to the power supply under sparking conditions by limiting the peak spark current, and reduce heat losses in the TR set. Larger reactors generally permit greater stability, but since the CLR is a primary circuit inductive impedance, larger reactors also reduce power input to the precipitator for a given TR voltage rating by limiting the TR-rated current. The net result is that to obtain optimum stability and TR performance, one normally has to design for a high impedance and low current capacity, and this usually limits the bus section collecting area that can be used. The methods of matching precipitator power supplies to actual operating loads are complex, but if a TR set is energizing over 25,000 ft of collecting surface and TR set reaction to sparking or arcing is extremely sluggish, or if the TR is operating at less than 50% rated secondary current and displays poor recovery after a spark event, the operator should investigate the benefits of better sectionalization. This may simply require a larger CLR if the current capacity of the modified energization equipment is still sufficient. In some cases, this may mean additional TR sets or modifying internals to obtain smaller bus sections. Of course, the reverse of this situation can occur, where the current capacities of the TRs for their respective bus sections are too small. In any event, if trouble is verified, correction will probably involve the original precipitator manufacturer, who is also in the best position to recommend modifications.

Sluggish reaction to sparking may also be caused by a design inability of the automatic voltage control. The automatic controls must minimize interruptions in the applied voltage caused by sparking or arcing. Complete bridging of the gap between discharge and collecting electrode by a spark or arc represents a momentary short to ground, during which time the precipitator applied voltage will tend toward zero, with a resulting loss in charging and collecting electric fields. A control should be able to respond to a spark and restore full voltage within 10–80 m sec. Sparks of low current are usually of natural short duration and extinguish themselves. Extremely fast response should be possible with this sort of spark. High-current sparking discharges, or arcs, generally do not extinguish readily. Or a burst of sparks may occur that can last several seconds if uncontrolled. Under these sorts of discharges, a slower response, up to 80 m sec may be required to quench the arc and prevent another immediate occurrence.

The primary debilitating effect of poor sectionalization or control is that the applied voltage to the precipitator is limited or interrupted too often to sustain high performance. The effect can be observed on any energization system by connecting an oscilloscope to a convenient secondary ground return current signal and setting the trace at a slow speed. Frequent reductions in the waveform to zero or long recovery times after a spark event are indications that the energization equipment itself can be improved. Methods and equipment are available to correct the deficiencies in energization design, and the necessary effort should be made prior to attempting additional optimization.

Problems with energization also frequently arise as a result of changing process, fuel, or by-product properties. In industrial processes, each process generally produces the same sort of energization characteristics from one plant to another. For power-industry applications, the situation is more complex, because fly ash is truly not a general process. All coals and the ashes they produce are different to some extent, and even coal from the same mine can produce vast differences in ash properties. Finally, an identical coal fired in different boilers can produce different energization results. The most frequent problems that arise are associated with ash resistivity, space charge, and corona quenching. Fortunately for the operator, there are techniques useful in diagnosing these problems and fine-tuning the power supplies for optimum precipitator performance.

The single most valuable diagnostic tool available to the operator is the V–I relationship. A strong indication of a high-resistivity situation can be obtained by comparison of a clean plate V–I curve to the operating V–I curve. Figure 15.4 illustrates this concept. The clean plate curve can be obtained by leaving a bus section deenergized during startup and recording the V–I relationship at normal operating temperature immediately upon energizing the TR set. Operating V–I curves can be obtained at any time after the clean plate curve, the one shown illustrating a resistivity around 5 × 10″ ohm-cm.

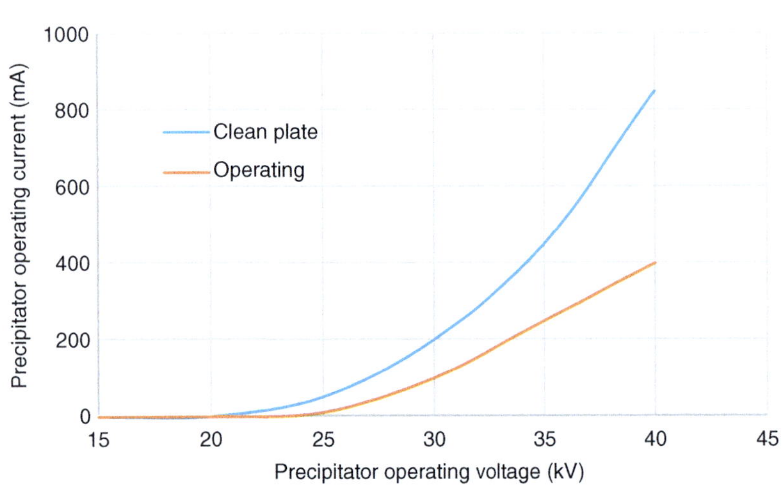

Figure 15.4 Voltage–current (V–I) characteristics: 9 in. gas passage on fly ash at 300°F.

There would be an infinite number of curves for intermediate and greater resistivities, and although they may be altered somewhat by changes in particle-size distribution, moisture, and gas temperature, all would show the trend toward lower current at sparkover and lower corresponding operating voltages. In other words, higher resistivities would shift the curve farther to the right without a large change in starting voltage. The detriment to energization is that the useful corona current density and maximum sparkover voltage are decreased with increased resistivity, as shown for a given plate spacing in Figure 15.5. The important thing to understand is that forcing the voltage or current above the critical point for a given application can be detrimental because excessive back corona may be generated. In fact, the operator may be surprised to find that performance improves when input power is reduced! Lower precipitator performance at higher power settings is caused by operating above the critical point where the back corona is generated. The energization level that corresponds to the lowest discharge opacity is the best verification that maximum useful power has been attained. The amount of trial and error needed to reach this point can be reduced by observing the minimum and peak value of the secondary voltage waveform with an oscilloscope connected to TR sets equipped with voltage dividers.

Space-charge effects and corona quenching can produce energization effects somewhat similar to high resistivity, the major differences being the voltage level that can be achieved prior to sparkover. Figure 15.6 illustrates the different types of V–I relationships characteristic of different conditions. Space-charge current suppression and corona quenching effects normally result in sparking at higher voltages than high-resistivity cases, typically near the maximum voltage attainable for a certain gas passage spacing, even though the current may be low. Space-charge suppression is caused by high dust concentrations, and corona quenching may be caused by a high concentration of fine particulate matter. These conditions, although not necessarily good for precipitator performance, are usually not as detrimental as

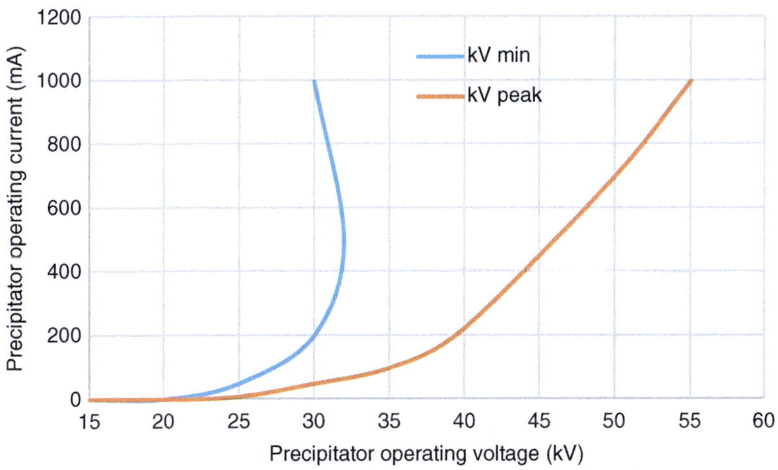

Figure 15.5 Effect of resistivity of attainable operating voltage and current. *Source:* Adapted from[7].

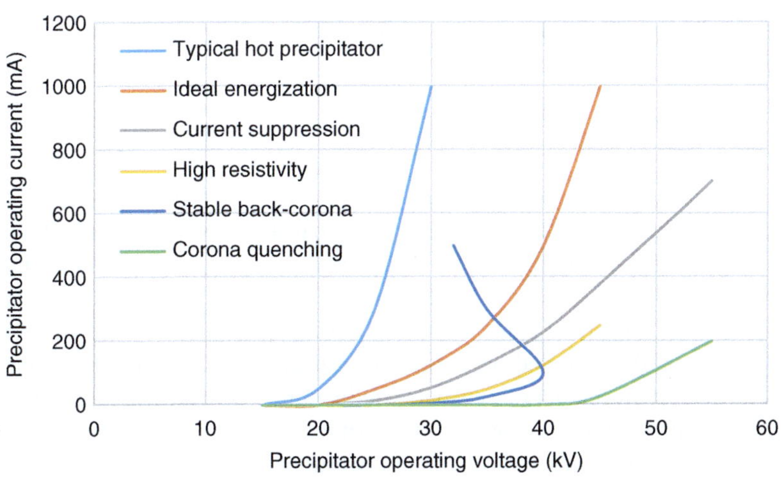

Figure 15.6 Fly ash voltage–current characteristics for various conditions. *Source:* Adapted from[8].

operations with high resistivity. Maximum performance is usually obtained by maximizing voltage while minimizing spark rates.

Full-wave rectification often results in better performance under conditions of corona quenching and operations not limited by high resistivity. For severe space-charge current suppression and high-resistivity situations, half-wave rectification is more beneficial. The reason for this is generally attributed to the longer decay period between half-wave voltage pulses. The operator who experiments for several hours with half- and full-wave energization may not detect any useful differences in performance, since as many as 4 or 5 days may be required for the effect to be evident.

15.6.2 Rapping

Energization is intrinsically related to the quality of rapping, and the most critical aspect is collecting electrode rapping. The inability to remove certain difficult particulates, commonly associated with dusts that exhibit high-resistivity characteristics, leads to reduced voltage and current input. Even low-resistivity dusts, if allowed to build up to a level where alignment is distorted, can cause premature sparking. For these reasons, rapping and energization often have to be optimized simultaneously.

The first step in rapping optimization is establishing the relative frequency of rapping by field. Since the inlet field in a reasonably well-behaved precipitator will collect the most ash, the inlet field should be rapped more frequently than downstream fields. The outlet field should require the least frequent rapping. The best pattern depends upon the number of precipitator fields in operation. A frequency ratio of 4–9 raps in the inlet field for every rap in the outlet field is a reasonable starting point, and intermediate values should be used for intermediate fields. The optimum ratio can be refined by trial and error, but precision is not critical for this aspect.

The next step is determining the interval between raps for each field. Experience has shown that longer cycle times are better than short ones, the rationale being that the long time interval permits natural particle agglomeration tendencies to take place, reducing the likelihood of reentrainment. This represents highly generalized cases, for cycle time is related to the reentrainment potential of the particulate. High-resistivity ashes, usually tenacious when collected and not prone to reentrainment, generally respond best to longer cycle times, on the order of one hour or so on the inlet fields to four or more hours between raps in the outlet fields. Low-resistivity ashes typically reentrain readily. They respond best to shorter cycle times, as low as 10–15 minutes on inlet fields to not more than several hours in outlet fields.

The final step in rapping optimization is finding the optimum intensity by field. High-resistivity and other tenacious ashes require higher-intensity rapping than do reentrainment-prone particulate. Although the actual rapping intensity required will depend upon the design of the precipitator, the best starting point for tenacious dusts is the maximum intensity for which the unit was designed, and variation between fields is not important. Final optimization is reached at the lowest intensity required to maintain power input while prolonging component life. For dusts sensitive to rapping reentrainment, intensity is critical, but again, there may be little required variation between fields.

If attempts to improve electrode cleanliness and energization by following the guidelines above are not fruitful, it may be necessary to improve the rapping hardware itself. This is especially true where the dust is known to be ordinarily difficult to remove and the rapping density is greater than 3,000 ft^2 of collecting electrode per rapper. Correction may entail additional rappers, modifying support systems, or cutting high-tension bus sections, but if the expected end result can justify the means, these options are considered valid optimization alternatives.

15.6.3 Gas Flow Distribution

Uniformity of the bulk flow entering the precipitator is extremely important to maximizing precipitator performance, and a customary practice in precipitator supply is the conduct of a complete model study prior to installation to define the flow control devices necessary to ensure compliance with stipulated uniformity criteria. Occasionally, it is found that the uniformity criteria is not attained in the actual field unit, and if subnormal collection efficiency is experienced, there is a natural inclination to suspect that maldistribution of the bulk flow is the cause. However, experience has shown that correction of the gas flow maldistribution rarely achieves the expected improvement in precipitator performance. This experience may reflect the empirical aspects of precipitator performance design, where reduction in theoretical performance is unconsciously accounted for in the sizing. Also, improvements in gas flow uniformity will not result in improved collection efficiency if the fundamental problem is unrelated, such as a limitation of useful energization caused by high dust resistivity. If improvement in flow uniformity in view of these precautions is still considered mandatory, the manufacturer or a qualified consultant should be involved before any decisions or modifications are made.

Elimination of flow-induced reentrainment losses is often more rewarding than small refinements in gross flow uniformity, and in most cases, the operator can detect the source of these losses on his own. The most sensitive area for reentrainment losses in a precipitator is the outlet field. Large-scale reentrainment can be detected visually by using a high-intensity lamp in the outlet plenum. Observations should be made with the gas flow backlighted, accomplished by shining the light through a special port on one side of the plenum and making the observations through a similar port on the opposite side. Precipitators performing as low as 80% efficiency provide a remarkably clear view of the exiting gas, and areas of dense reentrainment are readily detected, pinpointing the source of the reentrainment or indicating an area for closer observation. The course of this sort of investigation may lead to visual observation of flow in the hoppers. This can be performed in a manner similar to making plenum observations, but caution must be used to avoid accidental electric shock. No lighting or other device should be inserted through a port and into the hopper. Three categories of flow reentrainment can be detected by visual observation. Dense dust discharge at the junction of the hopper opening and the plenum floor indicates large-scale hopper turbulence. A similar dense discharge congregating at partition walls is a strong indicator of hopper cross flow. Localized dense discharges well above the floor, and directly adjacent to side or partition walls, indicate raw gas sneakage around collecting plates, caused by loose, inadequate, or completely missing anti-sneak baffles at the leading edges of the outboard collecting plates. Dense discharges not near walls or the plenum floor are indicative of high-velocity jetting, which may be caused by severely maldistributed flow at the precipitator inlet or by internal obstructions. Elimination of flow-induced reentrainment is critical to maximum precipitator performance, because the associated dust losses may account for up to 90% of the total emission. Raw gas sneakage is eliminated by properly designed and installed anti-sneak baffles around the perimeters of the collecting zones. Hopper flows and jetting can be reduced by additional hopper baffling or gas distribution plates at the precipitator outlet.

In summary, the best approach to improving operation and performance involves the following sequence of events:

1) Eliminate known mechanical/component deficiencies.
2) Take care in the proper operation of the precipitator, get to know its normal behavior, and maintain it regularly, preventing misalignment, large dust buildups on electrodes, and full hoppers.
3) Look for and eliminate sources of gas sneakage or large flow-induced reentrainment losses.
4) Get the energization equipment right, matching TR set capacity to actual load and obtaining fast-acting automatic voltage control.
5) Optimize power and rapping.

If the factors that influence precipitator performance are understood and accounted for in the operation of the precipitator, and the sequence above is followed carefully but does not provide the desired results, performance may be limited by precipitator size or inherent design limitations, requiring the use of more precipitator collecting surface or a contingent technology to meet emissions compliance.

15.7 Recent Developments

The basic design of ESP has remained relatively unchanged since first used in the early part of the twentieth century. Some modest equipment changes and new types of devices have appeared in the last 20 years, but all have essentially employed the same capture mechanisms used in the past.

Two recent significant technological developments were by-products of the early smelter installations. One discovery was that an electrical precipitator could be used to fractionally separate the constituents of the furnace fumes. By cooling the gases in steps, followed by collection with precipitators, it is possible to recover reasonably pure materials. This method has served as the basis for chemical processes to recover materials of commercial value.

Another early discovery was that the operation of ESP in collecting high electrical resistivity smelter dusts, such as lead oxide, improved when sulfur trioxide (from sulfide ores) was present in the gas. This discovery led to the practice of "conditioning" a high-resistivity dust by the injection of SO_3 and moisture into the flue gas to lower the resistivity of the dust and thus make it easier to collect.

It is important to know which conditioning agent is best suited for which type of precipitator application. Ammonia appears to have the widest range of use, having been successfully applied to cold-side ESPs with both low- and high-sulfur coals, and on some hot-side units. The conditioning mechanism is not well known, but appears to enhance particle agglomeration and increase space charge. Sulfur trioxide has been most widely used to increase the surface conductivity of low-sulfur coal ashes.

Pulse energization can enhance the collection efficiency of precipitators where performance limitations are caused by poor energization, either from high resistivity or fine particulate corona quenching. Pulse energization consists of electrical components that may superimpose high-voltage pulses of very short duration on an underlying, relatively constant electric potential, or the pulses may be solely applied and varied automatically with the pulsed parameters, such as voltage, frequency, and width, to tune the precipitator operation and minimize the debilitating effects of resistivity-induced sparking or back corona. The precipitator can then operate with the relatively high charging and collecting fields needed for effective particulate removal in the precipitator. Pulse energization produces a profuse corona along the discharge electrodes, resulting in uniform corona distribution, which plays an important role in enhancing particulate collection. Thus, performance enhancement is accomplished by altering the physics of the corona, not by raising the average level of the applied electric field. For optimum efficiency, the electric field strength should be as high as possible. This is accomplished by applying a high voltage to the discharge electrode with the consequent high corona current flow from the discharge electrode to the collection electrode.

15.8 Conclusions

ESP have demonstrated wide applicability in the control of air pollutants and recovery of valuable by-products from industrial plants, ranging from large units in power-industry generating stations to less familiar processes in the chemical and refining industries. Their almost universal acceptance is attributed to several of their inherent advantages. First, the corona charging mechanism is operable on essentially all particulate matter, regardless of size, consistency, or chemical composition. Second, since the capture forces work only on the particulate, the cost of energy consumption by mechanical filtration is avoided. Finally, precipitators have for many years demonstrated their performance capability on a wide range of particulate characteristics. They have become an established part of the pollution control equipment inventory worldwide.

The record of precipitator performance and reliability has not been without its problems. A simple listing of the major activities involved in the selection and use of precipitators reveals ample opportunities for something to go wrong:

1) Feasibility study and permitting
2) Specification
3) Proposal
4) Evaluation and selection
5) Engineering
6) Construction
7) Operation
8) Maintenance

Four of these activities are completed before the unit is even designed, and much of the pre-engineering activity depends upon sizing. Theories that have evolved to advance the state of the art are usually worked out in clean laboratories under controlled conditions, and the invaluable verification data that are obtained on field units is often incomplete and sometimes useless simply because the operators or technicians were unfamiliar with the theory. Despite these inaccuracies and the naturally complex nature of sizing, these theories have produced the means that have made larger and more efficient units possible, with collection efficiencies exceeding 99.9% actually achieved in the field on some difficult dusts. Much of the success of any precipitator also depends on operation and maintenance, but the importance of these activities is often overlooked in favor of specifying and designing units that are expected to operate automatically and trouble-free. However, a well-trained operator, given the responsibility and authority for optimizing performance and maintenance, has all the capability required to correct many of the field problems that occur. Most of these problems are no fault of nature, but rather a shortcoming of people who, driven by values not related to the nature of the process, are either unwilling or unable to exercise the knowledge in design and operation that the state of the art allows.

References

1 **L. Theodore**, *Air Pollution Control Equipment Calculations*, John Wiley & Sons, Hoboken, NJ, 2008.

2 **M. Robinson**, Electrostatic precipitation, in *Air Pollution Control* Strauss, W. Ed., Wiley Interscience, New York, NY, 1971.

3 **H. White**, *Industrial Electrostatic Precipitation*, Addison-Wesley, Reading, MA, 1963.

4 **J. Katz**, *The Art of Electrostatic Precipitation*, Precipitation Technology, Inc., Munhall, PA, 1979.

5 **P.L. Feldman**, *Effects of Particle Size Distribution on the Performance of Electrostatic Precipitators*, 68th Annual Meeting of Air Pollution Control Association, Boston, June 15–20, 1975. Air Pollution Control Association, Pittsburgh, PA, 1975.

6 **Air Pollution Control Association, ED**, *Proceedings of the Specialty Conference on Operation and Maintenance of Electrostatic Precipitators*, Dearborn, Mich., April 10–12, 1978. Air Pollution Control Association, Pittsburgh, PA, 1978.

7 **J. Casiglia and H. Fletcher**. *Unstable Vibration and Power Are Induced Failures in Electrostatic Precipitator Discharge Electrode Wires*. Unpublished paper, Detroit Edison Company, 1971.

8 **H. Hall**, *Solving problems in the electrical energization of electrostatic precipitators, JAPCA*, 28(9), 871–878, 1978.

16

Baghouses

Paul Farber

Consulting Engineer, Willowbrook, IL, USA

Fabric-filter dust collectors, commonly referred to as baghouses, are among the oldest and most widely applied particulate emission control devices. It is reported that there are currently over 200,000 fabric filters commercially operational in the United States. The universally employed principle of operation, simply stated, is the removal of dust from dust-laden gas by passing the gas through a filtration medium, normally a fabric. The cleaned gas emerges from one side of the medium while the dust is collected on the other side. Subsequently, the collected dust is removed from the fabric. Different collector designs, fabric types, configurations, and a variety of cleaning methods provide numerous hardware combinations. A number of the basic designs emerged dominant in the marketplace. These designs and their maintenance features will be discussed subsequently.

There were once over 100 US fabric-filter manufacturers. The larger companies usually offer more than one fabric-filter design type. The market was once extremely competitive, and therefore designs were constantly changed by the more aggressive companies. With so many manufacturers, product design possibilities, and a dynamic market, one realizes that hard and fast operating and maintenance rules are difficult to come by, and it is important for the user to establish where the fabric-filter dust collector fits into the design and application picture.

A fabric filter is capable of providing very high collection efficiencies for particles as small as 0.1 μm and will remove a substantial quantity of those particles as small as 0.01 μm. In its simplest form, the industrial fabric filter consists of a woven or felted fabric through which dust-laden gases are forced. A combination of factors results in the collection of particles on the fabric filters. When woven fabrics are used, a dust cake eventually forms; this, in turn, acts predominantly as a sieving mechanism. When felted fabrics are used, this dust cake is minimal or almost nonexistent and the primary filtering mechanisms are a combination of inertial forces, impingement, and so on. These are essentially the same mechanisms that are applied to particle collection on wet scrubbers, where the collection media is in the form of liquid droplets rather than solid fibers.

As particles are collected, the pressure drop across the fabric-filtering media increases. Owing in part to fan limitations, the filter must be cleaned at predetermined intervals. Dust is removed from the fabric by gravity and/or mechanical means. The fabric filters or bags are usually tubular or flat.

The structure in which the bags hang is referred to as a baghouse; the number of bags in a baghouse may vary from less than a dozen to several thousand. Quite often, when great numbers of bags are involved, the baghouse is compartmentalized so that one compartment may be cleaned while others are still in service. The basic filtration process may be conducted in many different types of fabric filters in which the physical arrangement of hardware and the method of removing collected material from the filter media will vary. The essential differences may be related, in general, to the following[1]:

1) Type of fabric
2) Cleaning mechanism(s)
3) Equipment geometry
4) Mode of operation

Additional details of each of the above areas are presented in the next section.

Optimizing Air Pollution Control Equipment Performance: Operation & Maintenance, First Edition. Jay Richardson and Louis Theodore.

16.1 Description of Control Device

To better understand one's fabric-filter dust collector (baghouse), it is helpful to determine where it fits among the various types of fabric filters. When one tries to group fabric filters into a number of categories, it soon becomes obvious that the task is not simple. There appears to be an exception to each of the rules. The creation of certain categories, though they are not rigid, is yet very helpful.

One such approach is to group fabric-filter designs by cleaning method (Table 16.1). There are three major cleaning methods: shakers, reverse air, and pulse jets. In addition to these three dominant cleaning methods, there exist a large number of other cleaning methods which are less often applied. Recently, combinations of the three primary methods have been occasionally employed. For example, reverse air and shake have been used in combination, and reverse air with a pulse assist. The three dominant cleaning methods are discussed below in the next three sub-sections.

Another approach to grouping fabric filters is by capacity (Table 16.2). Generally, the groupings are small volumes (i.e., below 10,000 acfm), medium volumes (i.e., 10,000–100,000 acfm), and large volumes (i.e., >100,000 to the multimillion acfm level).

The filter-media type and temperature capabilities provide two other ways to categorize and view fabric filters. Woven versus felted media are the media-type categories and high-temperature (>400°F), medium-temperature (200–400°F), and low-temperature applications (<200°F) are useful temperature groupings. Some of the fibers used for baghouse filter media are shown in Table 16.3.

Two additional ways to group filters are intermittent versus continuous service and inside versus outside bag collection.

One should be aware of these distinctions and attempt to find where the collector in question fits; thus, when considering operating or troubleshooting recommendations, one only applies recommendations that are suited to the type of collector being used.

Table 16.1 Fabric-filter cleaning methods.

Cleaning Method	General Characteristics Associated With the Cleaning Method
Shaker	Inside bag collection
	Manual & automatic cleaning
	Woven filter media
	Filtration by the filter cake
	Generally intermittent – small operation
	Large & small gas volumes
	Often single compartment design
Reverse air	Inside bag collection
	Automatic cleaning
	Woven filter media
	Filtration by the filter cake
	Continuous operation
	Medium & large gas volumes
	Multiple compartment design
Pulse jet	Outside bag collection
	Automatic cleaning
	Felted filter media
	Media filtration
	Continuous operation
	Small, medium, and large gas volumes
	Can be multiple compartment design

Table 16.2 Fabric-filter design difference based on gas volume capacity.

General Capacity	General Characteristics Associated With the Design Gas Capacity
Low capacity	"Off the Shelf" small units
	Treating gas volumes of hundreds to a few thousand actual cubic feet (of gas) per minute (acfm)
	Little or no assembly
	Bags installed at the factory (i.e., pre-bagged)
	Manual cleaning systems available with some types
	Cleaning may be shaker, cartridge, or pulse-jet
Medium capacity	Modular designs
	Treating gas volumes of a few thousands to approximately less than 100,000 acfm
	Little field assembly required
	Generally pre-bagged
	Pulse-jet – high gas-to-cloth ratio
High capacity	Shippable, modular design – 50,000 to over 1 million acfm
	Moderate field assembly
	Bags installed in the field
	Reverse-air, shake, and pulse-jet
	Structural considerations in assembly
	Shake and reverse air – low gas-to-cloth ratio
	Pulse jet – high gas-to-cloth ratio

Table 16.3 Characteristics of filter bag fibers.

		Maximum Temperature							
		Continuous		Surges					
Generic Name	Fiber	°F	°C	°F	°C	Acid Resistance	Alkali Resistance	Flex Abrasion Resistance	Relative Cost
Natural fiber cellulose	Cotton	180	82	225	107	Poor	Excellent	Average	0.4
Polyolefin	Polypropylene	190	88	200	93	Excellent	Excellent	Good	0.5
Natural fiber protein	Wool	200	93	250	121	Good	Poor	Average	0.8
Polyamide	Nylon	200	93	250	121	Poor to fair	Excellent	Excellent	0.6
Acrylic	Orlon®	240	116	260	127	Very good	Fair	Average	0.7
Polyester	Dacron®	275	135	325	163	Good	Fair	Excellent	0.5
Aromatic polyamide	Nomex®	400	204	425	218	Fair	Very good	Very good	2.0
Fluoro-carbon	Teflon®	450	232	500	260	Excellent (Except poor if fluorine is present)	Excellent except poor in the presence of trifluoride, chlorine, and molten alkali metals	Fair	6.7
Glass	Fiberglass® or Glass	500	260	550	288	Good	Poor	Poor to fair	1.0
Polymer	P84®	450	232	500	260	Good	Fair	Fair	2.5
Polymer	Ryton®	375	191	450	232	Excellent	Excellent	Good	2.5–4.0
Nextel	Ceramic	2,192	1,200	2,372	1,300	Excellent	Excellent	Fair	10

Source: McKenna and Turner, 1989, Greiner, 1993.

16.2 Cleaning Methods

16.2.1 Shakers

Simply stated, a shaker cleaning system removes the collected dust from the surface of the bag by mechanically shaking the bag. This can be done manually in the case of small dust collectors. Generally, however, it is accomplished by the use of a motor driving an eccentric, which in turn has a rod connected to the bags. It is probably the oldest bag cleaning technique and is currently employed over the full range of baghouse capacities from very small off-the-shelf units to extremely large structural design units. The bag is generally open at the bottom and closed at the top, being fixed in the tube sheet at the bottom and attached to the shaking mechanism at the top. With this configuration, the dust is collected on the inside of the bag. The bag normally contains no rings or cages. The flow of dirty gas to the bags is stopped during the cleaning process. Table 16.4 shows some of the typical characteristics of a shaker-cleaned baghouse.

In the United States, shakers normally employ woven cloth at gas/cloth (G/C) ratios below 4:1. Attempts thus far to use shake cleaning in combination with domestic felts have led either to ineffective cleaning and thus high-pressure drop or to the filter medium being shaken apart. The important parameters affecting the efficiency of cleaning are frequency, oscillation, and amplitude.

To achieve a continuous shake-clean baghouse, the unit needs to be subdivided into a number of isolatable compartments, such that one compartment at a time may be taken off-stream for cleaning. The normal procedure for the intermittent-shake cleaning process is to first close the inlet flow with a damper and, if necessary, the outlet flow of dirty gas. This should almost immediately result in zero pressure drop across the bags. Once this condition is achieved, shaking of the bags can be initiated. Shaking continues for as many as 100 shake cycles. The cleaning is then stopped and the compartment is momentarily kept dormant until the dust settles. The dampers are then opened and the filtration process resumed. If reverse air is employed to assist the cleaning process, the cleaning cycle may consist of a series of repeated shakes, then reverse-air cycles before the unit is brought on-stream.

16.2.2 Reverse-Air/Reverse-Flow Cleaning

Reverse-air cleaning is perhaps the gentlest cleaning method. In this method the dust is removed from the bags by back-flushing with low-pressure (a few inches water gauge) reversed flow. In the case of high-temperature applications, the just-cleaned hot gas is employed to back-flush rather than ambient-temperature air. Woven-filter media are generally employed in conjunction with reverse-air cleaning. The cleaning is motivated by a separate cleaning fan which is normally much smaller than the main system fan since only a fraction of the total system is cleaned at any one time. In the case of a negative pressure system, one often can clean without a reverse flow fan. The flowrate of cleaning gas is normally about equal to that of the dirty gas (see Table 16.5).

Most often, reverse-flow systems are comprised of a number of isolatable compartments. The G/C ratios usually employed are less than 4:1. Normally, the dust is collected on the inside of the bag, the bag being open at the bottom and closed at the top. The bag will contain rings to keep it from collapsing completely during flow reversal. Complete collapse would, of course, prohibit cleaning since the dust particles would be unable to fall down within the bag to the hopper.

Table 16.4 Shaker baghouse parameters.

Item	Parameter
Frequency	Usually several cycles/minute; adjustable
Motion type	Simple harmonic or sinusoidal
Peak acceleration	1–10 g's
Amplitude	Fraction of an inch to several inches
Mode	Off-line (compartment isolated) usually
Duration	10–100 cycles, 30 seconds to few minutes
Common bag diameters	4–12 in.

Table 16.5 Reverse air filter parameters.

Frequency	One compartment cleaned at a time, sequencing one compartment after another. Cleaning may be initiated either automatically by system maximum pressure drop or continuously.
Motion	Gentle collapse of the bag concave inward upon deflation. Compartment slowly depressurized after completion of a back-flush. NOTE that anti-collapse rings are sewn into the bag at intervals along the length to ensure that the bag is re-inflated after cleaning.
Mode	Compartments are isolated off-line during cleaning
Duration	1–2 minutes, including valve opening and closing and time for dust settling periods. Reverse-airflow itself is normally 10–30 seconds
Common bag dimensions	8–12″ diameter, up to 30 ft long
Bag tension	50–75 pounds typical; optimal values are determined after initial on-line operation

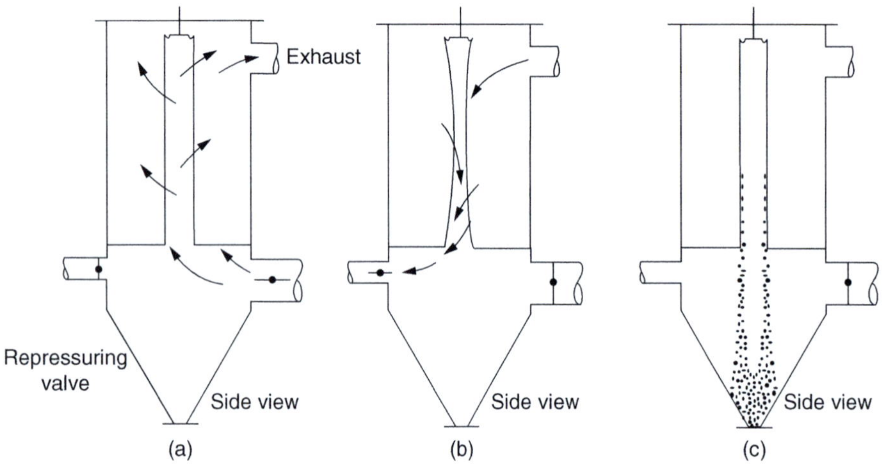

Figure 16.1 Reverse air cleaning (a) filtering; (b) collapsing; (c) cleaning.

Cleaning is accomplished both with and without flexing (partial collapsing) of the bag. Reverse flow without flexing can be, employed when the dust is very easily dislodged from the bag surface. Figure 16.1 shows a typical cleaning cycle. In the on-stream-gas filtering mode, the compartment and outlet dampers are open and the dirty gas enters the bag at the bottom. The dust is collected on the inside of the bag and the cleaned gas exits the outlet damper. During the bag cleaning cycle, the flow is reversed by closing the outlet plenum and opening a third damper, which allows cleaned gas to enter the compartment on the clean side of the bags, thus back-flushing the bags and exiting the compartment through the inlet damper. This now-dirty gas progresses to the balance of compartments that are on-stream. It should be noted that this process increases the system G/C ratio by adding to the total gas volume, the volume of gas employed in the back-flushing process. The system described here, although generally applicable, is normally associated with large gas volumes. Low- and medium-volume reverse-air systems can be found; however, these often employ felted media with a moving reverse-flow plenum.

16.2.3 Pulse-jet Cleaning

Pulse-jet cleaning is a method that employs high-pressure (60–120 psi) compressed air, with or without a venturi, to back-flush the bags vigorously. This method creates a shock wave that travels down the bag, knocking the dust away from the filter medium. Normally, this method is employed in conjunction with felted-filter media and the G/C ratio is generally higher than that of the shake and reverse-air cleaning methods. The duration of cleaning is lower than that of the other two methods; generally, the pulse lasts only a fraction of a second. The house is often not subdivided when pulse-jet cleaning is employed.

Table 16.6 Pulse jet filter parameters.

Frequency	Usually, one row of bags at a time in the compartment being cleaned. Sequenced one row after another. Cleaning can be controlled such that every other row is cleaned. Cleaning may be continuous or activated by maximum baghouse pressure drop.
Motion	Shock wave of cleaning pulse travels down the bag, momentarily distending the bag and releasing the dust cake.
Mode	Generally, bags are cleaned on-line. Off-line cleaning may be used with difficult-to-remove filter cakes or to thoroughly clean bags prior to compartment maintenance
Duration	Compressed air pulse generally approximately 0.1 seconds. The row of bags being cleaned is effectively "off-line" during the pulse.
Common bag dimensions	5–8″ diameter, bag length up to 20 ft.

The usual configuration has the bag closed at the bottom and open at the top, a metal cage is employed within the bag to keep the bag from collapsing. In the normal mode of operation, the dirty gas enters the hopper and proceeds to the bags. The dust is collected on the outside of the bags and the cleaned gas exits through the top of the bags and baghouse. Usually, a row of bags is cleaned simultaneously by introducing compressed air briefly at the top of each bag. The shock wave thereby created drives the dust off the outside of the bag and down into the hopper. Continuous discharge of the dust from the hopper is often employed. The importance of the venturi to provide an onrush of secondary air for cleaning is now being questioned, and some pulse systems operate with only the compressed-air manifold and without venturis over each bag. Pulse-jet bag parameters are given in Table 16.6. Notable is that this system has no internal moving parts and allows for removal of the bags from the clean side of the baghouse compartment, since the bags are usually connected only at the top.

16.2.4 Miscellaneous Cleaning Methods

In addition to the three most common cleaning methods, a number of other methods have been employed. These methods were either made obsolete by one or more of the three major methods or else never achieved the commercial significance of the three major cleaning methods.

One method that achieved some degree of commercial success is "reverse jet" cleaning, sometimes referred to as "blow ring" cleaning. This method employed a jet ring surrounding each filter tube; the rings were carried up and down the bags on a carriage driven by chains and sprockets or a cable system. Today, this method is found only in a very limited number of applications.

In addition to the reverse jet, high-frequency agitation has been employed. One such approach is to achieve fluttering of the cloth by vibration and rapping. This is normally associated with envelope filters. In the case of tubular bags, fluttering has been achieved by the use of an "air shake."

Even-higher-frequency agitation is achieved by the use of sonic and ultrasonic cleaning. Sonic cleaning has also been employed in combination with reverse flow and pulse-jet filters. In this method, sonic horns may be installed in baghouse compartments and activated during the compartment cleaning cycle. These sonic horns are often a "post-operation" installation when it is found out that there is a hard-to-remove dust from the bags. In this way, the sonic horns provide extra cleaning, when necessary.

One of the simplest cleaning methods used in the case of very small collectors is manual cleaning, where the bags are struck, shaken, or even removed and turned inside out.

A variation of the pulse jet which is commercial today but not as common as the pulse jet is the plenum pulse method of cleaning. Like the pulse jet, this method employs compressed air, but instead of delivering the air to each bag via a manifold, the compressed air is delivered to a compartment of bags via a compressed air plenum and damper system, cleaning the entire compartment simultaneously.

16.2.5 Cleaning Method Versus Capacity

As shown in Tables 16.1 and 16.2, fabric filters can be grouped by both cleaning method and gas capacity. There is a loose correlation between the two groupings. The extremely small collectors are generally found to be shakers. This is especially true if an intermittent collector is suitable. Recently, compact cartridge filters (similar to truck air filters) employing simple pulse cleaning have been finding success in the small-volume marketplace. The small and especially the medium-size collectors, where continuous operation is required, are most often found to be pulse-jet collectors.

The large continuous collectors of recent vintage are most often pulse-jet or reverse-air collectors, although large shakers are found as well. In all cases, continuous operation is achieved by subdividing the unit into a number of compartments so that when one compartment is off-line for bag cleaning, the total process gas volume can be handled by the balance of the baghouse system.

16.2.6 Bags and Cages

As previously mentioned, each fabric-filter dust collector consists of a number of fabric-filter elements. These elements normally consist of textile material and are constructed as tubes, envelopes, or most recently, cartridges.

The textile can be either felted or woven with a bag material (Table 16.3) chosen due to the chemical and thermal conditions of the gas stream being filtered as well as the chemical and physical characteristics of the dust being filtered.

If woven textiles are used, there are numerous constructions that can be employed, such as sateen, 2×2 twill, or crow-foot. In addition to various yarn constructions, the filter element can be varied further by the finishing process applied. The most commonly employed finishing steps include glazed, silicone graphite, Teflon, napping, calendaring, and singeing. Details of these finishes are provided in Table 16.7.

Finally, when reviewing bag construction, one must also consider stitching and seam types. The most commonly employed filter element design is that of a tubular-shaped "bag." This is employed to collect dust either on the outside or the inside of the tube. The envelope-constructed filter element is not as common in the United States as the tubular-shaped. The flat-enveloped baghouse normally collects on the outside of the filter element and provides more filtering area per volume of housing. Finally, the cartridge design, which is a pleated construction, provides maximum filtering area, but at this time seems to have temperature and filtering velocities limitations more restrictive than the other two constructions.

When reviewing the filter element and its construction, one must also consider the filter element support structure. In the case of shakers, as mentioned, the filter element normally has no support other than attachment at the top and bottom.

Reverse-air filter elements normally have support rings sewn into the bag. In the case of pulse-jet cleaning systems, the bag is usually supported by a cage placed within the bag. The standard construction has been 6–20 vertical wires (stringers) spot-welded to a number of rings that are generally located about 6–12 in. apart.

The use of polymer and glass bags in conjunction with pulse-jet cleaning has led to the use of close-mesh cages with either a 2×1 in. or a $1/2\times1/2$ in. mesh to provide greater fabric support. Variations in cage design beyond mesh considerations include differences in the bottom cage pan, top collar, and top flanges. Usually, these variations are related to bag – cage attachment schemes.

To summarize, baghouses are usually characterized and identified according to the method used to remove collected material from the bags and particle removal can be accomplished in a variety of ways, including shaking the bags, blowing a jet of air on the bags from a reciprocating manifold, or rapidly expanding the bags by a pulse of compressed air. In general, the various types of bag cleaning methods can be divided into those involving flexing and those involving a reverse flow of clean air. Figure 16.2 illustrates four (4) of the fabric-flex cleaning methods.

Table 16.7 Fabric finishing.

Calendaring	High-pressure pressing of the fabric; pushes the surface fibers down onto the body of the fabric
Napping	A scraping of the fabric surface that raises the surface fibers
Singeing	Passing of the filter medium over an open flame, thereby removing the independent surface fibers
Glazing	High-pressure pressing of the fabric at elevated temperatures which fuses the surface fibers to the body of the fabric
Coating	Immerses the fabric into a solution that will provide the fabric with a coating that will lubricate and thereby reduce self-abrasion; in the case of many woven fabrics the preferred coating is Teflon and/or silicon-graphite

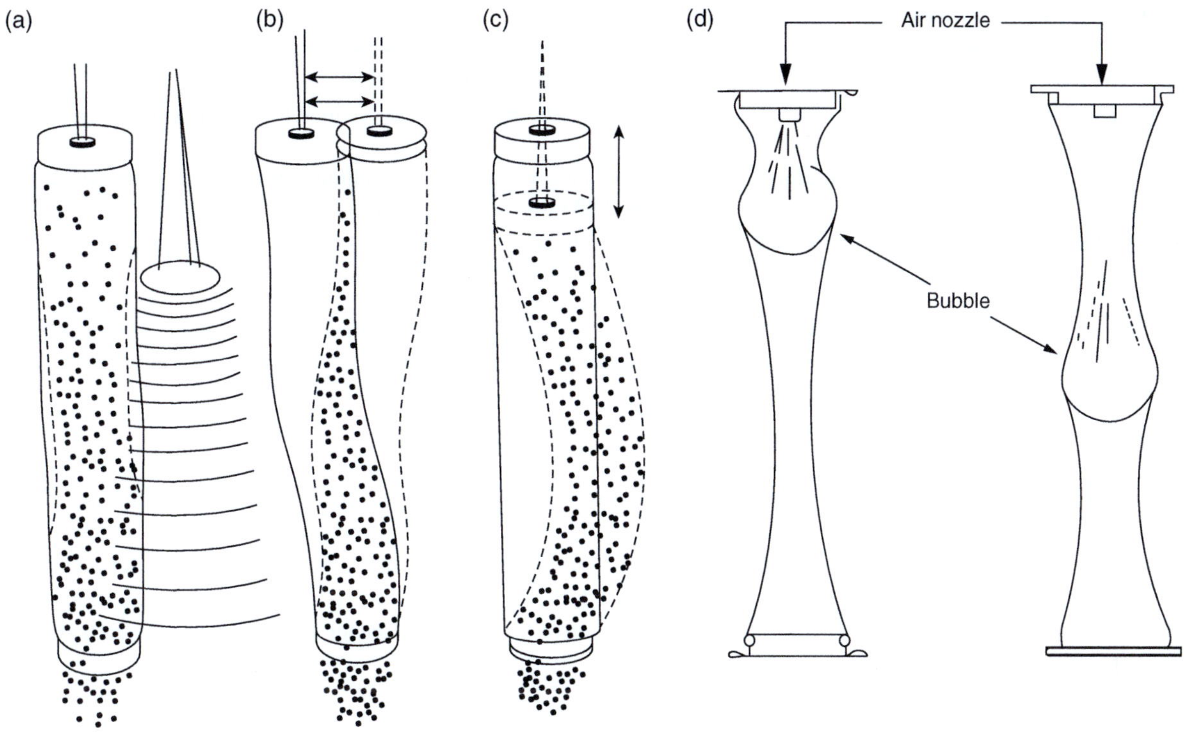

Figure 16.2 Fabric-flexing cleaning methods. (a) Sonic cleaning; (b) oscillating; (c) shaking; (d) pressure-jet cleaning.

16.3 Design Considerations

When selecting a baghouse, two major areas of design need to be considered in conjunction with each other. The first is that of the total dust collection system design. This includes all ductwork, fans, heaters, dampers, dust-conveying equipment, and controls that will directly affect and are intimately associated with the dust collector itself. The second area of concentration is that of the flange-to-flange dust collector, its cleaning method, bag type, materials selection, and so on. The flange-to-flange collector should not be selected without concurrent specification of the total dust collection system, and this in turn requires full consideration of the dust source operation (i.e., normally the production process). It can be that such an overview will lead to some modification of the production process or process controls in order to attain trouble-free baghouse operation.

The specification of the dust collector and its auxiliaries must begin with a definition of the emission problem. The better and more complete the problem definition, the better the chances that the dust collection system will function in a trouble-free, reliable manner. The dust properties of primary interest are the density, particle size, and dust loading. The gas properties essential to proper problem definition are volume, temperature, moisture content, and acid gas concentration. A general description of the source or production operation is helpful, particularly in defining whether the operation of the collector is to be continuous or intermittent.

It is not uncommon to find a dust collector selected simply on the basis of estimated gas volume, temperature, and a brief description of the production operation; however, performing a selection based on such limited criteria enhances the risk of selecting a baghouse system requiring constant maintenance and bag replacement, high energy (fan) use, and a potential deleterious effect on both downstream and upstream systems.

The size of the fabric-filter dust collector is tied directly to a parameter known as the air/cloth (*A/C*) or G/C ratio. This parameter is the apparent velocity of the dust-laden gas stream as it approaches the filter elements. It can be calculated by dividing the total gas volume, usually in ft^3/min, by the total area in ft^2 of cloth or filter medium available for filtration. Variations of the air-to-cloth ratio to take into account compartments (i.e., cloth) off-line for cleaning and/or maintenance are shown in Table 16.8

The *G/C* ratio is a measure of the amount of gas passed through each square foot of fabric in the baghouse. It is given in terms of the number of cubic feet of gas per minute passing through one square foot of cloth. In other words, the *G/C*

Table 16.8 Air/cloth ratio definitions.

Gas (air)-to-cloth ratio	Gas/cloth ratio = gas volume/cloth area
Gross gas (air)-to-cloth ratio	Gross gas/cloth ratio = total inlet gas volume/total filter cloth area in collector
Net gas (air)-to-cloth ratio [compartment off-line for cleaning and/or maintenance]	Net gas/cloth ratio = (total inlet gas volume + cleaning air volume)/total on-stream cloth area
Net-net gas (air)-to-cloth ratio [1 compartment off-line for cleaning and 1 compartment off-line for maintenance]	Net-net gas/cloth ratio = (total inlet gas volume + cleaning air volume)/ total on-stream cloth area
Units usually employed	Gas/cloth ratio = gas volume/cloth area = cubic feet per minute/square feet = feet/minute (essentially this is the velocity of the gas through the cloth)

ratio = gas volume rate/cloth area. Also note that this velocity is not the actual velocity through the openings in the fabric, but rather the apparent velocity of the gas approaching the cloth. As the *G*/*C* ratio increases, pressure drop (ΔP) also increases. In the United States, pressure drop in baghouse applications is generally measured by inches of water (WC).

Despite several sophisticated formulas that have been developed, there is no satisfactory set of published equations that allows a designer to accurately calculate the efficiency of a prospective baghouse. However, there are three more heuristic formulas worth mentioning that can help baghouse designers:

$$\text{Gross gas / cloth ratio} = \text{total inlet volume rate / total filter cloth area in collector} \tag{16.1}$$

$$\text{Net gas / cloth ratio} = (\text{total inlet gas volume rate} + \text{cleaning volume rate}) / \text{on stream cloth} \tag{16.2}$$

$$\text{Units gas / cloth ratio} = (\text{gas volume rate / cloth area}) = \text{ft}^3/\text{min}/\text{ft}^2 = \text{ft/min} \tag{16.3}$$

As mentioned earlier, baghouse design is still very much an artform and nowhere is this more evident than in the selection of *G*/*C* ratios. Factors influencing the G/C selection include the cleaning method, filter media, dust size, dust density, dust loading, and other factors. Quantifying the general relationship between these parameters has been attempted with very limited success.[2] Models specific to a given source operation and dust collector type have been proven more accurate for *G*/*C* selection for this more limited application.[3]

Most commonly, two design approaches have been used successfully. The first has been to collect all of the empirical data available for the source in question. If there is none in one's industry, one goes to an analogous industry that is using baghouses and determines the *G*/*C* range successfully employed in that industry and conservatively applies it. The risk here is that there is a subtle but essential difference between the two industries which prohibits the *G*/*C* translation. The alternative to this is to operate a pilot unit on a slipstream and vary the G/C ratio. Then one designs the full-scale unit around the *G*/*C*, which appeared to provide the lowest emission and pressure drop levels within predetermined ranges.

Some earlier work that appears in the earlier literature by one of the authors is available.[4–8]

16.4 Installation Procedures

Depending on the type of system chosen, installation of baghouse systems can take a few days to install or several months to erect and start up. In either case, proper planning and installation procedures will save time and money, during both the installation and future operation and maintenance periods.

This section discusses recommended installation practices, typical installation errors and oversights, as well as their potential immediate and long-term consequences. A typical pre-startup inspection checklist is also reviewed. The installation stage of a project plays an important role in the future operation and maintenance of a dust collection system. If poorly done, it can cause operation and maintenance (O&M) problems for years or require large expenditures to correct. If properly executed, it has the potential to ensure the initiation and continuation of a successful O&M program. How? This stage allows for the first time, on-site assessment of the compatibility of the system design and O&M procedures and requirements. Usually, and unfortunately, it is the first meeting of the system designers and the O&M personnel. Not enough emphasis is generally placed on a complete combined designer/O&M engineering review and inspection during the installation phase. This is understandable because of the usual separation of responsibilities and company affiliation; the design engineer's role terminates with the completion of paper design and analysis; the erection engineer is primarily concerned with meeting a schedule and often never understands the dust collector's function; plant O&M personnel, overburdened with existing production equipment O&M, have not yet found time to understand that this new equipment will provide no additional operating capacity and is generally regarded as a maintenance issue.

If, however, these obstacles are overcome and effective coordination of the system designer, installer, and O&M personnel is accomplished, the payoff in future dollars saved will be substantial. Some key areas to reevaluate during the installation period are:

1) Is there easy access to all potential maintenance areas (i.e., fans, motors, conveyors, discharge valves, dampers, damper operators, pressure and temperature monitor ports, and mechanical linkage)?
2) Is there easy access to all inspection and test areas?
3) How will weather affect the above?
4) Are manual operators available with respect to location and access? Does a worker have to climb a 50-ft ladder in winter to open a bypass damper to ensure plant equipment or improve personnel safety or continued operation – the design should be changed.
5) Can the system design tolerate all anticipated operating conditions?

Although most of the above should have been considered in the original system design, often they are not. This is primarily because the designer and real O&M personnel have never gotten together or O&M procedures have changed. Equally important is that the erection engineer and/or O&M personnel do not make field changes without the approval of the system designer. Most items of a system are interrelated, and modifying one part without regard for the whole is often costly.

Typical installation errors and their effect on O&M are shown in Table 16.9.

16.4.1 A Properly Designed and Installed System Will Provide

1) Uniform air and dust distribution to all filters.
   a) Duct design, turning vanes, and deflection plates all assist in obtaining this. Often, they arrive loose and are field-installed. If improperly installed, they can induce high-airflow regions that will abrade the duct or bag filters or cause reentrainment, high-dust-concentration regions that can produce uneven hopper loading and uneven filter bag dust cake.
2) Total seal of the system from dust pickup to stack outlet.
   a) In-leakage at flanges or collector access points either adds additional airflow to be processed or short-circuits the process gases. In-leakage to a high-temperature system is extremely damaging, as it creates cold spots and can lead to dew-point excursions and corrosion. If severe, it can cause the entire process gas temperature to pass through the dewpoint and result in condensation on the bags. Early bag failure and high-pressure drop will generally result. The best check for leaks is to inspect the walls from inside the system during daylight. Light penetration from outside isolates the problem areas. It is particularly important to seal the dust discharge points in negative systems. In-leakage here will result in incomplete or no discharge, which can lead to reentrainment problems, yielding high-pressure drop and hopper fires.
3) Effective coatings and paint.
   a) Most systems are painted on the exterior surfaces only. Take extra care to touch up damaged areas with a good primer and if equipment is not delivered finish-painted, apply it as soon as possible following erection. Unprotected primers will soon allow corrosion and require sandblasting plus many dollars to repair. If your system has been internally protected with a coating, thoroughly inspect for cracks or chips, particularly in corners, and repair them before operating. A poor interior coating can be worse than none at all because it will trap corrosive elements between the coating and the surface it was intended to protect.
4) Properly installed filter bags.
   a) The filter bags are the heart of any fabric-filter collection system. Improper installation can result in early bag failure, loss of cleaning effectiveness, and thus high-pressure drop and operating costs or increased stack emission. Each manufacturer provides instructions on the proper filter bag installation and tensioning (where required).

Table 16.9 Typical installation errors and their effect on O&M.

Item	Immediate Potential Effect	Long Term Effect
Improper installation, or omission of turning vanes and baffle lates	Uneven gas flow and dust distribution leading to higher pressure drops and uneven hopper loading	Filter bag wear and shortened bag life, excessive erosion of ductwork
Poor seal at flanges and access areas	In-leakage resulting in reduced gas volume, higher fan volume, lower gas temperatures (with potential condensation issues), higher operating costs	Localized cold spots resulting in component corrosion and bag degradation, including fabric caking, and lower bag life
Poor seal at dust discharge flanges	Incomplete dust discharge, hopper fires, reentrainment of dust	Reentrainment, steadily rising pressure drop
Cracked or chipped paint or coatings	Esthetics	Corrosion
Improper bag tensioning	Ineffective bag cleaning, bag collapse. Potential bag failure due to over-tensioning	Bag wear, high-pressure drops
Improper bag seating	Increased stack emissions	Compliance failure, high-pressure drop, increased bag wear
Incomplete insulation	Cold spots	Corrosion
Seal between dirty side and clean side	Increased stack emissions, dust buildup on clean side of plenum	Compliance failure
Duct damper alignment	Loss of flow control, increased pressure drop	Poor maintenance
Screw conveyor direction reversed	No discharge	Plugged compartment, bent screw; full hopper, hopper fires
Fan mount	Noise – vibration	Broken components
Fan belt alignment	Vibration, improper fan volume	Broken belts, increased fan bearing wear
Compressed air not dried	Freeze-up and condensation	Water in control systems, damaged components
Lack of inspection access	Lack of early indication of potential component failures	No notification of operating and equipment issues
Lack of maintenance access	Lack of regular preventive maintenance	Major component failure

These must be explicitly followed. Very often, early bag failures can be traced to improper installation. Remember, it is much easier to check and recheck bag connections, tensioning, locations, and so on, in a clean, cool, dry collector than it will be one day after startup. Bag maintenance usually accounts for 70% of annual maintenance time and money. Extra efforts in this area during installation can have a significant effect.

5) Proper insulation installation.
   a) High-temperature collector systems require special consideration in order to prevent O&M problems. When handling high-temperature gases, it is important to maintain the temperature of the gas and all collector components coming in contact with it above the gas dew point. Insulation is usually employed. Much of the time, all or a part of the insulation is field installed. Check to see that all surfaces and areas of potential heat loss are adequately covered. In particular, check to see that field flashing also has insulation beneath it. Cold spots cause local corrosion. Gross heat loss may cause excessive warm-up time or lower the gas temperature below the dew point.
6) Total seal between the dirty side and clean side of the collector.
   a) Remember, the primary purpose of the dust collector is to separate the particulate from the gas by means of fabric filtration. This means that all the gas must pass through the fabric. Any leaks bypassing the fabric filters will directly emit dust to the stack and therefore reduce the collection efficiency of the system. The time to inspect "bypass leaks" is before startup, when everything is clean and accessible. The best technique is to utilize a bright light on one side of the plenum and visually observe for light penetration on the other. This is the most effective in total darkness. Take that extra time to check this important area. Tracking down stack emissions not associated with bag failures can be extremely difficult after startup.

7) Properly installed and operating dampers.
   a) Most systems employ several dampers to isolate areas of the system or control the volume of airflow. Proper alignment is important both in the case of internal blades and the operating linkage. In high-temperature applications, special care must be taken to allow for proper operation and sealing at the operating temperatures. Some dampers may require readjusting after reaching high-temperature operation. In modular systems, single modules are normally isolated for bag cleaning and maintenance. Leakage through these isolation dampers can cause improper bag cleaning. It will also create a very poor ambient condition for maintenance workers to work in. This, in some applications, can pose a health hazard, and in all applications results in lower-quality workmanship or incomplete maintenance.
8) Properly operating mechanical components.
   a) Most mechanical components are designed with a normal operating direction. Cylinder rod location, motor rotation, and so on, must be checked. Remember, when hooking up an AC motor, one has a 50% chance of being correct on the first try. Not only will a backward-moving conveyor produce no discharge, but it can pack material so that it bends the screw. Left uncorrected, a reversed screw conveyor will result in a full hopper. The industry abounds with horror stories where full hoppers have led to burned bags, or dust set up, requiring jackhammers to remove it.
9) Smoothly running fans.
   a) Fans must be checked for proper rotation, drive component alignments, and vibration. Fans should be securely mounted to sufficient mass to eliminate excessive vibration.
10) Clean, dry compressed air.
   a) Most systems employ compressed air to operate dampers, controls, instruments, and so on. Probably more systems suffer shutdowns and maintenance problems due to poor-quality compressed air than for any other reason. Clean, dry air is necessary to maintain proper operation of the pneumatic components. In installations where the ambient temperature drops below 32°F, a desiccant dryer system is generally employed. Sometimes, insulation of air lines and pneumatic components will be required. Often, these considerations are not included in the dust collector system, with "clean, dry compressed air to be supplied by owner." Remember, the air must be clean and dry when it reaches the pneumatic component.

16.4.2 The Checklist

Each installation should have its own checklist reflecting its unique design and components. The important thing is to prepare one before beginning your final inspection and initial startup. Startup is normally a hectic time, so a checklist will help in the final inspection to ensure that nothing is missed. Table 16.10 contains the items for a typical checklist for a fairly simple system.

16.4.3 The Formula for a Successful Startup

Implementation of the following steps will go a long way toward providing a successful startup:

1) Advance planning – create a checklist
2) Good communication with operating personnel both before and during startup
3) Be certain of the system's initial state through a thorough inspection
4) Define and communicate the operating conditions requiring an abort and the emergency steps required to accomplish it to all personnel involved
5) Activate each part of the system carefully and in proper sequence
6) Observe and log all pertinent data at frequent intervals until the system has "settled" out
7) Reinspect key areas under operating conditions
8) Identify any special action required of operating personnel

16.5 Operation

The operation phase of a fabric-filter collection system should start long before the actual startup. Prior to bringing the collector system on-stream there should be familiarization and training sessions for all O&M personnel. In these sessions, the following subject areas should be covered:

1) System design
2) Component functions

Table 16.10 Example checklist for inspections and startups.

Item No.	
1	Visually inspect: 1) Structural connections for tightness 2) Ductwork flange alignment and proper seal 3) Filter bags for proper installation (tensioning, fit into tube sheet, etc.) 4) Dampers for operation and sequence 5) Motors for system fans, reverse-air fans, conveyors, etc. for proper rotation 6) Electrical controls for proper operation 7) Instrumentation for proper installation 8) Automatic valves such as rotary valves and slide gate valves for proper operation
2	Remove inspection doors and check conveyors for loose items or obstructions
3	Adjust ductwork dampers to ensure full operation and set them to their proper settings
4	Remove any temporary baffles
5	Test all alarm systems
6	Test all conveyors for proper operation, especially direction of motion for mechanical conveyors
7	If included in design, start reverse-air fan
8	Start system fan
9	Record system pressures, temperatures, and other pertinent variables
10	Check to see that reverse-air dampers (if part of system design) are cycling
11	Check compartment and system pressure drops, especially pressure drop settings for compartment/system cleaning cycles
12	Determine system gas volume and adjust dampers as needed
13	Check compartments for dust leaks
14	Check for proper dust discharge from compartments to conveying system(s)

3) Control system
4) Critical limits and alarms
5) Key parameters to monitor
6) Good operating practices
7) Specific no-no's
8) Preventive maintenance program
9) Startup and shutdown procedures
10) Emergency shutdown procedures
11) Safety considerations

This training phase is too often left to on-the-job training. Such an approach can be both dangerous and costly. It should be the plant manager's and/or plant engineer's responsibility to see that all responsible O&M personnel are both qualified and trained in the operation and maintenance of the system prior to startup.

16.6 Startup and Shutdown

Operating a baghouse is like flying a plane – the most critical moments are usually at the beginning and the end of a run. Improper startup can lead to long-term negative effects. For example, partial or total blinding of the bags and residual pressure drop can result from allowing too high a filtering velocity or dew-point excursions. Improper shutdown can result in dewpoint excursions, over- or under-cleaned bags, inadequate hopper discharge, and so on. High-temperature applications are particularly sensitive to startup and shutdown periods. Each application has its special concerns and the baghouse manufacturer should be consulted as to the proper do's and don'ts for each system.

16.6.1 Initial Startup

The initial startup is potentially the most dangerous time for permanently damaging the fabric. The clean filters have not developed a protective dust cake and are particularly sensitive to dust abrasion and penetration of fine particles that can lead to permanent residual pressure drop. For some applications, the bags are precoated with a protective dust prior to the introduction of flue gas. In all cases, the filtering velocity should be kept very low until a sufficient dust cake is developed. This will be indicated by a pressure drop of 1–2 in. water gauge. The gas flow can then be slowly increased until design volume is achieved.

A system flow control damper is usually available to regulate the gas flow. If there is none, the pickup lines should be dampered back by using the balancing blast gates or restricting the pickup hoods with a temporary baffle. Some general rules for routine startup and shutdown are listed below:

16.6.1.1 Startup

Make sure all collector components are in working order and in proper mode. Do not allow higher-than-design filtering velocities or airflow. Avoid passing through (below) the dew-point within the baghouse when dirty gases are present. The system should be preheated to above the dew point with clean, hot air before the introduction of flue gas. During normal operation, maintain the temperature above the dew-point level. Operate the bypass system to assure its readiness in an emergency situation. Check all indicating and monitoring devices for proper operation.

16.6.1.2 Shutdown

Purge the collector with clean (hot when necessary) dry air before allowing the temperature to descend below the dew-point. Even ambient applications will pass through the dewpoint when the collector cools at night. Do not store dust in the collector. Maintenance workers have complained after spending a day with a pick and shovel inside a dust collector hopper. Additionally, this extra labor leads to excessive maintenance costs. Allow bags to clean down after the dust flow ends, but do not overclean. Check to see that all components are in the proper shutdown mode.

All O&M personnel should become familiar with each of the system components, the control system, and monitoring devices. The operator should monitor and keep records so that system or component deviations can be identified. These will serve as early warning signals for maintenance and assist in troubleshooting.

Normally, the principal parameters of concern are pressure drop, temperature, outlet emissions, and hopper discharge rates. Other information related to specific equipment may provide early warning signs of problems that are not easily detected or understood through the major parameter indicators. The object is to identify and correct problems before they reach the magnitude requiring shutdown. It is also recommended that alarms be included within the system sufficiently below shutdown levels to provide the necessary lead time in correcting the problem before shutdown. After "living" with the new system for a while, the operating and/or maintenance personnel may define additional monitoring or control equipment to accomplish this task.

Typical monitoring and indicating devices include:

- Pressure-drop indicators, magnehelics, photohelics, or manometers to define pressure drop across the baghouse or at other points of the system. Recorders of this parameter are very useful and often employed.
- Opacity meters which monitor the outlet and indicate the percent opacity.
- Gas flowmeters to indicate actual flow or percent of design flow rate.
- Temperature measuring and recording devices measuring inlet, outlet, and key temperature locations.
- Indicating lights defining operation condition of all motors, solenoids, dampers, and so on.
- Adjustable timers to control the frequency and duration of the cleaning mechanism.
- Fan-current indicator, fan-bearing-temperature indicator, and fan-vibration detector, all aimed at identifying early warning signs of maintenance needs.
- Heat-sensitive materials placed in key areas such as baghousings, cylinder heads, and motor housings. These change color when exposed to temperature.
- Corrosion chips placed inside the ductwork, collector areas, and so on, to monitor the degree of internal hardware corrosion.

16.6.2 Safety

All O&M personnel must know the safety limits of the system and when to abort the system. Normally, there will be an emergency shutdown button for this purpose, and often the abort system is automatic. Usually, the important thing to do in such an emergency situation is to stop the flue gas from entering the collector and that means shutting off the fan, opening the bypass damper, and shutting any inlet or outlet dampers, where available.

Whenever personnel enter a collector hopper or bag plenum, protection from dust, temperature, and flue gas contaminants and entrapment must be provided. All entry doors should include an electrical interlock to ensure compartment isolation and a positive mechanical method for ensuring that the door cannot close accidentally.

When performing work inside a collector, it is advisable that one member of the team always remain outside. Reports of personnel becoming trapped inside collectors have been alarmingly frequent. If adequate ventilation cannot be achieved, it will be necessary for maintenance and inspection personnel to wear life support systems and proper protective clothing.

Personnel should be trained as to emergency procedures in case of a fire, explosion, or electrical hazard in or around the collector. A dust collector is a relatively safe piece of equipment, but caution and preparedness are still required.

16.6.3 Maintenance

The award of the year for least-maintained equipment again goes to the fabric filter! This unfortunate lack of care not only costs industry millions of dollars each year, but also reduces the potential impact of air pollution control on air quality levels. Normally, a dust collector has been or will be installed at the encouragement of the local, state, or federal regulatory agency or union, not a production department. Even though it is often tied directly in series with your production process, it is often treated like a stepchild. The first step toward efficient, trouble-free operation of a fabric-filter system is to treat it the same as would any other piece of production equipment. It is a machine, and machines need "tender loving care." It requires routine inspection, preventative maintenance, and quick response to malfunctions.

Routine inspection of key components and operating parameters are the key ingredients to normal maintenance. Table 16.11 shows a typical periodic maintenance checklist. A preventative maintenance program is a must for systems requiring continuous duty. The preventative maintenance procedures must be continually updated to reflect actual experience, not some typical guidelines. The extent of preventative maintenance during scheduled shutdowns must be evolved, starting with the equipment supplier's recommendations, supplemented by plant staff routine inspection and maintenance records.

- Records – the heart of any maintenance system! No one wants to keep them, but everyone wants to see them. Log keeping of actual inspections, observations, maintenance performed, and so on, is invaluable for:
- Updating the maintenance program as a whole, modifying the preventive maintenance schedule or procedures,
- Defining problem areas to be corrected
- Defining maintenance costs and personnel/workforce needs
- Maintaining spare-parts control

Each plant normally has its own method of maintenance record keeping, but the following items should be included:

1) Activity date and time of day
2) Type of activity (i.e., routine inspection, emergency maintenance, and preventative maintenance)
3) Specific action(s) taken (i.e., replacement modification and adjustment)
4) Components affected – by part number and manufacturer
5) Problem cause, if known
6) Manhours expended
7) Materials used
8) Spare-parts usage
9) Any recommendations to prevent reoccurrence

The time used to record the details listed above will be well worth it. The maintenance section that recognizes the value of these items is diligent in obtaining and utilizing them, will be under control, and not have to respond to constant emergencies.

16.6.4 Troubleshooting

Detecting and correcting small problems before they grow into large and expensive crises is the goal of any maintenance program.

Table 16.11 Typical routine maintenance schedule.

		Frequency				
Item	**Check**	**Daily**	**Weekly**	**Monthly**	**Quarterly**	**Annually**
Dust pickup areas	Dust capture effectiveness	x				
	Monitoring flue gas volume	x				
	Test hood face velocity				x	
	Rebalance system					x
System dampers	Proper operation		x			
	Proper valve seating				x	
	Wear or corrosion					x
Bags	Observe stack/exhaust (either visually or with opacity meter)	x				
	Spot-check bag tensioning (inside collectors)		x			
	Spot-check bag condition and seating			x		
	Thoroughly inspect bags				x	
Cleaning system	Monitor cleaning cycle	x				
	Check compressed air	x				
	Inspect mechanical components for wear			x		
	Replace high-wear parts					x
Control system	Observe all indicators on panel/screen	x				
	Log ΔP	x				
	Blow out any manometer lines		x			
	Check compressed air system, including filters		x			
	Activate key shutdown and/or bypass controls		x			
	Verify accuracy of temperature indicting/monitoring equipment		x			
	Check accuracy of all other monitoring/indicating equipment			x		
System fan	Check drive components		x			
	Inspect for corrosion and material buildup			x		
	Check for vibration		x			
Outlet stack/ exhaust	Check emissions visually by monitoring opacity monitor	x				
	Calibrate opacity monitor				x	
Discharge system	Monitor discharge rate	x				
	Check all moving parts for wear and alignment			x		
General	Walk through entire system to check for normal and abnormal visual and audible conditions	x				
	Inspect system for wear and corrosion			x		
	Inspect door gaskets				x	
	Check for dust buildup in ducts				x	
	Inspect paint				x	
	Inspect baffles, hopper duct, etc., for wear					x
	Inspect general structural integrity of system					x

Trouble symptoms: A major malfunction in your collector system will result in one or more of the following identifying symptoms:

1) Abnormally high-pressure drop across the collector
2) Visible emission of dust in the exhaust stack
3) Inadequate face velocity or "puffing" at the pickup points and hoods
4) Lower-than-normal dust discharge
5) Loud or unusual noises
6) Severe corrosion of material

A general guide is shown in Table 16.12. If unable to identify and correct a problem in a reasonable period of time, do not hesitate to contact the system designer or the equipment manufacturer. Often, contacting the right person is all that is required. Remember, the manufacturer has the advantage of experience and feedback from many sources to assist in solving your problem.

16.6.5 Bag Maintenance

The highest source of maintenance and cost are generally the filter bags. All bag sets have a finite lifetime which will vary by application, installation, operating parameters, fabric type, and so on. Typical causes of bag failure are:

- High G/C abrasion
- Metal-to-cloth abrasion
- Chemical attack
- Bag-to-bag abrasion
- Inlet velocity abrasion (on inside-out cleaning)
- Accidents
- Upset conditions (e.g., temperature)
- Thread mismatched
- Cuff mismatched

In addition, each bag in a set may have a different life as a result of fabric quality, bag manufacturing tolerances, location in the collector, variation in the bag cleaning mechanism. Any one or a combination of these factors can cause bags to fail. This means that a baghouse will experience a series of intermittent bag failures until the failure rate requires total bag replacement. Typically, a few bags will fail initially or after a short period of operation due to installation damage or manufacturing defects. The failure rate should then remain very low until the operating life of the bags is reached, unless a unique failure mode is present within the system.

The importance of when to correct a broken bag will depend on the type of collector and the resultant effect on outlet emissions. In "inside bag collection" types of collectors, it is very important that dust leaks be stopped as quickly as possible to prevent adjacent bags from being abraded by jet streams of dust emitting from the broken bag. This is called the "domino effect" of bag failure.

"Outside bag collection" systems do not have this problem, and the speed of repair is determined by whether the outlet opacity has exceeded its limits. Often, it will take several broken bags to create an opacity problem and a convenient maintenance schedule can be employed instead of emergency maintenance.

In either type of collector, the location of the broken bag or bags has to be determined and corrective action taken. In a non-compartmentalized unit, this requires system shutdown and visual inspection. In inside collectors, bags often fail close to the bottoms, near the tube sheet. Look for accumulation of dust on the tube sheets, the holes themselves, or unusual dust patterns on the outside of the bags.

Other probable bag failure locations in reverse-air bags are near anti-collapse rings or below the top cuff. In shaker bags, also inspect the area below the top attachment.

Improper tensioning can cause early failure.

In outside collectors, which are normally top-access systems, an inspection of the bag itself is difficult; however, the location of the broken bag or bags can normally be found by looking for dust accumulation on top of the tube sheet, on the underside of the top-access door or on a blowpipe.

If the system is compartmentalized, the search for broken bags can be narrowed by monitoring the stack while isolating one known compartment at a time. Through an opacity monitor, and sometimes visual observation of the stack, the

Table 16.12 Troubleshooting guide.

Symptom	Possible Cause	Remedy
High collector pressure drop	Malfunction of bag-cleaning system	Check all cleaning system components
	Ineffective cleaning	Modify cleaning cycle, review with system designer
	Reentrainment of dust in collector due to low-density material or in-leakage at discharge	Check discharge valves Lower (if possible) air-to-cloth ratio
	Wetting of bags	Control dew-point excursions Dry bags with clean air Clean bags with vacuum or Wet wash Check insulation of ductwork leading to collector
	Too high A/C ratio either through added capacity or improper original design	Verify gas volume Reduce inlet volume if possible Review with designer
	Change in inlet leading or particle distribution	Test for inlet loading and particle size distribution Review with designer
Abnormally low-pressure drop	Manometer/pressure sensor line(s) plugged	Blow back through lines Protect sensing point from dust or water buildup
	Manometer/sensor line(s) broken or uncoupled	Verify with local/portable instrumentation Inspect and repair
	Overcleaning of bags	Reduce cleaning energy and/or cycle time
Stack emission	Broken bag	*See Bag Maintenance (Section 16.6.5)*
	Bag permeability increase	Test bag Check cleaning energy/cycle and reduce if possible
	Clean-to-dirty plenum leakage	Inspect and repair
	Change of inlet conditions	Test and review
Puffing	High pressure drop across baghouse	*See above*
	Low system fan speed	Check drive system Increase speed
	Improper duct balancing	Rebalance system
	Plugged duct lines	Clean out
	Poor hood design	Evaluate temporary modifications and implement
	Improper system fan damper position	Check and adjust
Low dust discharge	In-leakage at discharge points	Inspect and repair seals or valves
	Malfunction of discharge valve, screw conveyor, or material transfer equipment	Inspect and repair
	Reentrainment of dust within collector	Lower the A/C ratio
	Reentrainment of dust on filter bags	Increase cleaning
Loud or unusual noises	Vibrations	Inspect and repair
	Banging of moving parts	Inspect and repair
	Squealing of belt drives	Inspect and repair

(Continued)

Table 16.12 (Continued)

Symptom	Possible Cause	Remedy
Corrosion	Improper paint material or application Improper insulation Emission of nearby equipment Dew-point excursions Improper shutdowns	

compartment containing broken bags can be identified, since the emissions will be reduced when a compartment containing broken bags is isolated from the system. Once the compartment is identified, find the specific broken bag as detailed above.

One of the latest techniques for locating failed bags is the use of fluorescent or phosphorescent powder and an ultraviolet light. The powder is injected into the inlet gas stream and the ultraviolet light is used to scan the clean-air side of the collector. Very small leaks can be detected from the glow of the powder under the ultraviolet light where it has penetrated the clean air plenum.

Once a broken bag is found, it must be corrected. Individual bags within a compartment should not be replaced with clean bags, as they will be forced to filter a much higher volume of air than the older dust-laden bags, due to its lower-pressure drop. This higher filtering velocity could cause permanent "blinding" and high-pressure drop or early failure through dust abrasion. Instead, the broken bag should be tied off or plugged up. In inside collection bags where the failure has not occurred too close to the tube sheet, the bag should be cut, tied, and stuffed into its hole. For outside collection bags and inside collection bags with holes too close to tie, a hole plug should be used. The collector manufacturer can probably supply hole plugs or they can be made from steel plate with proper gasketing or heavy sandbags made from the same material as the filter bags.

Further action will be dictated by increased pressure drop. As the number of bags taken out of service rises, so will the pressure drop. At some point, the broken bags need to be replaced. When this occurs, it is recommended that all broken bags are replaced with used bags from one designated compartment. Change out all bags in the designated compartment with new bags and damper the flow into that compartment until a sufficient dust cake has been developed. Store extra "used" bags for future replacement.

When inspecting for plugging or changing out broken bags, care must be taken for the safety of the maintenance personnel. Protection from dust, temperature, and flue gas contaminants must be provided. This is especially important for inside collectors, where the maintenance personnel must enter the bag plenum.

Care should also be taken not to damage adjacent bags and to properly reinstall the new bag. This operation should not be rushed and a good working environment as possible should be provided. Errors here will only create more problems tomorrow.

16.7 Improving Operation and Performance

Comments in this section are directed toward improving the operation and performance of fabric-filter dust collection systems. It is assumed for the sake of this discussion that the user is at the point where the baghouse system has no major problems, and the dust collector is operating smoothly; however, there is a desire to improve operational reliability and reduce operating costs. As one orients to optimizing costs, it becomes clear that general cost rules can be very misleading because the gas volume capacity can vary greatly from one unit to another, and therefore operating, maintenance, and capital cost relationships can change dramatically. The approach to optimization can be uniform, however, and may include trading off additional capital investment or increased maintenance costs to reduce overall operating costs.

The initial cost objective when purchasing the system was to achieve a technically reliable system capable of meeting the outlet emission goals and achieving all of this at the minimum annualized cost.

The first and basic requirement for improved operation and operating costs is that of record keeping. Specifically, a daily log of any breakdowns, parts replacement, and operational problems needs to be kept. In addition, monitoring and recording of operational cost parameters is needed. For example, records of specific utility component costs need to be separately kept (e.g., main fan, reverse-air fan, and conveyors). Examples of their cost parameters are shown in Table 16.13.

After about six months of operation, these data should be reviewed and operational changes made based upon minimum operating cost considerations. It should be noted that data obtained in the early weeks may not be representative

Table 16.13 Installed and operating cost components.

Installed cost components
Design
FOB baghouse
Fans and motors
Ducting
Dust conveyors
Instrumentation
Freight
Foundations
Structural steel
Insulation
Installation labor
Start-up
Operating cost components
Electric power (fan motors, conveyor drives, compressors, lights, instrumentation)
Auxiliary heat
Bag and cage replacements
Maintenance
Dust disposal
Overhead

of long-term operational data, and therefore one should evaluate where the data appear to level out and more weight should be given to these data.

Analysis of the data obtained falls into two categories, the first being a review of any recurring operational problems and/or component failures. If this review identified any problems of a repetitious nature, corrective actions need to be defined and executed. These remedial actions can include component changes, modifications, additions, relocations, increased or modified preventive maintenance procedures, and additional monitoring equipment and controls. The second area of data review is a cost study to determine whether there are hardware or operational mode changes that can result in operation-cost savings. This analysis should be done in a rigorous fashion and probably put in graphical form. Points to be considered here are discussed below.

Can the system be run at a higher G/C ratio and allow for increased gas volume, and therefore increased production or the addition of another source operation?

Can the cleaning system be modified to allow for either pressure-drop reduction or longer bag life? Bag cost versus utility costs need to be weighed here. It is at this point that the user recognizes that the original selection of the *G*/*C* and cleaning cycle was purely an empirical selection and now, based on the actual operation, this can be modified to arrive at a lower-cost operation. Bag types and finishes are evolving rapidly in the current market, and it may be worthwhile to initiate one's own bag screening program. This can be done by installing one or more test bags and monitoring their performance. Such a program may be simply oriented to monitoring bag life or if conditions allow, provide pressure drop and emission data as well. In cases of very poor bag life (i.e., less than I year), excessive pressure drops (e.g., greater than 6–8 in. water gauge), or very costly bags, it is recommended that a bag screening program be instituted.

16.8 Recent Advances

A review of the literature of the past quarter century has produced little in terms of improvement regarding the state of the art for fabric-filtration systems. The following can be offered at this point in time;

1) Longer bag lengths for pulse-jet filters – Better pulsing systems and off-line cleaning modes have enabled manufacturers of pulse-jet baghouses to be able to offer bags that are 8–10 meters in length. This allows for, with the same air-to-cloth (A/C) ratio, a smaller footprint for the baghouse.

2) Sonic cleaning systems – The use of sonic horns, in both pulse-jet and reverse-air baghouses has been found to be useful in enhancing bag cleaning. This has been especially useful when trying to remove hard-to-dislodge dust layers from a bag. This is very similar to, and an outgrowth from, the sonic horns used in the electric utility industry for hard-to-remove deposits on boiler tubes, and for fly ash removal/cleaning from Selective Catalytic Removal (SCR) systems for NO_x emissions control.[9]
3) Polytetrafluoroethylene (PTFE) membrane – The use of a PTFE membrane in a bag can aid through its "non-stick" ability to help with cake release of bags during cleaning.
4) Very high-temperature bags – Some filter bag manufacturers can now offer woven filter bags fabricated from silica-boria fibers. This unusual fiber results in a filter bag capable of continuous operation at 1,000°F with surges up to 1,250°F.
5) Activated carbon-impregnated filters – Although not an advance in particle removal technology the impregnation of activated carbon particles into filter bag fabrics has enabled fabric filters using this bag material to also reduce mercury emissions along with reductions in particulate emissions. The activated carbon is effective in the reduction of gaseous and ionic mercury emissions and has been tested by the US electric
6) Pleated filter bags – The new bags achieve a filtration efficiency as high as cartridge-style collectors, without having to modify the baghouse. Replacing standard filter bags with pleated filter bags will increase the filtration area of the baghouse and result in a lower air-to-cloth ratio for that baghouse.

16.9 Conclusions

One final note is directed toward the total production process. A review of operating cost will indicate that utilities are of ever-growing concern and the dust collection system fan is usually the major utility component of the dust collection system costs. Reduction of the system fan operating cost requires reduction of the system pressure drop. This can be accomplished by optimization of the duct run design, inlet and outlet damper designs, the bag cleaning system, filter media type, or reduction of the gas volume processed. Often, the fan speed can be adjusted to an improved fan curve operating point. It is important that constant vigilance be maintained to ensure that the process volume is no greater than required (e.g., boiler excess air control) and that process gas inleaks arc eliminated.

The manufacturer's operating manual is at best only a guide to proper operation. The optimization of operation and minimization of operating costs are left in the user's hands. Awareness and sensitivity to the operational cost parameters and a semiannual review of the operational data collected are the route to improved operation, performance, and costs.

References

1 **L. Theodore**, *Air Pollution Control Equipment Calculations*, John Wiley and Sons, Hoboken, NJ, 2008.
2 **C.E. Billings, and J. Wilder**, *Fabric Filter Systems Study – Handbook of Fabric Filter Technology*" Volume 2, US Dept. of Health, Education and Welfare, Washington, DC, 1970. App. 5.1.
3 **R. Denis, et al.**, "*Filtration Model for Coal Fly Ash with Glass Fabrics*". EPA-600/7-77-084, Research Triangle Park, NC, 1977.
4 **J. Nowak, M. Mangone, J. Reynolds, and L. Theodore**, *A Theoretical Model to Predict Pressure Drop for Flow Through Fabric Filters*, Proceedings of the Fourth Natl. Conference on Energy and the Environment, Cincinnati, Ohio, 141–150 1976.
5 **L. Theodore, and J. Reynolds**, Effect of bag failure on baghouse outlet loading, *J. Air Poll. Control Assoc.*, 29(8), 870–872, 1979.
6 **J. Reynolds, and L. Theodore**, Analysis of an APCA baghouse operation and maintenance survey, *J. Air Poll. Control Assoc.*, 30, 1218–1223, 1980.
7 **L. Theodore**, *Engineering Calculations: Baghouse Specification and Operation, Simplified*, pp. 22–23, CEP, 2006.
8 **L. Theodore**, Prediction of particulate loading in exhausts from fabric filter baghouses with one or more failed bags, *J. Air & Was. Manag. Assoc.*, Pittsburgh, October, 2006. Letter to the Editor (author response)
9 **L. Theodore**, *Personal Notes*, East Williston, NY 2007.

17

Hybrid Systems

Sean Dooley

Dual Resource Partners, Woodstock, IL, USA

The basic design of air pollution control equipment has remained relatively unchanged since first used in the early part of the twentieth century. Some modest equipment changes and new types of devices have appeared in the first 20 years of this century, but all have essentially employed the same capture mechanism of the past.

One area that has recently received some attention is hybrid systems equipment that can in some cases operate at higher efficiencies more economically than conventional devices. Tighter regulations and a greater concern for environmental control by society have placed increased emphasis on the development and application of these systems.

This chapter is concerned with hybrid systems, as they apply to air pollution control equipment. Unfortunately, the term hybrid system has come to mean different things to different people. The two most prevalent definitions employed today for hybrid systems are:

1) Two or more different air pollution control devices connected in series, e.g., baghouse followed by an absorber.
2) An air pollution control unit that utilizes two or more collection mechanisms to enhance pollutant capture.

With regard to item (1), equipment connected in series are usually treated, from an engineering and calculational point of view, sequentially, with the outlet of one device becoming the inlet of the next control device. As such, this topic does not require detailed additional treatment other than that provided earlier in this book. It should be noted that this approach has found application in the treatment of metal particulates. A dry scrubber followed by a wet scrubber has been employed in this country to improve the collection of fine particulate metals in hazardous-waste incinerators; the dry scrubber captures metal that condenses at the operating temperature of the unit, and the wet scrubber captures residual metals (particularly mercury) and dioxins/furan compounds. Another recent application in Europe involves the injection of powdered activated carbon into a flue gas stream from a hazardous-waste incinerator at a location between the spray drier (dry scrubber) and the baghouse (or electrostatic precipitator [ESP]). The powdered carbon mixing with the lime particles from the dry-scrubbing system and the gas stream itself adsorb the mercury vapors and residual dioxins/furnace compounds and then are separated from the gas stream by a particulate control device.

This chapter will focus on item (2). Although there are numerous hybrid systems available in the marketplace, the discussion that follows will examine only the following four control devices:

1) Dry scrubbers – DRYS
2) Ionizing wet scrubbers – IWS
3) Wet ESPs – WEP
4) Electrostatic stimulation of fabric filtration – ESFF

These have been chosen because of their widespread use and/or potential application in the future, or both. Tighter regulations and a greater concern for environmental control by society have placed increased emphasis on the development and application of these systems.

Each system above is treated separately in individual sections and includes (where applicable) a brief introduction, a description of the system, some operational considerations, and applications. Some of the material has been drawn from company literature, reports, and/or bulletins. (These have been acknowledged in the reference section).

Optimizing Air Pollution Control Equipment Performance: Operation & Maintenance, First Edition. Jay Richardson and Louis Theodore.

17.1 Dry Scrubbers

As environmental concerns continue to become more mainstream in everyday life, new emissions regulations have become tighter. One of the most studied and publicized effects of industrial pollution is that of climate change and acid rain. Hydrochloric acid (HCl), sulfuric acid (H_2SO_4), and sulfur dioxide (SO_2) are typical byproducts of common combustion processes and are major contributors to climate change and acid rain. To operate responsibly and within environmental compliance, emissions of HCl and SO_2 need to be controlled and monitored. Since the 1990s a common method of control has been to chemically react pollutant acid gases with an alkaline material to generate a relatively benign waste product; this process is known as dry-scrubbing or dry sorbent injection (DSI).

The DSI industry has expanded to include other pollutant treatment targets such as hydrofluoric acid (HF), stack opacity improvement, particulate matter (PM) collection efficiency increases (on ESP applications), and metal leachate stabilization for process byproducts. Dry scrubbing and in particular DSI is now used as a complete solution to solve many issues arising from post-combustion emissions and byproducts, rather than a complement to other scrubbing technologies in the past. When DSI is applied (with proper particulate collection devices like a baghouse, cyclone, or ESP) the resulting air pollution control system becomes robust and flexible in terms of pollution capture efficiencies.

DSI processes involve pneumatic introduction (injection) of a dry, powdery alkaline material (usually a calcium or sodium-based sorbent) into the flue gas stream with subsequent fabric-filter (or other PM control equipment) collection downstream of the injection point. The injection point in such processes can vary from the boiler/furnace section all the way to the flue gas entrance at the particulate collection device. The injection points are determined by process operating conditions and design criteria.

The method of operation of dry scrubbing via DSI is relatively simple, requiring only two major equipment items; (i) a pneumatic conveyor and storage system (the DSI system), and (ii) a particulate control system (baghouse, cyclone, or ESP) to collect the fly ash or entrained flue gas PM and the sorbent PM. Baghouses have become most common because baghouses are generally designed to "cake" or generate a layer of sorbent material on their bags to further increase sorbent interaction with pollutants. The solid-bearing gas passes through the particulate control devices where the dry product is collected, and where a percentage of the unreacted alkali reacts with acid gases for further removal. Usually, the cleaned gas is then discharged through the particulate control device's plenum to an induced draft (ID) fan and the stack.

The success of fabric filters in removing fine particles from flue gas streams has encouraged the use of combined dry-scrubbing/fabric-filter systems for the dual purpose of removing both particulates and acid gases simultaneously. Dry scrubber processes such as DSI offer potential advantages over their wet counterparts, especially in the areas of energy savings and capital costs; a dry system can cost up to 90% more than a wet system with comparable performance. Furthermore, the dry-scrubbing process design is relatively simple for construction and operating purposes in the future, especially in existing facilities. These dry-scrubbing systems generate a solid byproduct waste at the PM control locations. Waste removal and disposal is often an overlooked, yet expensive and critical component of these air pollution control designs. A dry scrubber generates a dry powder waste product that is more easily and economically disposed of by the users, rather than the more difficult and expensive wet sludge of a wet system.

Among the inherent advantages that the DSI enjoys over wet scrubbers are

1) Lower capital cost
2) Lower draft losses
3) Reduced auxiliary power
4) Reduced water consumption
5) Lower waste disposal cost

DSI is evolving from its roots, first commercially initiated in the coal-fired power utility market in the 1990s for sulfur dioxide and sulfuric acid treatment. The application is now finding use in new markets such as hydrogen production, glass, cement, and refining or any other industrial application where acid gas or metal pollutants are generated and in need of capture.

DSI can take on numerous different applications with varying different sorbent supplies. DSI can utilize the following different injection materials for distinct types of pollution control or remediation:

1) Hydrated lime $Ca(OH)_2$
2) Sodium bicarbonate $NaHCO_3$
3) Sodium sesquicarbonate $Na_3H(CO_3)_2$

4) Activated carbon C
5) Magnesium hydroxide $Mg(OH)_2$
6) Various blends of the above materials

Sorbent injection systems now are not only used to treat flue gas acid pollutants but also to treat process byproducts. In many cases, the process byproducts or PM that are generated from the process are metals that exceed local and federal guidelines for traditional waste classification and need to be disposed of by hazardous-waste guidelines. In a patented process facilities and users are instituting specialized sorbent injection methods solely for the treatment of the waste byproduct solids within the flue gas stream. In a sense, these DSI systems are preventing the formation of hazardous waste within the process. There can be many combinations of these hazardous-waste stabilizers and DSI configurations used on the same process and treatment system. Dual Resource Partners, an air pollution control technology provider outside of Chicago, IL initiated the first hazardous metals waste stabilization program in 2023[1]. The DSI system and methodology transformed a previously hazardous-waste-generating process into a stable byproduct solution, meeting nonhazardous waste criterion.

17.1.1 DSI Equipment Pneumatic Convey Supply and Design

Preserving reliable operation of the sorbent injection equipment involves routine maintenance and anticipation in the initial design. Sorbents can be erosive and reactive products that can plug lines and limit or even cease injection if not properly stored and conveyed. Appropriate equipment selection is crucial to the system's performance. The overall goal of the systems is to capture as much pollutants as possible, so systems must be robust enough to convey the material – often 24 hours a day, 7 days a week, 365 days a year, and flexible enough to adapt to process changes in their respective facilities. Because these pneumatic convey systems are moving a reactive/unstable product (the sorbent), the convey air and convey routing must be engineered to be considerate of the unique conditions required of that sorbent.

DSI systems are often designed with silos or other storage devices that use conditioned air to ensure the product or sorbent is not reacting with ambient air and conditions. Convey systems will often utilize air conditioning pieces of equipment like heat exchangers and dehumidifiers to make sure the convey air used as the motive force to move the sorbent from the storage to the injection points is an environment that will not initiate a reaction of the sorbent within the convey line. Routing the convey lines to minimize convey distances and turns or "elbows" will also reduce the friction or kinetic energy placed on the sorbent. The less energy transferred to the sorbent, the less chance of reaction and buildup in the convey lines. Managing these pneumatic conveying principles can optimize the flow of the material while also avoiding costly and difficult-to-remediate plugging within the system.

17.1.2 DSI Chemical Interaction

Residence time, dispersion, and distribution of the sorbent are the biggest factors in a DSI system's overall performance and ability to capture pollutants efficiently. Residence time allows the sorbent material to mix, react, and capture their targeted pollutants. Short residence times will not allow the sorbent to initiate its chemical reaction thoroughly and effectively with the targeted pollutant. Distribution and dispersion are important criteria to ensure there is as close to 100% duct or flue gas coverage by the sorbent material. The more sufficient dispersion will yield the most improved results toward getting targeted pollutants to interact with that sorbent solution.

Traditionally the sorbent enters the flue gas duct from a pipe lance connected to the convey lines that travels from the main storage section, usually a silo or a bulk bag system. The lances are arranged in a way to get maximum duct coverage and to encourage interaction between the flue gas and the sorbent. As of 2010, sorbent injection systems have been retrofitted or designed without traditional pipe lances, opting instead to use nozzles or lanceless injectors to achieve better dispersion with less plugging. The lanceless systems provide better sorbent dispersion through the motive force of the conveying air that generates increased duct coverage. Lanceless systems are attached outside of the duct and are not exposed to harsh temperatures or corrosives, which known to be the root cause of plugs in the injectors. Lanceless systems have higher uptime and reliability than their lance predecessors, because of the lower likelihood of plugging.

A reliable DSI system provides important pollution control to meet imposed emission limits while not inhibiting the facility's production process. The progress of DSI has led to a more achievable cleaner environment.

17.1.3 DSI Expansion

There are new applications within the carbon sequestration and capture market that will need the dry sorbent application for improving their systems. Carbon capture systems utilize amine solutions to strip or remove carbon dioxide from the gas stream. Amine solutions are usually enormously expensive proprietary products that have been created over vast amounts of research and funding. Amine solutions are incredibly sensitive to acid gases like sulfur dioxide, prolonged exposure to acid gases will destroy amines that could otherwise be recycled and reused for further carbon dioxide stripping. The sensitivity to sulfur dioxide necessitates the use of intense and powerful dry scrubbing like a DSI system to remove the acid gases prior to the introduction of the amines.

As the world continues to expand pollution control demands, DSI flexibility will continue to be utilized as a part of pollution control systems for future processes.

17.2 Ionizing Wet Scrubber (IWS)

The term wet scrubber was first used by the Ceilcote Co., located in Berea, OH. The presentation to follow will key on the Seilcote system (Patent No.'s: 3,874,858 and 3,958,958), with information provided by the vendor.[2–3] This system has found wide application in the air pollution control field, particularly in controlling discharges from hazardous-waste incinerators. A short description of MikroPul's short-lived Elektrofil ESP is also included.

The Ceilcote IWS system is a proven means for the removal of pollutant from industrial process gas streams. The IWS combines the established principles of electrostatic particle charging, image force attraction, inertial impaction, and gas absorption to simultaneously collect submicron solid particles, liquid particles, and noxious and malodorous gases. The IWS system requires little energy and its collection efficiency is high for both submicron and micron-size particles.

The IWS utilizes high-voltage ionization to electrostatically charge particulates in the gas stream before the particles enter a Tellerette-packed scrubber section where they are removed by attraction of the charged particles to neutral surfaces. Larger particles of 3–5 μm and larger are collected through inertial impaction. As small particles flow through the scrubber, they pass close to the surface of the Tellerettes and scrubbing liquid droplets. The electrostatic charge on the particles causes them to be attracted to these neutral surfaces by image force attraction. All particles are eventually washed out of the scrubber with the exit liquor. Noxious and malodorous gases are absorbed and reacted in the same scrubbing liquor.

The IWS system utilizes Tellerette packing as its collection surface to achieve particulate removal. Scrubbing liquor droplets also act as collection surfaces. Particles of any size or composition are collected by the IWS. Fine particles (0.05–2 μm) are collected with high efficiency as well as coarse particles (2 μm or larger), regardless of their composition (organic or inorganic, with either high or low resistivity). Particle collection efficiency over long-term service remains consistently high.

The IWS system acts as a fractional collector. The percent of particulate removed varies little with load and particle size distribution over a wide range. As the particulate load increases, the percent removal remains nearly constant. The collection efficiency for fine particles is nearly as great as for coarse particles. The IWS system also simultaneously absorbs gases. Noxious gases are removed through physical absorption and/or absorption that is accompanied by a chemical reaction. The pressure drop through a single-stage IWS is only 0.5–1.5 in H_2O. The energy for particle charging is low – approximately 0.2–0.4 kVA/1,000 CFM. The shell and most internal parts of the IWS are commonly fabricated of Duracor (fiberglass-reinforced plastic) and thermoplastic materials. The predominance of plastic construction assures corrosion-free operation in the presence of acid gases such as HCl, HF, Cl NH_3, SO_2, and SO_3. For noncorrosive applications, metallic construction is also available. Factory-assembled modules, available in standard capacities from 900 ACFM to 54,000 ACFM, can also be grouped together to handle virtually any gas volume.

Regarding IWS system components, particles are first electrostatically charged within an ionizer section that utilizes a high-voltage DC power source. Discharge electrodes have negative polarity and wetted "mini-plates" serve as grounded electrodes. High-voltage DC power for the ionizing section is provided by a high-voltage transformer/rectifier. This is a sealed, watertight, oil-filled unit. Rectification is accomplished by silicon diodes which are suitably protected against transient voltages for efficient and long-term, trouble-free service. All control and remaining power functions for the IWS are combined in a single, standard, factory-prewired control cabinet which can be located at any convenient location, either adjacent to or remotely located from the IWS. Field hook-up is simple. The control cabinet is connected to the high-voltage transformer/rectifier by five leads attached to cabinet terminals. The maximum voltage on the interconnecting wiring is 480 V. Voltage control for the high voltage is achieved by thyristor control through phase-angle variation. An ultrafast response, solid state, plug-in type standard control module automatically controls high-voltage levels and spark rate for optimum performance. A manual control is also provided for backup and troubleshooting.

Operation is simple and straightforward. All functions are automatic. Necessary meters and indicator lights are provided to monitor the operation. Protective interlocks and circuits safeguard the equipment against overloads or malfunctions. The grounded electrode plate surfaces of the IWS are continuously flushed by a flow of water or other suitable liquid. This assures that optimum electrical performance is maintained at all times by washing away any resistive solids or residues that could collect on the "mini-plates" and cause reduced performance. A second flushing system is also provided which periodically deluge flushes all internal areas of the ionizer section. The deluge flushing is completely automatic and is set to provide a flushing cycle every eight hours or as experience dictates. Water for the periodic deluge flush is recycled liquid from the scrubber recycle system. The IWS incorporates a cross-flow scrubber section following the ionization section. The scrubber section employs Tellerette packing to absorb gases and collect both liquid and solid particles from the gas stream. The contaminated gas stream moves horizontally through a Tellerette packing bed that is irrigated by scrubbing liquid flowing vertically down (creating cross-flow contact) through the packing.

The inherent advantages of the cross-flow design result in high operating reliability, high airflow rate, low-pressure drop, and low operating cost. The IWS contains an integral sump to recirculate the scrubbing liquid directly after it passes through the packed-bed section. Use of an integral sump reduces cost and minimizes floor space requirements. Pump(s) and complete recirculation piping for the IWS are normally provided for continuous wetting of the ionizer plates, deluge flushing of the ionizer section, and continuous irrigation of the cross-flow scrubber. Alternative design approaches can be adapted for use in the IWS liquid distribution system. The method employed is contingent upon the customer's requirements and job site restrictions.

As described earlier, the IWS shell and most internal parts are normally fabricated of Duracor fiberglass plastic (FRP) and other polymeric materials. The preponderance of plastic as a material of construction assures corrosion-free operation in the presence of gases such as HCl, HF, NH_3, SO_2, and SO_3. Additionally, structural support requirements are minimized by the unit's light weight. Extensive use of proven, long-service life plastics throughout the IWS is possible because of its unique design, construction, and operating principles. The only metal required is for the electrical components employed within the ionizing section. These metallic components are fabricated of alloy materials chosen for each application to provide maximum corrosion resistance in the service environment. The IWS can also be constructed of metallic components where corrosive conditions are not a problem and the use of FRP materials is not economical or practical in a specific application.

There are several important operating characteristics for an IWS. Results from earlier pilot tests and more than 50 actual full-size operating systems verify that particles of virtually any size and composition are collected by the IWS. Submicron solid/liquid particles as minute as 0.05 μm, coarse particles having organic or inorganic structures, and conductive or highly resistive particles are collected at high efficiency.

The predominant influence acting within the IWS to collect submicron-size particulate is the mechanism of "image force attraction." Image force attraction occurs whenever an electrostatically charged particle comes within the boundary layer (usually within 1 mm) of a neutral surface. As the charged particles come close to the neutral surface, an electrostatic charge of opposite polarity is induced at the neutral surface. The force of attraction is similar to that which exists between a charged particle and an imaginary particle of equal and opposite polarity located at an equal distance behind the surface. It is as though the charged particle was seeing its "mirror image."

Particles of 3–5 μm and larger are collected by inertial impaction. Particles impinge on a Tellerette packing surface, are captured, and flushed away. Tellerette packing is used exclusively in the IWS. The unique geometrical design configuration of the Tellerette makes it an excellent impingement target. Highly efficient collection is due to the large number of small target areas. Greater particulate collection efficiency, higher gas absorption efficiency, and greater flow capacities are all characteristics of Tellerette packing, as opposed to extended surface packing. The IWS collection efficiency in the fine particle range decreases only slightly as the particles become smaller. In conventional ESPs collection efficiency drops in the submicron area, especially in the 0.2–0.6 μm range. For cloth collectors, collection efficiency also decreases sharply when particles are 0.5 μm and smaller. The IWS is a fractional collector. A single-stage IWS unit will remove a nearly constant percentage of incoming particles regardless of particle size distribution. Where a higher collection efficiency is required than is possible with a single-stage IWS, a second IWS stage can be used. The percent collection efficiency of the second-stage IWS will be approximately the same as the first-stage IWS despite the reduction in loading. The IWS can remove the light loadings of submicron particulates that cause a "blue haze" emission. Generally, an emission having an opacity of 20% (Ringelmann 1)[4–6] or less is acceptable. The IWS can be designed to obtain opacities of 5% or less to give clear or nearly clear stack discharges.

Removal of gaseous contaminants in the scrubber is achieved by the principle of mass transfer whereby gaseous molecules from the airstream are transferred to the scrubbing liquid. Transfer is achieved by a combination of diffusion, physical absorption, and/or chemical reaction. Gaseous collection efficiency is directly dependent upon the irrigation rate, the horizontal depth of packing, and the gas velocity through the scrubber. Depending on the depth of the bed, the liquid flow rate, and composition of the scrubbing liquid, efficiencies up to 99.9% gas absorption can be obtained. In an application where two or more gaseous contaminants are to be removed, it is often desirable to scrub with chemically different liquids. This can readily be achieved with cross-flow scrubber geometry, which easily accommodates two or more separate scrubbing beds.

Pressure drop through a single IWS is only 0.5–1.4 in H_2O and is controlled primarily by the pressure drop through the wet scrubber section. Additional small amounts of energy are required for high-voltage charging and for the recycle pump. Total system energy usage is approximately 2.0–2.5 Bhp/1,000 ACFM for a single-stage IWS and 4.0–5.0 Bhp/1,000 ACFM for a two-stage IWS installation.

The energy-efficient IW can be applied to a wide variety of air pollution control problems. It offers the greatest advantages in applications requiring the collection of submicron particulate, mist, and gases, especially where corrosive conditions exist. The Ceilcote Company has designed, constructed, and installed numerous IWSs. They can provide a listing of these applications.[2–3]

MikroPul's Elektrofil ESP utilizes a continuous flow of finely divided liquid droplets, which become electrically charged and are attracted to the collecting plates. The precipitated water droplets create an evenly distributed water film over the entire collector surface. The solid particulates are charged and simultaneously migrate to the collecting surfaces and are captured by the continuously downward-moving film of water. The water film on each of the collecting plates and discharge electrodes flows by gravity into the slurry trough for disposal.

The operation of the Elektrofil system is somewhat unique. The dirty gas stream is preconditioned to achieve saturation. It then enters the inlet and loses much of its velocity in the inlet expansion chamber. Strategically located spray nozzles in the inlet section ensure complete saturation and maintain a steady water flow down the housing sides and baffles to remove any particulates collected by impingement. The water sprays in the inlet expansion chamber help break up the airstream, and transverse baffles, double- or triple-staggered rows of channels, and assure even distribution of the gas flow over the entire cross-section of the precipitator. After passing through the transverse baffles, the dirty gas stream then passes into the electrostatic field. Water sprays located in front of and behind the transverse baffles, and above the electrostatic field section, introduce the proper amount of water droplets to the gas stream. The solid particulates and the water droplets in the electrostatic field pick up a charge and migrate to the collecting plates or to the discharge electrodes. The water droplets form a continuous downward-moving film over all the collecting surfaces, and help to entrap the droplets and particulates as they migrate to the electrostatic elements. The water film and collected particulates slowly flow down the collecting elements and fall to the discharge trough below. The collected slurry is then withdrawn through a drain and piped to a treatment system.[7]

Where efficiency specifications require the use of more than one field, a row of transverse baffles between each field redistributes the gas flow. Each field has its own spray system which introduces the proper amount of water to assure effective cleaning. After passing through all the electrostatic fields, the gas stream again is subjected to another set of transverse baffles where any water droplets or dust particles not deposited on the plates are eliminated. All baffle systems are arranged so that a walkway runs across the front or back of each of the electrostatic fields. The discharge electrode frames are mounted on collar-type high-voltage support insulators. Insulator compartments are heated and pressurized to prevent moisture and particulate leakage.

17.3 Wet Electrostatic Precipitators (WEPs)

The two major types of high-voltage single-stage ESP configurations are plate and tubular. As the term single stage implies, particles are both charged and collected in a single stage. The plate type configuration was examined in detail in Chapter 15. This section examines the wet ESP – a unit that has a tubular configuration and is normally operated wet. This unit has appropriately been called a wet ESP. It should be noted that its use has declined in recent years. In addition, plate ESPs are used more often than tubular ESPs in industrial applications.

Tubular precipitators consist of cylindrical collection electrodes with discharge electrodes located in the center of the cylinder. Dirty gas flows into the cylinder where precipitation occurs. The negatively charged particles migrate to and are collected on grounded collecting tubes. The collected dust or liquid is removed by washing the tubes with water sprays located directly above the tubes. These precipitators have also been referred to as water-walled ESPs. Tubular precipitators are generally used for collecting mists or fogs. Tube diameters typically vary from 0.5 ft to 1 ft (0.15–0.31 m), with lengths ranging from 6 ft to 15 ft (1.85–4.6 m). The gas to be cleaned flows around the outside of the cylindrical electrodes and up through the inside of the cylinders where precipitation takes place. The collected dust or liquid is removed from the bottom of the chamber. On occasion, hexagonal tubes (enabling closer packing) may be used to increase the amount of available collecting surface in a given volume, a problem treated in the literature.[8]

17.4 Electrostatic Stimulation of Fabric Filtration

The use of fabric filters to remove particles from gas streams is a well-established industrial practice and its importance is growing for electrical utility boilers. The fabric filter is roughly equivalent in price to other utility control devices and provides increased particle removal efficiency. However, fabric-filter operating experience is limited, and enough questions have been raised because of unexpectedly high baghouse pressure drops to cause concern in the industry.

A concept to improve fabric filtration through the use of electric fields (ESFF) was developed on a laboratory scale by researchers at the Textile Research Institute (TRI), Princeton, NJ, whose work was partially supported by EPA.[9–10] TRI showed that this concept could significantly improve filter performance by reducing filter drag and increasing filter efficiency. In work supported by EPA, the Research Triangle Institute (RTI), Research Triangle Park, NC. and ETS, Inc., Roanoke, VA transferred the ESFF technology to the field environment by successfully operating a pulse-cleaned ESFF baghouse on a slipstream from an industrial boiler.[11]

The ESFF effect can best be understood as the net effect of fundamental forces on particles – either charged or uncharged – in the presence of an electric field. While only one primary electric field is applied to the filter, that field interacts with the filter fibers and collected particles to form many localized gradients in the electric field. A particle approaching the filter is thus exposed to a complex field before it is collected. The electric field continues to have an effect on the particle after it is collected.

There are several changes in the pattern of dust deposition on a filter that might account for the reduced pressure drop effects observed with ESFF. These include:

1) An increased fraction of the collected dust being deposited on or near the upstream filter surface.
2) Changes in the pattern of collection on a filter fiber as more dendritic collection and bridging occurs.
3) Formation of a highly nonuniform dust deposit – at the scale of the electrode spacing – on the filter surface.

An increased fraction of the dust being deposited near the upstream surface of the filter has been observed in laboratory studies of ESFF. This observation is important because the upstream side of a fabric filter has a lower packing density than the body of the fabric. Collection in the less dense surface region, which is enhanced by the electric field, leads to reduced pressure drop for a given quantity of dust and reduces the amount of dust that penetrates the fabric. Reduced penetration into the depth of the fabric accounts for the reduced residual pressure drop observed.

Changes in the pattern of dust deposition on fibers have been observed by a number of researchers. In general, how the presence of an electric field leads to increased dendritic deposition on filter drag is not clear. If the dendrites bridge between fibers and pack tightly, the resistance should increase; if they remain in an open structure, the flow resistance should be reduced. Visual observation within a filter is difficult, and the effects of particle deposition on fibers remain uncertain.

If electrostatic forces cause the formation of a dust deposit with a nonuniform areal mass density, a reduction in pressure drop should result. This will be true, for instance, if coulombic forces cause most of the dust to collect in the immediate region of the electrodes, leaving reduced deposits on the balance of the filter.[10] Highly nonuniform deposits have been observed with redispersed fly ash in a laboratory ESFF filter apparatus.[11]

Stabilization of the deposited dust layer by the electric field has also been studied as a possible ESFF mechanism. The theory was developed in conjunction with fluidized beds and adhesion forces within ESP dust layers, but its application to the ESFF situation is fairly direct. It is postulated that the initial dust cake formation in an ESFF filter is quite porous due to the electrical gradient forces, and that the electric field prevents the collapse of this porous cake as the pressure drop increases. The electrical stabilization preserves the initial high porosity, resulting in prolonged operation at lower drag than in the more compressible conventional operation.

17.5 Recent Advances in Control Equipment Technology

Recent advances in this field have been primarily involved in the treatment and control of metals. A dry scrubber followed by a wet scrubber has been employed in the United States to improve the collection of fine particulate metals in hazardous-waste incinerators; the dry scrubber captures metals that condense at the operating temperature of the unit and the wet scrubber captures residue metals (particularly mercury) and dioxin/furan compounds. Wastes from the wet scrubber are recycled into the dry scrubber resulting in a no-liquid discharge system.

Another recent application in Europe and the United States involves the injection of powdered activated carbon into a flue gas stream from a hazardous-waste incinerator at a location between the spray dryer (the dry scrubber) and the baghouse (or ESP). The carbon, mixing with the lime particulates from the dry-scrubbing system and the gas stream itself, adsorbs the mercury vapors and residual dioxin/furan compounds and is separated from the gas stream by a particulate control device.

17.6 Conclusion

The two most prevalent definitions employed today for hybrid systems are (i) two or more different air pollution control devices connected in series, e.g., a baghouse followed by an absorber, and (ii) an air pollution control unit that utilizes two or more collection mechanisms to enhance pollutant capture.

There are two major types of so-called dry scrubber systems: spray drying and dry injection. The first process is often referred to as a wet-dry system. When compared to a conventional wet scrubber, it uses significantly less scrubbing liquid. The second process may be referred to as a dry-dry system since no-liquid scrubbing. is involved. The spray drying system is predominantly used in utility and industrial applications.

The IWS incorporates a cross-flow scrubber section following an ionization section. The scrubber section employs Tellerette packing to absorb gases and collect both liquid and solid particles from the gas stream.

Wet tubular ESPs consist of cylindrical collection electrodes with discharge electrodes located in the center of the cylinder. Dirty gas flows into the cylinder where precipitation occurs. The negatively charged particles migrate to and are collected on grounded collecting tubes.

A concept to improve fabric filtration through the use of electric fields (ESFF) was developed. The ESFF effect can best be understood as the net effect of fundamental forces on particles – either charged or uncharged – in the presence of an electric field. While only one primary electric field is applied to the filter, that field interacts with the filter fibers and collected particles to form many localized gradients in the electric field.

References

1 **S. Dooley**, *Personal Notes, Dual Resource Partners*, Chicago, IL, 2023.
2 Ceilcote Co., *Bulletin 12–19, IWS System*, Berea, Ohio, not dated.
3 Ceilcote Co., *Private Communications*, J. MacDonald, Berea, OH, not dated.
4 **L. Theodore**, *Air Pollution Control Equipment Calculations*, John Wiley & Sons, Hoboken, NJ, 2008.
5 **L. Theodore, and R. Allen**, *Air Pollution Control Equipment, A Theodore Tutorial*, Theodore Tutorials, East Williston, NY, 1993.
6 **J. Mycock, J. McKenna, and L. Theodore**, *Air Pollution Control Engineering and Technology*, CRC Press/Taylor & Francis Group, Boca Raton, FL, 2010.
7 MikroPul Bulletin, *The Elektrofil Electrostatic Precipitator*, NJ, 1972.
8 **J. Louie Kovach**, *Private Communication: Dr. J. Louie Kovach, President and CeEO, Nucon*, Columbus, OH, 1980.
9 **D. Van Osdell and D. Furlong**, *Electrostatic Augmentation of Fabric Filtration*. USEPA Report No. 68-02-3186, E.S. EPA, Research Triangle Park, NC, 1986.
10 **G. Lamb, and P. Costanza**, A low-energy electrified filter system, *Filtr. Sep.*, 17, 319–322, 1980.
11 **C. Wang and C. Ho**, *Inertial and Electrostatic Deposition of Solid Particles on Isolated Fibers*. Invited paper presented to the Fiber Society and the Filtration Society Joint Symp. Fibers, Electrostatics, and Filtration, Princeton, NJ, November 14–15, 1979.

18

Controlling the Oxides of Nitrogen

The oxides of nitrogen consist of the elements nitrogen and oxygen, both of which are briefly described below.

1) *Oxygen*. Oxygen is a gaseous element with the symbol O and is located in group VI of the periodic table, with atomic number 8, and atomic weight of 16. Oxygen is the most abundant of all of the elements on planet earth. It represents 21% by volume (23% by mass) of the atmosphere; 85.8% of the oceans; and as a constituent of most rocks and minerals, 46.7% of the solid crust of the earth. Oxygen comprises 60% of the human body. It is a constituent of all living tissues; almost all plants and animals, including all humans, require oxygen in the free or combined state to maintain life.

   Oxygen is a colorless, odorless, and tasteless nontoxic gas. It can be transformed into a pale blue liquid by compression. Three allotropic forms of oxygen are known to exist: ordinary oxygen containing two atoms per molecule (O_2), ozone containing three atoms per molecule (O_3), and a pale blue nonmagnetic form (O_4) containing four atoms per molecule which readily breaks down.
2) *Nitrogen*. Nitrogen is a gaseous element with the symbol N and located in group V of the periodic table with atomizing number 7 and atomic weight of 14. The free gas represents about 78% of the volume of the atmosphere. The element occurs in the combined state of several minerals.

   Nitrogen is a colorless, odorless, tasteless, and nontoxic gas. It can be transformed into a colorless liquid, which can be compressed into a colorless, crystalline solid. Nitrogen exists in two natural isotopic forms and four radioactive isotopes.

18.1 The Oxides of Nitrogen

Molecular nitrogen (N_2) is a relatively inert gas. However, the element nitrogen can be reactive as a single atom. Thus, nitrogen can form several different oxides. The stable gaseous oxides of nitrogen include N_2O (nitrous oxide), NO (nitric oxide), N_2O_2 (dinitrogen dioxide), N_2O_3 (nitrogen trioxide), NO_2 (nitrogen dioxide), and N_2O_5 (nitrogen pentoxide). NO_3 also exists but is unstable. Of these, the only ones present in the atmosphere in any significant amount are N_2O, NO, and NO_2. Thus, these three are potential contributors to air pollution and are referred to as NO_x.

1) Nitrous Oxide (N_2O) is an inert gas with anesthetic characteristics. Its ambient concentration is approximately 0.5 ppm, which is considerably below the threshold for biological effects. In addition, it has a balanced environmental cycle which is independent of the other oxides of nitrogen.
2) Nitric Oxide (NO) is a colorless gas, and its ambient concentration is usually significantly less than 0.5 ppm. At these concentrations its biological toxicity in terms of human health is insignificant. However, nitric oxide readily oxidizes in the atmosphere to form nitrogen dioxide (NO_2) and is an active participant in photochemical smog formation as well. It initiates reactions that produce air pollutants.
3) Nitrogen dioxide (NO_2) is a reddish-brown gas and is quite visible in sufficient amounts. A concentration of 1 ppm of NO_2 is detectable to the naked eye. The NO_2 ambient air quality standard in California was initially set at 0.25 ppm (on an hourly average), primarily on the basis of visibility concerns, but the toxicology and epidemiological effects of NO_2 on human beings are not completely known.

Optimizing Air Pollution Control Equipment Performance: Operation & Maintenance, First Edition. Jay Richardson and Louis Theodore.

Over 90% of all man-made NO_x that enter the atmosphere are produced by the combustion of various fossil fuels. In the United States, approximately one-half of the NO_x is from stationary sources, while the remainder is from mobile sources. Over 95% of NO_x formed from industrial and utility combustion is in the form of NO, with the remaining 5% as NO_2 and a trace amount of N_2O. Environmental regulations typically consider all NO_x as NO_2 because all the NO in the atmosphere is eventually converted to NO_2. N_2O is considered a greenhouse gas (GHG) as it can remain in the atmosphere for hundreds of years. By comparison, NO_x only lasts a matter of days. Because of its long lifespan, N_2O will eventually reach the upper atmosphere where it decomposes under UV light and the products of this reaction remove ozone from the upper atmosphere. The man-made emissions rate of NO_x is of increasing concern as NO_x reacts with moisture and other compounds to form nitric acid and create acid rain. To form smog, NO_x undergoes a photochemical reaction with volatile organic compounds (VOC's) and sunlight that result in airborne particles and ground-level ozone.

NO_x formation from a specific stationary combustion source (boiler, turbine, incinerator, etc.) is determined by the interaction of chemical and physical processes occurring within the boiler furnace. There are three principal chemical processes for NO_x formation:

1) Thermal NO_x, which is the oxidation of atmospheric nitrogen from the combustion air supplied to the burners and boiler furnace;
2) Prompt NO_x, which is formed by chemical reactions between hydrocarbon fragments and atmospheric nitrogen from the combustion air. Prompt NO_x tends to form quickly; often at the lower temperature stages of combustion; and,
3) Fuel NO_x which is formed from NO_x chemical reactions involving nitrogen atoms chemically bound within the fuel. Fuel NO_x is generally a more active pathway with solid fuels (such as coal and heavy oil) due to the nitrogen present in the material.

18.1.1 Thermal Oxide of Nitrogen Formation

Thermal NO_x results from the oxidation of atmospheric nitrogen from the combustion air in the high-temperature post-flame region of the boiler. During combustion, oxygen radicals are formed and attack atmospheric nitrogen molecules to start the reactions that comprise the thermal NO_x Zeldovich formation mechanism:[1]

$$N_2 + O \leftrightarrow NO + N \tag{18.1}$$

$$N + O_2 \leftrightarrow NO + O \tag{18.2}$$

$$N + OH \leftrightarrow NO + H \tag{18.3}$$

The first reaction is generally assumed to determine the rate of thermal NO_x formation because of its high activation energy of 76.5 kcal/mol. Because of this reaction's high activation energy, NO_x formation is slower than other combustion reactions causing large amounts of NO to form after combustion is "complete."

The major parameters that influence thermal NO_x formation are: (i) temperature; (ii) oxygen concentrations; (iii) nitrogen concentrations; and (iv) residence time. If temperature, oxygen concentrations, or nitrogen concentrations can be reduced quickly after combustion, thermal NO_x formation is suppressed. Of these four factors, temperature is the most important, and as such, suppression of NO_x formation is generally referred to as "quenching."

Thermal NO_x formation is an exponential function of temperature and is related to flame temperature by the Arrhenius rate relationship.[2–4] Generation begins at approximately 2,200°F and increases exponentially with an estimated doubling of NO_x for every 100°F of flame temperature increase (see Figure 18.1).

One of the fundamental parameters affecting temperature is the local equivalence ratio, ϕ, within the boiler furnace. Flame temperature peaks at equivalence ratios near 1.0 as shown in Figure 18.2. If localized areas within the boiler furnace are considered fuel-rich, then there is insufficient oxygen to burn all the fuel, and peak temperatures decrease. If localized areas within the boiler furnace are considered fuel-lean, this will result in decreasing peak temperatures. A non-premixed flame, or a flame that exists where the reactants must diffuse into each other during chemical reaction will generally react near an equivalence ratio of one, causing high peak temperatures. Thus, reducing peak flame temperature through combustion modifications can significantly reduce thermal NO_x formation.

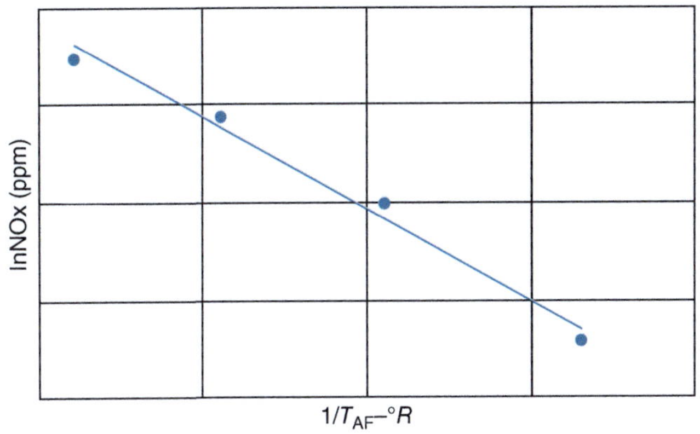

Figure 18.1 Relationship of adiabatic flame temperature (AFT) to thermal oxides of nitrogen production.

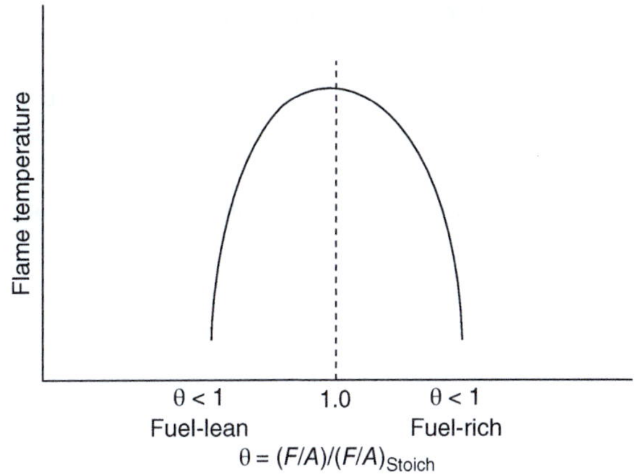

Figure 18.2 Variation of flame temperature with equivalence ratio.

18.1.2 Prompt Oxide of Nitrogen Formation

Prompt NO_x formation occurs in the boiler furnace through the reactions of hydrocarbon fragments and atmospheric nitrogen from the combustion air. As opposed to slower thermal NO_x formation, prompt NO_x formation is rapid and on a time scale comparable to the energy release reactions (i.e., within the flame). It is not possible to "quench" prompt NO_x formation by decreasing the temperature as is the case with thermal NO_x formation. However, the contribution of prompt NO_x to the total NO_x emissions of a system is rarely significant and accounts for a relatively small percentage of the total NO_x production (typically <10%). Although there is some uncertainty in the detailed mechanisms for prompt NO_x formation, it is generally believed that the principal product of the initial reactions is hydrogen cyanide (HCN) or CN radicals, and that the presence of hydrocarbon species is essential for the reactions to take place.

The HCN radical is further reacted to form NO_x. Measured levels of prompt NO_x for a number of hydrocarbon compounds in a premixed flame show that the maximum prompt NO_x is reached on the fuel rich side of stoichiometry. On the fuel-lean side of stoichiometry, few hydrocarbon fragments are free to react with atmospheric nitrogen from the combustion air to form HCN, the precursor to prompt NO_x. With increasingly fuel-rich conditions, an increasing amount of HCN is formed, creating more NO_x. However, above an equivalence ratio of approximately 1.4, there are not enough O_2 radicals present to react with HCN and form NO, so NO levels decrease, and as such is the preferred methodology employed by many low NO_x burners (LNB) or combustion staging technologies to reduce prompt NO_x formation.

18.1.3 Fuel Bound Oxide of Nitrogen Formation

The oxidation of fuel-bound nitrogen (FBN) is the principal source of NO_x emissions in the combustion of coal and some oils. All indications are that the oxidation of FBN compounds to NO is rapid and occurs on a time scale comparable to the energy release reactions during combustion. Thus, as with prompt NO_x, the reaction system cannot be quenched as it can be for thermal NO_x.

Although some details of the kinetic mechanism for conversion of fuel nitrogen to NO_x are unresolved at the present time, the sequence of kinetic processes is believed to be a rapid thermal decomposition of the parent fuel-nitrogen species to low molecular weight compounds such as HCN, and subsequent decay of these intermediates to NO or nitrogen (N_2).

In stoichiometric or fuel-lean situations, the intermediates will generally react to form NO over N_2, whereas in fuel-rich systems, there is evidence that the formation of N_2 is competitive with the formation of NO. This may, in part, be the cause of high NO_x emissions in fuel-lean and stoichiometric mixtures and lower NO_x emissions in fuel-rich systems.

For high nitrogen fuels such as No. 6 oil, fuel NO_x may be as high as 70% of the total NO_x emissions. For low-nitrogen fuels such as natural gas, fuel NO_x is insignificant.

18.2 NO_x Control Methods

Factors that affect the formation of NO_x in combustion processes are described in this and the next two sections. A brief discussion of each factor is also included. Much of this material is driven by the principles associated with the 3 T's of combustion: time, temperature, and turbulence.[2–4]

There are essentially six principles of NO_x control:

1) Reducing peak flame temperature
   a) Flue gas recirculation (FGR)
   b) Fuel reburning
   c) LNB or ultra-low NO_x burners (ULNB)
   d) Combustion optimization
   e) Burners out of service (BOOS)
   f) Reducing excess air
   g) Water/steam injection
   h) Over-fired air
2) Reducing residence time at peak temperature
   a) Air injection
   b) Fuel injection
   c) Steam injection
3) Chemical reduction of NO_x
   a) Selective catalytic reduction (SCR)
   b) Selective non-catalytic reduction (SNCR)
4) Removal of nitrogen
   a) Using oxygen instead of air
   b) Low-nitrogen fuels
5) Use of a sorbent
6) Combination of multiple methods

Details on each method follow:

1) Reducing peak flame temperatures often involves avoiding the stoichiometric ratio (the exact ratio of chemicals that enter into reaction). Essentially, this technique dilutes with excess air, flue gas, or steam. Combustion controls use different forms of this technique and are different for fuels with high and low-nitrogen content. The basic technique is to reduce the temperature of the combustion products. This method keeps the vast majority of nitrogen from reacting to form NO and is limited by the kinetics at the temperature in question.
2) Reducing residence time at high combustion temperatures can be accomplished by ignition or injection timing with internal combustion engines. It can also be implemented in boilers by restricting the flame to a short region in which the combustion air becomes flue gas. This is immediately followed by injection of fuel, steam, air, or recirculating flue gas. This residence time at peak temperatures keeps the vast majority of nitrogen from reacting and NO formation is limited by the kinetics at the operating temperature.
3) Chemical reduction of NO_x provides a chemically reducing (for example: reversal of oxidization) substance to remove oxygen from nitrogen oxides. Examples include SCR which uses ammonia, SNCR which uses ammonia or urea, fuel reburning (FR), non-thermal plasma, and other emerging technology.
4) Removal of nitrogen from combustion is accomplished by removing nitrogen as a reactant either by using oxygen instead of air in the combustion process or using low-nitrogen content fuel to form less fuel NO_x. Eliminating nitrogen by using oxygen tends to produce a rather intense flame that requires special refractory chambers and must be subsequently and suitably diluted.
5) Treatment of flue gas by injection of sorbents (such as ammonia, powdered limestone, aluminum oxide, or carbon) can remove NO_x and other pollutants (principally sulfur). There have been successful efforts to make sorption products a marketable commodity. This kind of treatment has been applied in the combustion chamber, flue, and particulate control devices. This method uses either absorption or adsorption followed by filtration and/or electrostatic precipitation to remove the resultant sorbent.
6) Many of these methods can be combined to achieve a lower NO_x concentration than can be achieved alone by any one method.

18.3 Reducing NO_x Generation via Pollution Prevention

Some information on NO_x pollution prevention methods are described below.

1) Reducing excess air
Excess air flow for combustion has been correlated to the amount of NO_x generated. Limiting the net excess air flow to under 2% can significantly limit the NO_x content of flue gas. Although there are fuel-rich and fuel-lean zones in the combustion region, the overall net excess air is limited when using this approach.
2) BOOS
Multiple-burner equipment can have part of an array of burners with some "BOOS" (not feeding fuel but supplying air or flue gas). This allows the burners around them to supply fuel and air to the air or flue gas flowing from the BOOS. The result is combustion in stages with the temperature always lower than when all burners are in service.[5]
3) Over-fired air (OFA)
The OFA approach removes a portion of the air from the burner and reinjects it downstream. Similar to fuel staging, OFA allows combustion to operate outside of the peak flame temperature zone and operate rich at the burner front and lean when OFA is introduced downstream. The amount and location of OFA is designed to ensure complete burnout of the fuel before exiting the boiler.
   The supply for the over-fire-air is typically taken from the burner wind box or supply ducting where sufficient pressure is available. In general, the higher the available pressure, the better the penetration and mixing, and the better the OFA performance. In cases where a combustion air source or pressure is limited, boosted over-fire air systems can be used to enhance NO_x performance, but the cost-benefit analysis should be considered for these applications.
   Proper modeling of the furnace provides calculation and visualization of OFA mixing with the furnace gases, CO produced by incomplete combustion, and NO_x reduction as a result of port location, flow, and sizing.
   OFA ports can include design features that include flow straightening and a split port design option to provide better mixing and performance throughout the load range.
4) Low NO_x burners (LNB)
Advancements in burner design have led to the advancement of burner nozzles and hardware to reduce NO_x emissions through fuel/air staging, low excess air operation, and aerodynamic analysis. An example of a typical low-NO_x burner is shown in Figure 18.3. Fuel/air staging is a proven technique used to reduce thermal NO_x emissions. Gas nozzles can be designed to establish fuel "rich" and "lean" zones at the burner outlet (staged combustion). During combustion, the fuel burns in the "rich" reducing environment and secondary air mixes further downstream to complete burnout. Operating the burner in fuel-rich/lean zones avoids peak flame temperatures and thereby reduces thermal NO_x emissions.
   LNB are able to operate in the 2–3% excess O_2 range. Low excess air operation not only reduces peak flame temperatures but has been shown to reduce the FBN conversion rate as well as prompt NO_x generation.
   The ULNB expands upon fuel staging by injecting the staged gas outside of the combustion air stream. This allows the staged fuel to mix with inert flue gases before combustion, mimicking the effect of external FGR. This mixing reduces the peak flame temperature and further reduces thermal NO_x emissions.
5) Flue gas recirculation (FGR)
FGR is a very effective method of reducing thermal NO_x emissions on industrial gas firing applications. FGR works by circulating relatively cool, low excess O_2 exhaust gas back into the combustion air. This dilutes the burner excess O_2 from the ambient concentration of 20.9% to 16–19% depending on the application. This reduces available O_2 for fuel or prompts NO_x formation in the primary combustion zone while also lengthening and cooling the flame for significant reductions in thermal NO_x (see Figure 18.4).

Figure 18.3 STEP Combustion axial flow, staged-fuel, low NO_x burner (LNB).

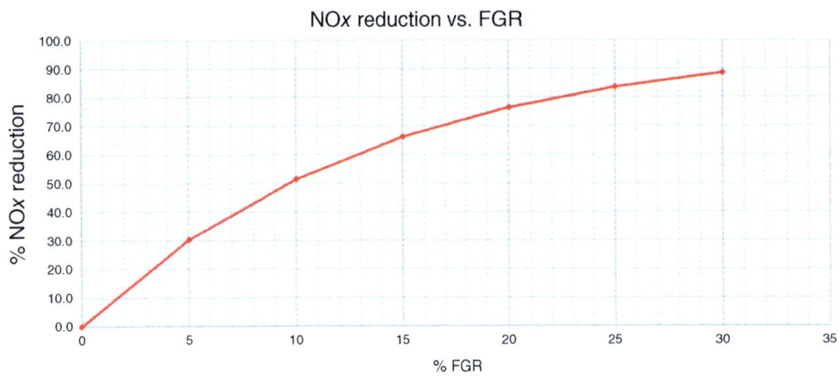

Figure 18.4 Percent flue gas recirculation (FGR) versus NO_x reduction for gas-fired applications.

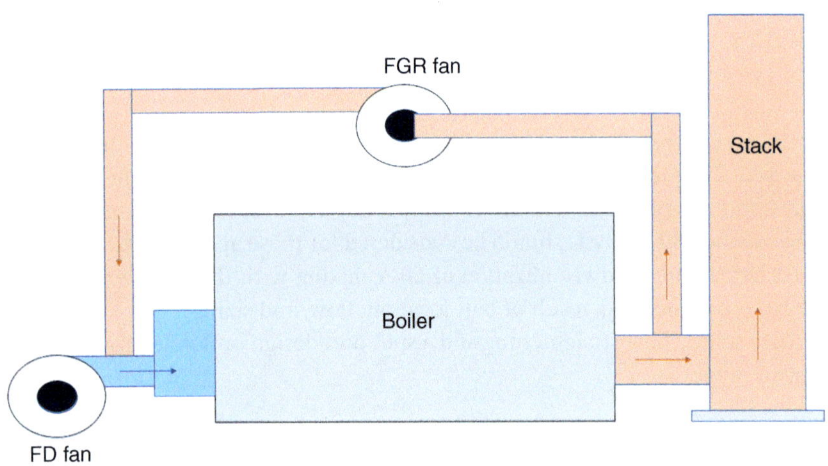

Figure 18.5 Diagram of forced FGR system.

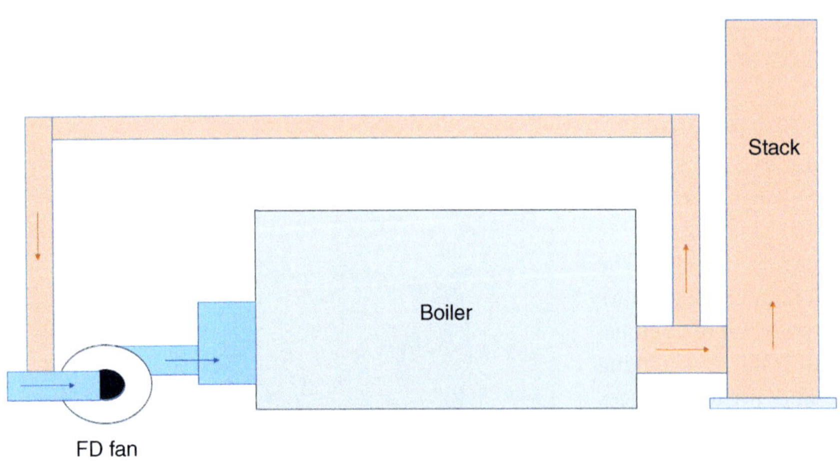

Figure 18.6 Diagram of induced FGR system

There are two basic methods of implementing FGR: forced; and, induced. Forced FGR (see Figure 18.5) will utilize an additional fan (booster fan) to recirculate the flue gases back into the combustion air downstream of the forced draft fans. A damper controls the recirculation rate, based on the firing rate of the burners. Induced FGR (see Figure 18.6) utilizes the existing combustion air fan to recirculate the flue gases back into the combustion air stream. A portion of the flue gases are routed by ductwork to the inlet of the existing FD fans, where they are premixed with the combustion air and introduced into the flame through the burner. The FD fans will need to be assessed to determine whether forced FGR is required, or if the existing fans can accommodate the additional FGR flow. The assessment of the existing FD fans is completed to determine if the boiler nameplate-rated capacity can be derated and still meet the normal operating performance of the boiler.

6) Water/steam injection

Injection of water/steam dilutes the mixture and cools combustion. If flame temperature is sufficiently reduced, thermal NO_x will not be formed in as great a concentration. If fuel oil is used as a backup fuel in gas fired application, steam injection is readily available through the oil burner.

7) Reduced air preheat

Some boilers implement an efficiency device called an air preheater. The air preheater is a gas–gas heat exchanger that transfers heat from the exhaust flue gas into the combustion air. Primarily an efficiency device, the air preheater is designed to retain heat destined to leave the stack unutilized, and recirculate it back into the unit. These devices were primarily utilized before NO_x emissions were of significant importance. It has since been discovered that the elevated combustion air temperature significantly increases NO_x emissions. Combustion air temperature effects NO_x emissions on different types of boilers differently. For tangentially fired boilers the relationship is roughly described in Figure 18.7.

Removing the air preheater, will not only reduce NO_x emissions, but also reduce boiler efficiency. Therefore, removing the air preheater is typically accompanied by adding an economizer. An economizer performs a similar function to the air preheater relative to boiler efficiency. The economizer is a gas–liquid heater exchanger that transfers heat from the flue gas to the boiler feed water. The economizer can then be placed in the same location as the removed air heater so there are minimal ductwork modifications. The economizer is more efficient than the air heater, so the overall boiler efficiency will be improved.

8) Fuel reburning

Recirculation of cooled flue gas with added fuel (this can be natural gas, pulverized coal, or even oil spray) causes a reduction in energy content similar to FGR, and primary combustion temperature can be lowered.

9) Combustion optimization

"You don't get NO_x out of the box." The first step in any NO_x reduction process is to take a complete set of baseline readings and perform combustion optimization tuning. As mentioned previously, thermal NO_x emissions account for essentially all of the NO_x emissions, and thermal NO_x is primarily generated by elevated flame temperatures. The systems used to control industrial and utility combustion equipment are complex and sophisticated. They require in-depth tuning by qualified professionals to optimize performance and minimize emissions.

Gaseous fuel balancing can be achieved using burner-mounted pressure gauges and manual balancing valves or orifices. If the gas spuds are in good condition, they can be effectively used as metering orifices and balancing can be achieved through providing uniform supply pressures at the burner fronts.

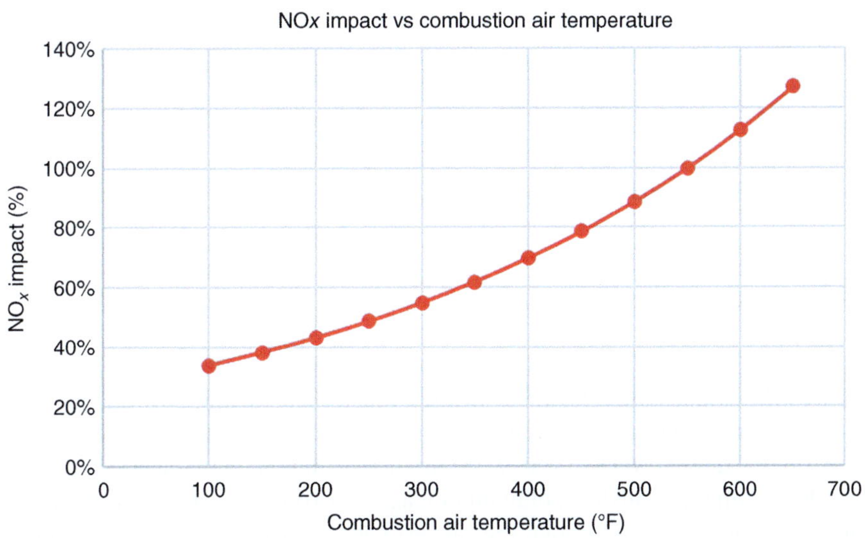

Figure 18.7 Combustion air temperature on thermal NO_x.

Airflow balancing is paramount to combustion optimization. Roughly 95% of the mass flow through the burner is combustion air. Poor airflow can decrease efficiency, increase emissions, and result in poor stability. Airflow balance testing can be conducted online with installed burner airflow measurement devices, but it is also recommended to perform periodic offline airflow balancing. The offline balancing is conducted while running combustion air fans only and can be done while the unit is hot. Balancing is typically completed in a single shift using an articulating probe. The airflow measurement probe is inserted into the burner and velocity measurements are taken at multiple circumferential points. Based on the calculated flows, balancing is performed using the existing air dampers.

10) Oxygen instead of air for combustion
As mentioned previously, NO_x is formed by the breakdown of diatomic nitrogen in the air. If pure oxygen is used and an oxidant for the combustion process, all nitrogen (and therefore NO_x) is eliminated from the reaction.

11) Catalytic combustion
Use of a catalyst or a ceramic mesh to stabilize combustion can occur below NO formation temperatures and can provide a suitable means of reducing combustion temperature. This technique is not used often because it is very sensitive. However, where it is used, catalytic combustion can achieve less than a 5 ppm concentration of NO_x in the flue gas.

18.4 Control of Flue Gas NO_x

In addition to the pre-combustion or in-combustion techniques mentioned above, NO_x can also be reduced in post-combustion processes. This allows burners to operate in a more efficient, high-intensity manner while still maintaining low NO_x emissions.

18.4.1 Selective Catalytic Reduction (SCR) Overview

Nitrogen oxides (NO_x) in combustion flues gases can be reduced by the process of SCR. In an SCR, NO_x reacts with a reducing agent, typically ammonia (NH_3), in the presence of a catalyst to form molecular nitrogen (N_2) and water (H_2O). The NH_3 is injected into the flue gas prior to the catalyst in either aqueous ammonia, anhydrous ammonia, or urea (which thermally decomposes into ammonia). Proper dosing and mixing are critical to SCR performance. Temperature and velocity must also be considered for proper NO_x reduction. The reaction occurs most effectively at temperatures between 500°F (260°C) and 1,050°F (566°C), with optimal performance based on catalyst design/makeup.

The reaction occurs as NH_3 diffuses into the catalyst surface pore structure and is adsorbed. This active catalyst site then reacts with NO_x in the flue gas. The rate of and effectiveness of the reaction depends on the number of available active sites, flue gas temperature, and pollutant concentration. A well-balanced process will maintain the desired NO_x emissions with minimal NH_3 slip.

When designed and operated properly, the SCR system should have a limited impact on upstream processes. The system will perform reliably and achieve high NO_x reduction under varying loads. Furthermore, a well-operating SCR system has limited impact on the operation of downstream equipment.

18.4.2 SCR Chemistry

The SCR reactions that occur to reduce NO and NO_2 with NH_3 are simplified as follows:

$$4NO + O_2 + 4NH_3 \rightarrow 4N_2 + 6H_2O \tag{18.4}$$

$$NO + NO_2 + 2NH_3 \rightarrow 2N_2 + 3H_2O \tag{18.5}$$

$$6NO_2 + 8NH_3 \rightarrow 7N_2 + 12H_2O \tag{18.6}$$

The first reaction is the primary reaction in this application. This reaction shows that, theoretically, only one mole of ammonia is required to remove each mole of NO from the flue gas. However, in practical applications, the process requires more ammonia to be injected than necessary for the desired NO removal. With variations in flue gas and imperfect mixing, more NH_3 must be available to react. Also, due to improper mixing, pollutant or reagent distributions, as well as temperature variations (to name a few), can lead to ammonia passing through the catalyst bed unreacted. This is referred to as ammonia slip.

Secondary reactions may also occur depending upon the operating conditions and changes in flue gas composition. Examples of possible secondary reactions are as follows:

$$4NH_3 + 5O_2 \rightarrow 4NO + 6H_2O \tag{18.7}$$

$$SO_2 + O_2 \rightarrow 2SO_3 \tag{18.8}$$

$$2NH_3 + SO_3 + H_2O \rightarrow (NH_4)_2SO_4 \tag{18.9}$$

$$NH_3 + SO_3 + H_2O \rightarrow (NH_4)HSO_4 \tag{18.10}$$

Equation 18.7 illustrates that ammonia oxidation can result when normal operating temperatures exceed 800°F. As the flue gas temperature increases, the NH_3 oxidizes more readily, increasing the concentration of NO and decreasing the available NH_3 for the SCR process.

Equation 18.8 is prevalent with higher sulfur fuels and not with low sulfur fuels or natural gas fuels. CO oxidation catalysts may also result in increased SO_2 to SO_3 conversion. Sulfur in the fuel generates SO_2 (predominantly) and SO_3 (SO_x) during combustion; the presence of both SO_3 and NH_3 in the flue gas can lead to the formation of ammonium sulfate and or ammonium bisulfate as indicated in Equations 18.9 and 18.10. Once formed, these substances may deposit on the catalyst and other system equipment, impacting and reducing the system's performance.

18.4.3 Selective Non-Catalytic Reduction (SNCR)

Similarly to SCR, SNCR is a method of post-combustion NO_x reduction where an ammonia-based compound is injected into the furnace to react with NO_x to form N_2 and H_2O. SNCR systems can use anhydrous ammonia, aqueous ammonia, or aqueous urea. There are pros and cons to each reagent that should be evaluated for each specific site. The SNCR reaction occurs within a temperature range of approximately 1,600–2,100°F. The optimum temperature within that range is dependent on residence time and carbon monoxide levels at the injection location, thus making the completion of a computational fluid dynamics (CFD) model of the boiler a requirement while evaluating the application of this technology. Lower temperatures require longer residence times. Elevated carbon monoxide levels shift the temperature curve such that better efficiencies are achieved at lower temperatures.

While NO_x conversion rates of 90% are possible in the laboratory, practical application of SNCR in conventional industrial and utility boilers currently limits this reduction to roughly 35–55%. This is due to temperature gradients, stratification, as well as droplet evaporation and distribution. Hotter furnaces will have lower NO_x reduction efficiencies, and boilers with furnace exit gas temperatures (FEGT) of 2,100°F are not good candidates for SNCR (see Figure 18.8).

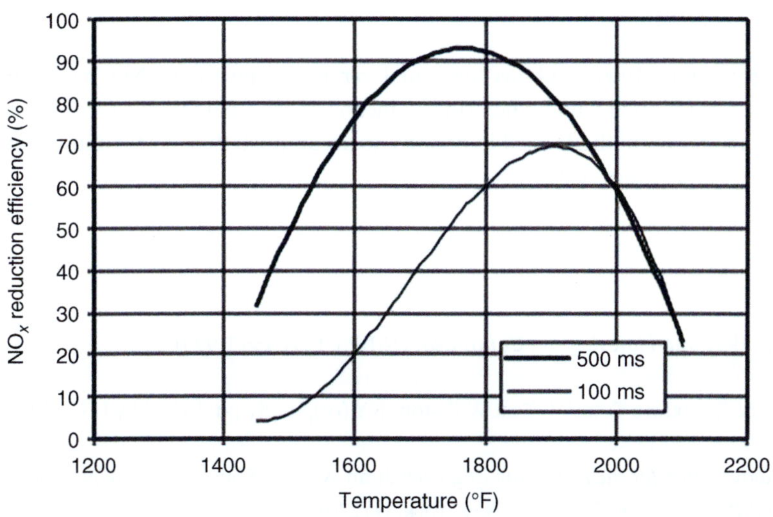

Figure 18.8 Selective non-catalytic reduction (SNCR) NO_x reduction as a function of temperature and residence time.

18.4.4 Absorption

While not commonplace practice in industry, the oxides of nitrogen can be absorbed by water, hydroxide and carbonate solutions, sulfuric acid, organic solutions, and molten alkali carbonates and hydroxides. When aqueous alkaline solutions such as NaOH and MgOH are used, complete removal requires that one-half of the NO must first be oxidized to NO_2. The best absorption occurs when the NO/NO_2 molar ratio is 1:1, which indicates that absorption of the combined oxide, N_2O_3, is the most favorable.

Absorption of nitric oxides by strong sulfuric acid has also been available. The compounds formed in this case are violet acid ($H_2SO_4 \bullet NO$) and nitrosyl sulfuric acid, ($NOHSO_4$). The latter is very stable in concentrated acid. The reaction is:

$$NO + NO_2 + 2H_2SO_4 = 2NOHSO_4 + H_2O \tag{18.11}$$

Aqueous scrubbing systems using aqueous alkaline solutions of concentrated sulfuric acid appear to offer the best potential for the control of both sulfur and nitrogen oxide emissions. Equimolar concentrations of NO and NO_2 (equivalent to N_2O_3) are required.

18.4.5 Adsorption

The use of char (activated carbon) to adsorb oxides of nitrogen has been studied.[6] Char has a high adsorption rate and capacity, compared to other materials. However, regeneration may be a problem. A potential fire and explosion hazard may be another difficulty with this material since O_2 is usually present in most stack gases. The technique does not show much promise at the present time. Adsorption may provide some limited means for the reduction of NO_2, emissions from flue gases associated with stationary combustion sources. Common adsorbents such as silica gel, alumina, molecular sieves, charcoal, and ion exchange resins show some capacity for oxidizing NO to NO_2 and adsorbing the resultant NO_2. However, the capacity of these adsorbents is quite low at typical NO flue gas concentrations. Manganese and alkalized ferric oxides show some potential.

18.5 Operation, Maintenance, Inspection, and Optimization Considerations

The number of processes available for NO_x control is extremely large. In addition, the number of factors that can impact the processes is also very large. Because of this, it is not feasible to include any meaningful discussion on operation, maintenance, inspection, and optimization considerations for the various processes.

18.6 Conclusions

The initial push for large-scale NO_x reduction started in the 1990's and has continued to ramp up post 2010. NO_x has been a significant concern due to its GHG effect and the formation of smog. The primary means of NO_x reduction is focused on reducing in furnace flame temperature. There are extensive methods used in practice to accomplish this. Post-combustion NO_x reduction is also a viable alternative, though typically more costly.

References

1 USEPA Technical Bulletin, *Nitrogen Oxides (NO_x), Why and How they are Controlled*. DARPS, EPA 456 1-99-006F, RTP, NC, 1999.

2 **L. Theodore**, *Thermodynamics. A Thedore Tutorial*, Theodore Tutorial, East Williston, NY (originally published by USEPA/APTI, RTP, NC), 1983.

3 **L. Theodore, F. Ricci, and T. VanVleet**, *Thermodynamics for the Practicing Engineer*, John Wiley & Sons, Hoboken, NJ, 2007.

4 **L. Theodore**, *Chemical Engineering: The Essential Reference*, McGraw Hill, New York, NY, 2014.

5 **L. Theodore, R. Dupont, and K. Ganesan**, *Pollution Prevention: Sustainability, Industrial Ecology, and Green Engineering*, CRCPress Taylor & Francis Group, Boca Raton, FL, 2017.

6 **L. Theodore, and R. Nelsen**, *Pollution Prevention. A Theodore Tutorial*, Theodore Tutorial, East Williston, NY (originally published by USEPA/APTI, RTP, NC), 1994.

19

Carbon Capture and Storage

Gases have performed varied functions in the economy. This is particularly true of carbon dioxide (CO_2). This chemical has been used in the liquid and solid phases for nearly two centuries. A half-century ago, over one million tons of CO_2 were produced in a solid, liquid, and gaseous forms. The largest use of solid CO_2 was for refrigeration purposes; it also helped reduce bacteria spoilage and for carbonating some beverages. The main use of liquid CO_2 is for carbonating liquid beverages and as a fire extinguishing source. The gaseous state has been used in the production of urea, certain acids, and soda ash.

There are three main sources of CO_2 that have been deployed in the chemical process industries:

1) Flue gases (10–18% by volume) which result from burning fossil fuels.
2) By-products from fermentation industries (99%) produced from the reaction of dextrose breakdown into alcohol and carbon dioxide.
3) By-product of lime kiln operation (10–40%) where carbonates are calcined to oxides.

However, it appears that anthropogenic emissions of CO_2 may be causing global heating and climate change. It has been estimated that two-thirds of global CO_2 emissions are from the utility and transportation sectors. Since fossil fuels remain the least expensive source of energy, combustion of fossil fuels continues and will remain the energy source in the *near future*. Carbon capture and storage (CCS) is the process of capturing CO_2 from large point sources and producing concentrated and pressurized CO_2 for storage. However, it should also be noted that approximately 60% of CO_2 emissions can be attributed to dilute, distributed sources.

19.1 Properties of Carbon Dioxide

Carbon Dioxide is a colorless, odorless, and lightly acid-tasting gas, sometimes called carbonic acid gas. The molecule consists of one atom of carbon joined to two atoms of oxygen (CO_2). Carbon dioxide is about 1.5 times as dense as air. It is soluble in water, with 0.9 volume of the gas dissolving in one volume of water at 20°C (68°F). The atmosphere contains carbon dioxide in variable amounts, usually 3–4 parts per 10,000. It is used by green plants in the process known as photosynthesis, by which carbohydrates are manufactured. It does not burn and does not support ordinary combustion, and because of these properties, it is used for extinguishing fires.

The CO_2 extinguisher is a steel cylinder filled with liquid carbon dioxide, which, when released, expands suddenly and causes so great a lowering of temperature that it solidifies into a powdery "snow." This snow volatilizes (vaporizes) on contact with the burning substance, producing a blanket of gas that cools and smothers the flame. Solid carbon dioxide, known as dry ice, is widely used as a refrigerant. Its cooling effect is almost twice that of ice water; its special advantages are that it does not melt as a liquid but turns into gas, and that it produces an inert atmosphere that reduces bacterial growth. The presence of carbon dioxide in the blood simulates breathing. For this reason, carbon dioxide is added to oxygen or ordinary air in artificial respiration and to the gases used in anesthesia.

Optimizing Air Pollution Control Equipment Performance: Operation & Maintenance, First Edition. Jay Richardson and Louis Theodore.

19.2 Global Carbon Cycle

Carbon dioxide comprises only a very small portion of the Earth's atmosphere but has risen from about 280–380 ppmv (parts per million by volume) recently (approximately 0.03% by volume). However, it is undoubtedly one of the most essential molecules for life on Earth. Without this colorless, odorless gas, animals and plants would not be able to survive. Nevertheless, an excess of carbon dioxide in the atmosphere arising due to the burning of fossil fuels might prove to be harmful. The exchange of carbon throughout the world, termed the "global carbon cycle," is of particular importance for studying the greenhouse effect.

The transfer of carbon occurs naturally through both the terrestrial and oceanic cycles. The terrestrial cycle starts with photosynthetic plants, which use sunlight, water, carbon dioxide, and a pigment called chlorophyll to form glucose and oxygen. It has been estimated that this consumes approximately 500 billion tons of carbon dioxide each year, converting it into organic compounds and oxygen. Carbon dioxide is naturally returned into the atmosphere by respiration and decay. The oceans, which cover about 70% of the globe, consist of all of the gases contained in the atmosphere. There is an exchange of carbon dioxide between the oceans and surrounding air until equilibrium is reached. Carbon dioxide is then transferred to deeper water by convective transport cycles, which lower the concentration at the surface, thereby allowing more of the gas from the atmosphere to diffuse in the ocean.

Surprisingly, about 25% of the increase in atmospheric carbon dioxide over the past century and a half has come from changes in land use, such as deforestation and soil cultivation. During the last 150 years, about 155 billion metric tons of carbon were released into the atmosphere from such actions. The amount released each year generally increased over the period, and by the 1990s the rate of release averaged about two billion metric tons of carbon per year.[1] Deforestation increases carbon dioxide levels in two basic ways. First, when trees decay or are burned, they release carbon dioxide. Second, without the forest, carbon dioxide that would have been absorbed by photosynthesis remains in the atmosphere. For example, a rainforest can hold 1 or 2 kg of carbon/m^2/yr as compared to a field of crops, which can absorb less than 0.5 kg/m^2 every year.[2]

In addition to deforestation and other such actions, the burning of hydrocarbons has caused a severe man-made influx of carbon dioxide into the global carbon cycle. The International Panel on Climate Change (IPCC) used knowledge of this natural process plus data on fuel consumption and land use in order to predict the concentration of carbon dioxide in the atmosphere. According to their model, approximately 3.3 gigatons more carbon dioxide is entering than leaving. This abundance of carbon dioxide has caused a disruption of the global carbon cycle. Without a "quick fix," some fear that this increase will result in catastrophic events, detailed in a later section.

Recently, the Environmental Protection Agency (EPA) appropriately replaced global warming with the term "climate change" to account for all of the environmental changes that are occurring in the world, not just the increase in temperature. Along with the federal government, the EPA has created a number of initiatives and incentives to encourage major polluters, particularly large corporations, to go "green" and reduce their environmental impacts. The US government has also enacted many climate change programs to help curb greenhouse gas (GHG) emissions as well as prevent the release of other pollutants. A program introduced over a decade ago (2009) was the Climate Change Technology Program (CCTP), which will be discussed in more detail.

19.3 The Greenhouse Effect

The "greenhouse effect" is a phrase properly used to describe the increased warming of the Earth due to increased levels of carbon dioxide and other atmospheric gases, called GHGs. Just as the glass in a botanical greenhouse traps heat for growing plants, GHGs trap heat and warm the planet. The greenhouse effect, a natural phenomenon, has been an essential part of Earth's history for billions of years. The greenhouse effect is the result of a delicate and non-fixed balance between life and the environment. Yet, the greenhouse effect may be leading the planet to the brink of disaster. Since the Industrial Revolution, the presence of additional quantities of GHGs threatens to affect global climate and the predicted effects of this increase are still debated among scientists.

The greenhouse effect works as follows. The energy radiated from the sun to Earth is absorbed by the atmosphere and is balanced by a comparable amount of long-wave energy emitted back to space from the Earth's surface. Carbon dioxide molecules (and GHGs) absorb some of the long-wave energy radiating from the planet. Because of the greenhouse heat-trapping effect, the atmosphere itself radiates a large amount of long-wave energy downward to the surface of the Earth and makes the Earth warmer than if warmed by solar radiation alone. The GHGs trap heat because of their chemical makeup and, in particular, their triatomic nature. They are relatively transparent to visible sunlight, but they absorb long-wavelength infrared radiation emitted by the Earth.[3]

As an environmental problem, climate change/global warming must be considered on an entirely different scale than that of most other environmental issues: The effects of climate change are long-term, global in magnitude, and largely irreversible. Because of the enormity of the problem and the uncertainties involved – it may take decades to determine with absolute certainty that global warming is underway – the difficult questions faced today are how and when one should react.

Fossil-fuel burning and forestry and agricultural practices are responsible for most of the man-made contributions to the gases in the atmosphere that act like a greenhouse to raise the Earth's temperature; hence the term "greenhouse effect." Most of the processes that produce GHGs include common everyday activities such as driving cars, generating electricity from fossil fuels, using fertilizers, and using wood-burning stoves. Because so many of these activities are so ingrained in society, reducing emissions could be a difficult task.

The search for solutions has begun. However, there is a growing concern that the costs of reducing emissions may be too high. But to put cost concerns in proper perspective, one must ask what kind of future one wants on this planet and how much does one value the environment and the cultural heritage that depends on it.[4]

A consensus has emerged in the scientific community that global warming has occurred. Most (but not all) scientists are certain that the concentrations of carbon dioxide and other GHGs in the atmosphere are increasing and these individuals generally agree that these gases will warm the Earth. Four questions remain to be answered:

1) How will the temperature rise?
2) When will the temperature rise?
3) Can the cause be attributed to man-made activities?
4) What are the potential consequences?

Recent estimates indicate that if the concentrations of these gases in the atmosphere continue to increase, the Earth's average temperature could rise by as much as 1.5–4.5°C in the next century. While this may not sound like a tremendous increase, one must keep in mind that during the last ice age 18,000 years ago, when glaciers covered much of North America, the Earth's average temperature was only 5°C cooler than today.

Certainly, global cooperation is an important consideration when addressing climate change issues. No single country contributes more than a fraction of GHGs, and only a concerted effort can reduce emissions. In the future, as developing nations grow and consume more energy, their share of GHG emissions will steadily increase. It is important for other nations to offer technological assistance so that these developing nations can grow in an energy-efficient manner.[2]

The sources of GHGs are so numerous and diverse that no single source contributes more than a small fraction of total emissions. Similarly, no single country contributes more than a fraction of emissions.

Unlike other environmental problems that the EPA could address with the stroke of a regulation, potential climate change is a problem that needs innovative global solutions. Future trends of emissions will depend on a wide range of factors, from population and economic growth to technological development and policies to reduce emissions. Past trends show that all countries have been producing GHGs at a increasing rate, and many countries will continue to do so for years to come. Based on a careful study of the sources and trends of greenhouse emissions around the globe, countries can begin implementing measures for slowing down emissions while increasing economic development.

In the coming years, the technical community must reevaluate how emissions are likely to change. But given this preliminary picture of the future, it is important to take the next step of assessing the specific technologies and policy measures that can reduce emissions now at low costs. Each country will have to examine its unique situation and determine appropriate responses. However, only by acting together will the global community slow the trend toward high emissions in the next century.

19.4 Effects of Global Warming/Climate Change

A rise in average global temperature is expected by many to have profound effects. The need to study future situations has led the EPA and other environmental scientists to investigate what would happen to the planet after a 3–8°F warming. These climate changes will have a significant effect on weather patterns. There will be changes in precipitation, storms, wind direction, etc. Rising temperatures are expected to increase tropical storm activity. The hurricane season in the Atlantic and Caribbean oceans is expected to start earlier and last longer. Storms will be more severe. Some researchers believe that the planet will be a wetter place. Global circulation models (GCM) predicted that a doubling of carbon dioxide could increase humidity by 30–40%. However, such increases will not occur uniformly around the world. Perhaps humid tropical areas will become wetter while some semiarid regions will become drier. It is possible that some farmers will experience a longer growing season, while others would suffer from more frequent droughts.

As the Earth gets warmer, the IPCC predicts that there will be a 0.5–3 ft rise in the average water level of oceans. As the ocean water is heated, it expands, or increases in volume. In addition, the polar ice caps will continue to diminish in size and contribute to the rising sea levels, even after taking into account the accumulation due to an increase in precipitation. Eventually, low-lying coastal areas are expected to flood. Beach erosions will be an increasing problem. The EPA has estimated that if the sea levels rise 3 ft (0.9 m), the nation will lose an area the size of Massachusetts, even if it spends more than $100 billion to protect critical shorelines. In addition to lost beaches, houses and other buildings that sit close to the water's edge will be engulfed and destroyed. This could bring havoc to the real estate and insurance industries, as well as the lives of many people.

If polar and temperate zones become warmer, there will be a poleward shift of ecological zones. Animal and plant species that now live in a particular area may no longer be able to survive there. The EPA predicts an increase in extinction rates as well as changes in migration patterns. As the ecological zones shift poleward, there may be a decrease in the amount of area suitable for forests, with a corresponding increase in grasslands and deserts. This means an overall loss in productive land, both for agriculture and for habitats of a broad range of plants and animals. For example, the United States' major crops grown – corn, wheat, and soybeans – are strongly affected by precipitation and temperatures. The warming trend may also cause changes in water quantity and quality in some areas. This will affect drinking supplies as well as the water needs of industry and agriculture. Rising sea levels may also contaminate water supplies as seawater migrates up rivers. For example, a ft. (0.6 m) rise in sea levels would inundate Philadelphia's water intakes along the Delaware River, making the water too salty to drink.

The IPCC stated that "climate change is likely to have wide-ranging and mostly adverse impacts on human health, with significant loss of life." Heat waves take a physical toll on people. The heat puts a strain on the heart, as the body tries to cool itself. Studies have shown that the number of cases of heart and lung disease increases when temperature rises. Contagious diseases, such as malaria, influenza, and pneumonia, and allergic diseases, such as asthma, are also affected by the weather and become more prolific and are expected to spread among the public. The life cycles of mosquitoes and other disease-carrying insects are also extended in warmer weather. Warming is likely to allow tropical diseases, such as malaria, to spread northward in some areas of the world.

Climate changes, an increase in sea levels, and other direct effects of global warming can already be seen in many parts of the world. These changes are expected by many to have major social, economic, and political consequences for the future.

19.5 Carbon Dioxide Control Technologies

Currently, scientists are working hard to develop new technologies to safely reduce and dispose of carbon dioxide. There are a number of proposed advancements, many of which are still in the research and development stage. Absorption, adsorption, and reaction chemistry are the processes upon which these technologies are based. Some of these are discussed below.

One existing technology used to harvest carbon dioxide is a scrubber. Amine solutions extract carbon dioxide from flue gases through the processes of absorption. The spent amine solution is then sent to a regeneration process, where the CO_2 is collected, compressed, and liquefied.[5]

Another process uses semipermeable membranes or catalyst-coated monoliths to collect carbon dioxide through adsorption. Membranes can either function according to the molecular size of the entrained molecules or ease of diffusion through the membrane. The latter poses an effective solution for CO_2 capturing. The membrane selectively allows certain gases that diffuse easily to pass through, while certain gases such as carbon dioxide are captured. In this case, the ease of diffusion of the flue gas through the membrane is dependent on the material of construction, which is polymer-based. Polyimide has some potential for this application.[6]

A somewhat recent patent uses reaction chemistry for CO_2, removal. The process called accelerated weathering of limestone typically involves reacting CO_2 in a flue gas stream with water and calcium carbonate (limestone). Instead of releasing CO_2, a wastewater-rich stream of bicarbonate ions is produced, which can be directly released into the ocean. The coinventor of this project claims that carbon waste in the form of dissolved bicarbonate will have minimal adverse effects on the ocean and may even be beneficial for coral reefs. The Department of Energy has patented that technology, and several power plants along coastlines will serve as the testing grounds for this CO_2 sequestering method.[7]

In general, it is not very difficult to isolate carbon dioxide. However, safe disposal of the GHC has proven to be a daunting task. Many possibilities exist, but economic consequences are a major consideration when determining the feasibility of these technologies. Because industrial and commercial uses of CO_2 are minimal compared to amounts produced (approximately 1%), there needs to be safe, cost-effective disposal methods available. Such disposal methods that are likely to provide the first large-scale opportunity are direct injection to oil and gas reservoirs, deep, unmineable seams, or saline aquifers. These environments for injection have the ability to sequester large amounts of carbon dioxide as well as store and separate CO_2 from areas where injection would be harmful. Also, it could contribute to oil and methane recovery by the displacement method. Furthermore, ocean sequestering can be effective by the method of direct injection. This must be performed at 1,000–2,000 m deep, where the temperature gradient is great enough that it cannot mix with surface waters. It could also be injected over 3,000 m deep, where CO_2 becomes negatively buoyant which in turn would form a CO_2 lake. In this case, piping could be a major cost. For all of these methods, many technical, safety, liability, economic, and environmental issues remain unresolved.

Researchers in Australia conducted the biggest test of carbon disposal. They injected this material underground into a 1.3 mile hole in the ground, located at the Otway Basin. The base of this hole is a natural gas reservoir covered by several layers of impermeable rock. In 2008, 100,000 tons of carbon dioxide was introduced into the reservoir, and the hole was sealed. Scientists continue to monitor and test the surrounding areas to test for any leaks. A small leak could deem this project useless, while a large leak could potentially lead to disaster.[8]

The direct air capture (DAC) is developing technology for removing CO_2 from the atmosphere that can, in principle, be built anywhere.[9] Unlike other CO_2 removal processes, it does not require a relatively concentrated source of CO_2 such as with flue gas. One method for DAC uses hydroxide solutions. The process consists of two main cycles: an air contactor, the CO_2 capture, and a sorbent regeneration cycle. In the first cycle, CO_2 from the atmosphere is absorbed by means of a chemical to form a carbonate (e.g., K_2CO_3). The carbonate solution then reacts with calcium hydroxide ($Ca[OH]_2$) to form calcium carbonate ($CaCO_3$). In the next cycle, the $CaCO_3$ is separated from the solution and calcined to lime (CaO) in an oxy-fired fluidized bed that operates at approximately 900°C. During this stage, CO_2 is released while the CaO is recycled and rehydrated to Ca(OH). Finally, this hydroxide is used to react with the carbonate solution.

19.6 Carbon Dioxide Sequestration

Sequestration is to separate, seclude, and in this case, store something. Why the interest in sequestration? There is a simple answer: concern with carbon dioxide emissions into the atmosphere.

Preliminary estimates indicate that the amount of CO_2 captured from industrial processes, including combustion and chemical manufacturing, and produced from naturally occurring subsurface CO_2 reservoirs is approximately 44 million metric tons carbon dioxide equivalent.[10] Currently, more than 95% of this CO_2 supplied to the economy is injected underground for enhanced oil and gas recovery. However, CO_2 may be injected underground for geologic sequestration (GS). GS is the long-term containment of a CO_2 stream in subsurface geologic formations and is a key component of a set of climate change mitigation technologies known as carbon dioxide capture and geologic sequestration (CCS). CCS has the potential to enable large emitters of CO_2 such as coal-fired power plants to significantly reduce GHG emissions.

19.6.1 Carbon Capture Processes

19.6.1.1 Geologic Sequestration

Underground geologic formations that can be used for GS include deep saline formations, oil and natural gas reservoirs, and unminable coal seams. In addition. CO_2 may be injected into other types of subsurface geologic formations, such as basalt formations.

GS occurs through a combination of structural and stratigraphic trapping, residual CO_2 trapping, solubility trapping, mineral trapping, and preferential adsorption trapping. These mechanisms are functions of the physical and chemical properties of CO_2 and the geologic formations into which the CO_2 is injected.

19.6.1.2 Enhanced Oil and Gas Recovery

CO_2 is currently being injected into subsurface geologic formations in the United States for enhanced recovery (ER). The CO_2 currently being used in ER is primarily produced from naturally occurring underground CO_2 reservoirs but also captured from industrial processes, including combustion and chemical manufacturing.

ER involves injecting CO_2 into oil or natural gas reservoirs via injection wells for the purposes of increasing crude oil production or to enhance recovery of natural gas. The crude oil and CO_2 mixture is produced from production wells and sent to a two-phase separator where the crude oil is separated from the gaseous hydrocarbons and CO_2. The gaseous CO_2-rich stream then is typically dehydrated, recompressed, and reinjected into the oil or natural gas reservoir to further enhance recovery. If the concentration of hydrocarbons in the CO_2 stream from the dehydrator is significant, then an acid gas recovery unit is used to separate the hydrocarbons from the CO_2. Injection of CO_2 into unminable coal seams may result in displacement of methane (CH_4) from the coal seam and subsequent production of the methane as a product (or enhanced coal bed methane).

19.6.1.3 Other Industries that Use CO_2

EPA identified and considered a total of 22 commercial end-use sectors that use gaseous liquid, or solid CO_2, excluding ER and GS. These end-use sectors are available in the literature.[11, 12]

19.7 Final Editorial Thoughts (of One of the Authors)

Regarding future trends, there appears to be three approaches to the response to this problem.[13] The first approach is the "wait-and-see" approach, where countermeasures may be ineffective if too much time is wasted. The second course of action is the "adaptation to incurable changes" attitude, which is based on the assumption that there will be plenty of time in which to decide and act on climate change, and the change is unavoidable. The third line of attack is the "act now" approach, which is the only one that demands an immediate legislative (and industrial) response. But the problem lies in the legal and economic systems which normally respond only to immediate and certain threats.[14] Which course of action the world will adopt regarding the greenhouse problem is difficult to predict at this time. At this point in tune, the answer appears to be the third approach. Whether it will be wise or successful is another story.

The EPA earlier introduced the aforementioned CCTP in 2005. The purpose of this initiative was to attain the sustainable technology necessary to provide clean, safe, and reasonably priced energy sources on a global scale. The CCTP se a Strategic Plan that outlines six major goals to be accomplished using the $3 billion in federal spending that was set aside for this program. These objectives are summarized below[14]:

1) Reduce emissions from energy end use and infrastructure
2) Reduce emissions from energy supply, particularly by development and commercialization of no- or low-emission technologies
3) Capture, store, and sequester carbon dioxide
4) Reduce emissions of GHGs other than carbon dioxide
5) Enhance the measurement and monitoring of GHG emissions
6) Strengthen the contributions of basic science to climate change technology

"It has always been a challenge for me to convince college graduates to join an oil and gas company. There is a strong desire for young professionals and academics to work with 'green' technology. This is completely understandable, but what I always told them was that we can't forget to improve what exists while we keep an eye on the future. It would not be possible to turn off fossil fuels overnight, so the questions that need immediate action are: how do we improve coal combustion, what cleaner fuels such as natural gas can be substituted, how can we improve efficiency, how can we reduce emissions? Overhauling our energy system to renewable fuels is potentially feasible, but not in the short term. While it is important to look to the future, it is imperative to also give some attention to improving existing equipment and systems in a cost-effective manner."[13]

19.8 Final Editorial Thoughts (of the Other Author)

The literature that has appeared over the last decade on the influence of carbon on global warming has been extensive as noted earlier.[15] Most technical individuals strongly believe that human behavior is unalterably changing the world's climate by continuing to release carbon dioxide into the atmosphere, carbon that was previously bound for millennia on millennia in fossil fuels. These concerns have fostered new initiatives and concepts such as the Kyoto Protocol, carbon caps and trading, carbon sequestration, carbon footprint analyses, and restrictions on coal-fired utility boilers. Engineers and scientists believe there are lesser or insignificant impacts of human influence citing other phenomena such as sunspot activity, methane releases, and cosmic radiation, as examples of climate-forcing events. The difficulty with any opinion on climate change is that the subject is so broad and complex, that it is difficult for any one individual to be truly knowledgeable of all competing factors and analyses used to be able to reach conclusions on the impact of each process and their ultimate effect on climate. As an example, the Earth may be entering a period of significantly reduced solar radiation from sunspot activity in the immediate future. Such events have coincided with mini-ice ages in the past. Depending on the magnitude and duration of the reduction of solar activity in the near term, warming caused by increased carbon dioxide concentrations in the atmosphere may be counterbalanced by the cooling effects of minimal solar activity. The hypotheses on human's role in global warming are compelling but competing theories continue to emerge. Until independent, non-biased research and analyses are performed to resolve the current differences in the theories on climate change, it may be best for the subject to remain outside the scope of this book. The reader is directed to the current literature to form their own opinion on human's role in global warming but is cautioned because there are many biased opinions that are being promoted by individuals with specific agenda, making it difficult even for a trained engineer or scientist to separate reality from hyperbole.[16]

Here is an excerpt of part of a recent article by this same author[17]:

You are no doubt aware that lawyers are adept at creating problems while engineers are noted for their ability to solve problems. In the days of old, some lectures to my chemical engineering students emphasized how to solve problems. Here is basically what I had to say. First, clearly define the problem. Then apply my three Cs concept: Cause, Consequence, and Cost. In effect, what is the cause, what are the consequences, and what will it cost to remedy/eliminate the problem? Here is what we have when this simple concept is applied to global warming and climate change: We really don't know the cause and we really don't know the consequences. Some of my critics claim that I often complain about problems but never offer any solutions. Well, here are seven possible solutions to this potential environmental problem by your favorite author that will have little to no adverse effect on society.

1) Plant 500,000,000 trees. The prestigious journal Science claimed a trillion trees would do the job in the most efficient manner.
2) Convert totally from a coal energy economy to one totally based on natural gas. I have previously shown that it would halve (actually 46%) CO_2 emissions. This partial energy conversion in the United States accounts for our achieving the goals of the (ridiculous) 2015 Paris Climate Account.
3) Convert to a natural gas/oil energy economy. Oil emits approximately 25% less CO_2 than coal per unit of energy generated.
4) Embark on more aggressive domestic and industrial energy conservation programs. This can include converting to electric cars, capping unused wells, and controlling CO_2 and CH_4 (methane) emissions from fracking operations and landfills.
5) Convert to nuclear power. It's simple, safe, and economically attractive and there would be no CO_2 emissions. This is obviously the cheapest and most effective solution.
6) Halt World Bank funding to China for potential coal-fired power plants since China is no longer (2015) classified as a developing nation.
7) Any combination of the above.

Bottom line: For the past 4.5 billion years, the Earth's temperature has varied from year to year, decade to decade, century to century, millennium to millennium, etc., and the changes, at times, were really significant. The same can be said about Earth's climate. None of these changes were man-made, and only some of these changes have been satisfactorily explained. Yet here we are, not only having survived but also prospered. At the present time, the cause(s) for today's changes are unknown, the consequences are unknown, and therefore, any attempt to address these phenomena with massive economic expenditures and altering present lifestyles would be unreasonable.

References

1 Global Carbon Cycle. The Woods Hole Research Center. Available from: http://www.whrc.ore/carbon/index.htm.

2 **F. Lyman**, *World Resources Institute: The Greenhouse Trap*, Beacon Press, West Palm Beach, FL, 1991.

3 **S. Thompson**, *The Greenhouse Effect*, Lucent Books, New York, 1993.

4 **L. Theodore**, *Personal Notes*. East Williston, NY, 1991.

5 **J. Shao**, *Amine Purification System (AmiPur) – Continuous Heat Stable Salts Removal from Amine Solutions. Oil & Gas*, Eco-Tec. Inc., Canada. http://www.eco-tec.com/techpapers/TP%20166%20AmiPur.pdf.

6 **D. Dortmundt, and D. Kishore**, *Recent Developments in CO2 Removal Membrane Technology*, UOPLLC, Des Plaines, IL, 1999.

7 PhysOrg. Patented Technology Captures Dioxide from Power Plants. 2005. Available from: http://PhysOrg.com.

8 **P. Tesler**, Air repair, *Curr. Sci.*, 93, 6–7, 2008.

9 **M. Samari et al.**, *Direct Capture of Carbon Dioxide from Air via Lime-Based Sorbents*, Springer, 2019.

10 US EPA, *Inventory of U.S. Greenhouse Gas Emissions and Sinks*. 1990–2008. EPA 430-R-10-006 RTP, NC2010.

11 US EPA, *Vulnerability Evaluation Framework for Geologic Sequestration of Carbon Dioxide*. EPA 430-R-08-009 RTP, NC, 2008.

12 **R. Boyle**, *Dead Heat*, Basic Books, New York, 1990.

13 **J. Richardson**, *Personal Notes*. CT 2023.

14 U.S. EPA. *Climate Change Technology Program*. Available from: http://www.epa.gov/climatechange/policy/cctp.html. Accessed April 28, 2009.

15 **L. Theodore**, *Personal Notes*, East Williston, New York, 2001.

16 **K. Skipka, and L. Theodore**, *Energy Resources: Availability, Management, and Environmental Impacts*, CRC Press/Taylor & Francis Group, Boca Raton, FL, 2014.

17 **L. Theodore**, *On Solving the Greenhouse Gasses Problem*, Queens Gazette, Astoria, NY, 2020.

20

Flue Gas Desulfurization Systems

Note: Parts of this chapter on flue gas desulfurization systems are an edited and updated version of an earlier chapter written by Timothy W. Devitt for a 1981 Prentice-Hall publication titled "Air Pollution Control Equipment: Selection, Design, Operation, and Maintenance" and which was edited by L. Theodore and A.J. Buonicore. The rights of the book were later transferred to Theodore Tutorials during the 1985–1986 time period.

Flue gas desulfurization (FGD) is the process of removing sulfur oxides, primarily SO_2 from combustion gases. In wet process FGD systems, flue gases are contacted with an absorbent in a vessel called either an absorber or a scrubber. The SO_2 reacts with the absorbent or dissolves into the solution to produce a slurry or liquid that contains dissolved or solidified sulfur compounds. In most operating systems, water is used to dissolve or suspend the reacting chemicals, although a number of dry absorption systems are used in commercial applications. The advantages of dry scrubbing include potential cost and energy savings and relative ease of operation compared with wet FGD systems. Dry scrubbing, however, requires higher chemical consumption and may be restricted to low-sulfur coals. Because most of the commercial dry scrubbing installations are lesser in number, there is a lack of data on full-scale systems. Therefore, this chapter is limited to wet FGD systems.

20.1 Description of Control Device

Wet FGD processes are grouped into two general categories – regenerable and non-regenerable – depending on whether the sulfur is separated from the absorbent as waste. Non-regenerable processes produce a sludge that requires disposal in an environmentally sound manner. Regenerable processes include additional steps to convert the sulfur into by-products such as liquid sulfur dioxide, sulfuric acid, or elemental sulfur.

Six basic wet FGD systems are used in this country, primarily to treat flue gas from large, coal-fired boilers. Of these, the lime scrubbing, limestone scrubbing, sodium carbonate scrubbing, and dual alkali systems are classified as non-regenerable. The magnesium oxide scrubbing and Wellman – Lord systems are classified as regenerable processes.

Because most operating FGD systems are non-regenerable (a trend that is likely to continue because of their generally lower capital and operating costs), only the non-regenerable systems are reviewed here. Furthermore, only the wet lime- and limestone-based systems are discussed, although many of the design, operation, and maintenance considerations are similar for other systems.

20.1.1 The Lime Scrubbing Process

The first step of the process is the preparation of a lime slurry. Lime (CaO) is reacted with water in a slaker to produce a slurry of calcium hydroxide ($Ca[OH]_2$) and water, according to the following reaction: $CaO + H_2O \rightarrow Ca(OH)_2 + \text{heat}$. This material is diluted with recycled water to produce a scrubber slurry containing about 15–20% solids.

The lime slurry is added at a regulated rate to a recycle tank that contains the scrubbing slurry. Large pumps circulate the scrubbing slurry through the scrubber or absorber vessel. In the absorber, the slurry droplets are brought into contact with

Optimizing Air Pollution Control Equipment Performance: Operation & Maintenance, First Edition. Jay Richardson and Louis Theodore.

the flue gas containing SO_2, and the SO_2 passes from the gas into the droplets driven by a concentration gradient driving force. The dissolved SO_2 reacts with calcium from the lime according to the following reactions:

$$Ca(OH)_2(s) \leftrightarrow Ca(OH)_2(aq) \leftrightarrow Ca^{2+} + 2OH^- \tag{20.1}$$

$$2OH^- + CO_2 \leftrightarrow CO_3^{2-} + H_2O \tag{20.2}$$

$$CO_3^{2-} + CO_2 + H_2O \leftrightarrow 2HCO_3^- \tag{20.3}$$

$$Ca^{2+} + CO_3^{2-} \leftrightarrow CaCO_3 \tag{20.4}$$

$$CaCO_3(s) \leftrightarrow CaCO_3(aq) \leftrightarrow Ca^{2+} + CO_3^{2-} \tag{20.5}$$

$$SO_2(g) \leftrightarrow SO_2(aq) + H_2O \leftrightarrow H^+HSO_3^- \tag{20.6}$$

$$HSO_3^- \leftrightarrow H^+ + SO_3^{2-} \tag{20.7}$$

$$Ca^{2+} + SO_3^{2-} \leftrightarrow CaSO_3(aq) \tag{20.8}$$

$$Ca^{2+} + SO_3^{2-} + \frac{1}{2}H_2O \leftrightarrow CaSO_3 \bullet \frac{1}{2}H_2O(s) \tag{20.9}$$

$$CaSO_3 + H^+ \leftrightarrow Ca^{2+} + HSO_3^- \tag{20.10}$$

$$H^+ + HCO_3^- \leftrightarrow H_2CO_3 \leftrightarrow CO_2(g) + H_2O \tag{20.11}$$

$$CaSO_3 \bullet \frac{1}{2}H_2O + \frac{3}{2}H_2O + \frac{1}{2}O_2 \leftrightarrow CaSO_4 \bullet 1H_2O \tag{20.12}$$

These reactions take place to some extent in every lime scrubbing system; similar reactions occur in limestone systems. Factors such as ionic concentrations, pH, temperature, and retention time in the reaction tank influence the speed and completeness of the various reactions. The net effect, however, is the removal of SO_2 from the flue gas, depletion of calcium hydroxide in the slurry, and generation of insoluble calcium sulfite and calcium sulfate. The calcium ions used in the reactions are replenished by the dissociation of calcium hydroxide. The use of an insoluble alkaline reagent is the primary difference between the lime and limestone slurry scrubbing systems and the so-called clear solution scrubbers (e.g., double alkali scrubbers), where the predominant reactive alkali (sodium) is more soluble.

Forcing most of the sulfite/sulfate precipitation to occur in the reaction tank requires a certain amount of retention time to permit the slurry to reach chemical equilibrium before it is recirculated to the scrubber. For this purpose, a reaction tank, sometimes an integral part of the scrubber vessel, is usually sized with a retention time of between 5 and 10 minutes.

The calcium sulfite and calcium sulfate that are formed in the reaction tank must be removed from the system. This is accomplished by bleeding a stream from the recirculating slurry to a thickener or clarifier. The calcium sulfite and sulfate settle out and are removed from the underflow of the clarifier in a slurry of about 30% solids. The overflow, a clear liquid of less than 1% solids, is brought back to the system for reuse in the process. The sludge from the underflow is disposed of by one of several means.

One method of disposal is to pump the sludge into a pond. In this case, the sludge is pumped from the thickener underflow (or occasionally directly from the reaction tank) to a large pond, where the solids settle. Clear water is recycled from the pond, as necessary, to keep the pond from overflowing. Return of all this clear liquor from the pond to the process is called a "closed-loop" operation.

Another method of disposal is to dewater the sludge in a vacuum filter, chemically stabilize it using generic or proprietary processes, and dispose of it in a landfill. When vacuum filtration is used, the filtrate is returned to the system to create a closed-loop operation.

20.1.2 Limestone Scrubbing Process

Limestone slurry preparation is the only step that is significantly different from the lime scrubbing process. Lump-size limestone is pulverized to ~200 to ~300 mesh in a ball mill and slurried with water. The fine size is needed because the dissolution of $CaCO_3$ is a slow reaction, the rate being directly related to the particle surface area and therefore to the particle size. For further limestone dissolution, the slurry is maintained at an acidic pH, in contrast to lime-based systems, in which scrubbing slurry is slightly alkaline. The overall process reactions are the same as for the lime slurry scrubbing process, except the reactant species is $CaCO_3$, not $Ca(OH)_2$.

Figure 20.1 illustrates a typical limestone process. The lime process is similar, except a lime slaker replaces the ball mill of the limestone process.

Other wet process FGD systems operate similarly. The SO_2 is scrubbed from the gases in a contactor, and reaction products are removed in a separate vessel. Several excellent texts and reports are available that describe the various systems, their chemistry, and the status of commercialization.[1–8]

20.2 Design Procedures

Proper system design is crucial to the successful operation of an FGD system. Although this is true for all air pollution control systems, it is especially important for FGD systems. In contrast to most other air pollution control systems, there are two design considerations; in addition to adequate mechanical design, an FGD system must receive careful chemical process design. Many operating problems have developed in these systems as a result of inadequate chemical design.

Table 20.1 presents a partial list of the factors outside the scrubbing system battery limits that influence system design, and Table 20.2 lists the mechanically and chemically related design parameters that must be considered within the scrubber battery limits. A significant body of knowledge has been established concerning proper mechanical design, but the factors bearing on chemical process design and their interrelationships with the mechanical equipment design have only begun to receive adequate consideration.

Chemically related problems, primarily scaling and corrosion, are due to improper consideration and/or inadequate control in the operational phase of process chemistry. Mechanically related problems refer to the malfunction or failure of mechanical components in the scrubber and auxiliary systems, such as pumps, fans, and agitators. Failure of field instruments, expansion joints, and piping may sometimes be caused by interrelated mechanical and chemical conditions. Indeed, the most persistent and significant problems with FGD systems have both chemically and mechanically related contributing factors. The design practices being used to minimize these problems are discussed below.

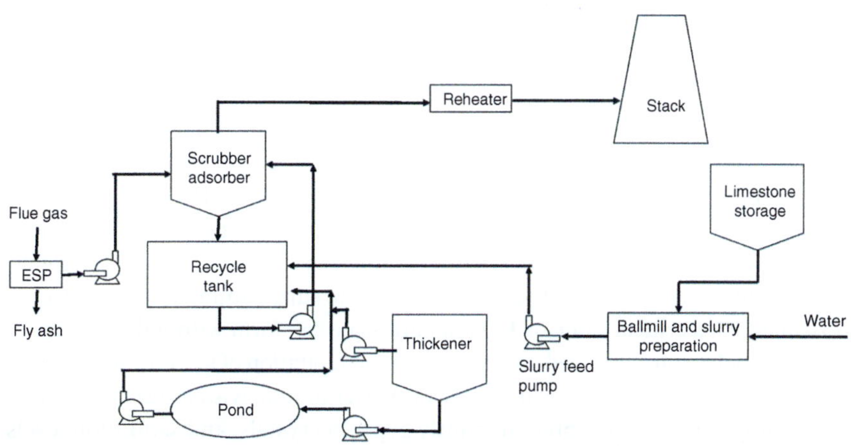

Figure 20.1 Flow diagram for a limestone scrubbing system.

Table 20.1 Factors influencing scrubber system design outside the battery limits.

Coal properties	Sulfur content Ash content Fly ash composition/particle size Chloride content Heating value Moisture content Composition variability
Boiler design	Type of boiler Size of boiler Age of boiler Flue gas flow Loading characteristics
Reagent properties	Percent inert Ca, Mg contents Reactivity Size
Size considerations	Land availability Soil permeability Rainfall Climate
Regulations	SO_2 emission/ambient standards Particulate standards Plume visibility standards PSD/nonattainment considerations Water quality standards
Makeup water	Source Chemical composition
Flue gas	Temperature Flow Dew point Particulate loading Particulate alkalinity SO_2 content O_2 level

20.2.1 Control of Scale Formation

Scale-free operation can be accomplished by operating in either of two basic modes: coprecipitation or control of supersaturation. Coprecipitation involves the removal of calcium sulfate from the system as part of a calcium sulfite/sulfate solid solution. If a system is operated so that the maximum oxidation in the slurry circuit is about 16% (i.e., 16% of the SO_2, on a molar basis, is oxidized completely to the SO_4^{2-}), the scrubbing liquor can remain subsaturated with respect to calcium sulfate (gypsum). Above this level, more calcium sulfate is formed in the slurry circuit than can leave the system in a coprecipitated form. This causes the sulfate concentration to increase and the system to operate supersaturated with respect to gypsum. When the relative saturation levels approach a critical value of 140%, a hard scale can form within the system.

Supersaturation can be controlled by providing nucleation sites (seed crystals) for the precipitation of gypsum and calcium sulfite/sulfate solid solution. This maintains the system below the critical (140%) level of saturation with respect to gypsum.

Alternatively, the slurry in a forced-oxidation system contains predominantly gypsum crystals, and saturation levels are maintained at 100–110% with respect to gypsum. The use of forced-oxidation systems has been especially beneficial for limestone systems treating flue gas with low-sulfur dioxide concentrations. For these applications, the low-sulfur dioxide levels, coupled with the long liquid retention times, result in high "natural" oxidation levels. This makes forced

Table 20.2 Factors influencing scrubber system design inside the battery limits.

Absorber type
Waste slurry disposal scheme
Degree of redundancy
Reheat type and amount
Degree of instrumentation
Mist eliminator
Raw material reactivity
Process layout
Materials of construction
Makeup water distribution
Process Chemistry
pH gradient
Sulfite oxidation
Chloride balance
Liquid/gas ratio
Point of fresh water addition
Reagent utilization

oxidation a particularly attractive method for improving sludge quality, minimizing scaling, increasing sulfur dioxide removal, and improving reagent utilization.

Another method for controlling scaling involves increasing the magnesium-ion concentration. This increases the liquid-phase alkalinity of the scrubbing slurry and the amount of sulfite and sulfate the scrubbing slurry can hold without exceeding solubility limits. The overall effect is a subsaturated operation that produces higher sulfur dioxide removal efficiencies and higher utilization. Organic acids, such as citric, benzoic, and (more recently) adipic, are also being evaluated for use in increasing the total alkalinity of the scrubbing liquor.

20.2.2 Mechanical System Design

20.2.2.1 Balanced Draft or Booster Fans

In addition to placing these fans upstream of the FGD system, another development is the use of variable-pitch, axial-flow fans. The main advantage of this design is its consistently higher efficiency (versus centrifugal fans) over the entire boiler operating range, which produces substantial energy savings. Other advantages are better flow control with variable frequency motor drives, easy access and maintenance, less severe construction requirements, and increased reliability.

20.2.2.2 Dampers

Bypass and isolation dampers are used to regulate the flow of flue gas into and around the FGD system. The primary purpose of isolation is continuance of unit operation while the scrubber modules are under maintenance. Common damper designs include slide-gate (guillotine), single-blade butterfly, and multi-blade parallel (louver). Corrosion and erosion of the various types of dampers and damper seals have been common. In some cases, dampers have failed entirely or have been so inefficient that the modules could not be maintained during bypass situations. The current trend is toward two-stage louver dampers having a pressurized seal-air system that maintains a positive pressure between them. Pressurized seal air increases the energy demand of the system, but it contributes significantly to successful damper operation.

20.2.2.3 Scrubbers

The trend in the design of lime/limestone slurry FGD systems is toward the use of simple, relatively open units, such as spray towers, or combinations of venturis and spray towers. Spray towers have few internal components in the gas – liquid contact zone and thereby offer fewer sites for the deposition of solids, collected fly ash and unused reagents, and for the formation of scale. Spray tower operation has been very successful, but SO_2 removal efficiency may be limited when high-sulfur coal is fired. In such circumstances, the use of more reactive absorbents such as lime or magnesium-promoted lime or limestone, or the use of combinations of scrubbers such as a venturi in series with a spray tower, may be required.

20.2.2.4 Reaction Tanks

Coinciding with gas-side staging via a venturi/spray tower or similar combination is liquid-side staging, in which the hold tanks are arranged in series to approach plug flow reactor design.[9]

20.2.2.5 Mist Eliminators

Chevron and baffle-type mist eliminators[10] continue to be the predominant designs used in US utility FGD systems. The popularity of these separators is due to design simplicity and flexibility, adequate collection efficiency for medium- to large-size drops, relatively low-pressure drops, open construction, easy access for maintenance, and relatively low cost.

20.2.2.6 Reheaters

A pronounced preference for stack gas reheat is still evident for systems in service and committed for future operation. In several systems, the use of wet stacks has been abandoned in favor of reheat because of problems with stack corrosion, plume dispersion, and localized "rain" caused by plume condensation. Stack corrosion is more pronounced where high-sulfur coal is burned because sulfur dioxide concentrations are higher.

20.2.2.7 Solids Separation (Sludge Dewatering)

The major development in this area is the increased emphasis on dewatering and stabilization of FGD wastes for managed landfill disposal and the corresponding decreased emphasis on ponding. Formerly, a pond was expected to fulfill three functions: clarification, dewatering, and temporary or final sludge storage. The increased emphasis on disposal in landfills and structural fills and on attaining closed water-loop operations further stimulated the use of clarifiers, centrifuges, and vacuum filters. In addition to using these techniques, several installations are using forced-oxidation strategies to enhance solids settling and filtration properties, to improve process chemistry and waste sludge quality, and to decrease land requirements for sludge disposal.

20.2.2.8 Process Control and Instrumentation

Virtually all full-scale operating systems regulate the reagent feed rate by controlling slurry pH. Dip-type pH sensors have been more successful than in-line sensors because they are easier to clean and calibrate; the in-line, flow-through sensors are subject to more wear and abrasion and generally require much more frequent maintenance. Other systems for control of reagent feed have also been evaluated on full-scale systems. One type involves feedforward reagent control based on the inlet flue gas flow rate and sulfur dioxide concentration, with trim provided by slurry pH. Another type involves control of reagent feed rate by use of the outlet sulfur dioxide as the control variable. Limited success has been reported for both systems, primarily because of the difficulty in obtaining accurate and consistent readings from sulfur dioxide gas analyzers; this has been especially true on high-sulfur coal applications. Nonetheless, this type of process control should enable much better control of process chemistry, particularly with limestone systems and variable boiler loading, and should continue to be developed.

Other critical instruments are those that measure and control the density of recycled slurry and the level of reaction tank liquid. In general, the more sophisticated instruments based on nuclear absorption or ultrasonic principles have proved more satisfactory than mechanical bubble-tube devices.

20.2.2.9 Construction Materials

The construction materials generally preferred for use in critical areas of the FGD system are 316 and 316L stainless steels. However, some designers prefer carbon steel with a surface lining or coating that physically shields the bare metal from the corrosive environment. These linings are usually resins applied in liquid or semiliquid form, by spray or trowel, and allowed to cure. The use of lined or coated carbon steel offers the potential benefits of being able to withstand low pH/high-chloride environments much better than the 316 stainless steels and being considerably less expensive. Operating experience, however, has shown that these materials are susceptible to damage by high-temperature excursions. In addition, improper application of linings has been the suspected cause of lining breakdown and subsequent base metal corrosion, especially in reapplications, where proper preparation of the metal surface is more difficult.

The use of natural rubber, neoprene, polyvinyl chloride (PVC), fiber-reinforced plastic (FRP), and flaked-glass polyester generally predominates in the liquid and thickener circuits of the FGD system (tanks, pumps, agitators, piping, and thickeners) where the metal parts are 316 stainless steel or Alloy 20. Neoprene, polypropylene, FRP, and Alloy 20 are generally used in filtration systems.

Many systems are also being constructed of more resistant alloys in trouble spots (e.g., a wet/dry, high-temperature, high-chloride environment such as a pre-scrubber or pre-saturator). Alloys such as Hastelloy C-276, Hastelloy G, Alloy 20, Inconel 625, Incoloy 825, 317 low-carbon stainless steel, 904 low-carbon stainless steel, Jessop JS-700, and E-Brite 26-1 are being selected for special problem areas.

The use of stack lining or coating materials has been troublesome, especially where high-sulfur eastern coal is burned. Lining failures have also been reported in systems that included an apparently adequate degree of reheat. Developments in this area indicate that elastomer (Coldbran's CXL-2000) has been applied with apparent success on the flue gases at the Bruce Mansfield station, which burns high-sulfur eastern coal without reheat. The use of acid brick also appears to be successful in a similar application. At installations burning low-sulfur coal, adequate linings have not been as great a problem; many installations can get by with unprotected carbon steel flues.

20.3 Installation Procedures

The installation of an FGD system is a major construction project, and the details that affect operation and maintenance are primarily a part of the engineering design. Unless considerations of operation and maintenance are included in the design, the system will be expensive to operate and maintain.

The most serious design errors result when an item of equipment that should be expected to require eventual replacement or major repair is buried deeply within the assembly. In particular, the rotating machinery, such as pumps, fans, reagent conveyors, and vessel agitators, should be installed with sufficient free space to permit their removal without an extensive disassembly of other parts of the plant. Large dampers also may require eventual replacement. Some compromise of this provision is usually necessary because open space in a process unit greatly increases the original construction cost. Design arrangements can usually be devised, however, in which those items most likely to require major service are located near the outside of the assembly.

Instruments present a special installation problem because the cost of maintaining field instrumentation has been especially high in some FGD systems. Since the instruments are often the most delicate items of equipment in the plant, they will usually require much more frequent maintenance than will the larger pieces of equipment. These instruments can usually be serviced easily and quickly, however, and the time required to service an instrument is often much less than that required to gain access to it and to remove it from the process line or vessel. It is especially important that the installation of such higher-maintenance instruments as SO_2 analyzers and pH electrodes permit simplified access and removal.

Installation procedures that simplify FGD system maintenance can usually also simplify operations. The system design must be reviewed to ensure that the test connections provided for routine operational use are conveniently accessible to the operators.

The continuing maintenance cost of an FGD system can be adversely affected by errors that occur during checkout or startup of a new system. Failure to flush piping thoroughly before the unit is placed in operation can cause damage to pumps or vessel linings that cannot be field-repaired to its original condition and therefore create continued maintenance problems. Improper startup can begin the process of corrosion or scale formation, which will show up later as a premature equipment failure. The first few days in the operation of a new system may form the basis for excessive maintenance that may require years to correct.

20.4 Operation

Two areas of FGD system operation are addressed here, in which the problems of operation are quite different. Startup of an FGD system involves not only a thorough checklist of the unit, but also a rapid approach to the conditions of normal operation. On completion of the startup phase, the slower-paced period of normal operation begins, in which a fixed routine of activities must be maintained to keep the system performing properly. The following paragraphs first describe the considerations that apply during normal operation and then discuss the activities associated with startup.

20.4.1 Normal Operation

The routine program conducted during the normal operation of an FGD system has two objectives. First, it is necessary to determine the conditions under which the system performs best and to hold the system to those conditions. Second, an organized procedure must be devised and executed to assess continuously the mechanical condition of the unit and each unit component. Although problems and minor emergencies are always changing during normal operation, the routine of operation is never-ending.

Especially with high-sulfur coal, operation of the reagent feed equipment necessitates close coordination with the operation of the power-generation equipment to permit the FGD system operator to anticipate a change in boiler loading. In such systems, it may be necessary to manually override the automatic instrumentation during excessive load changes to prevent an out-of-limits excursion in outlet SO_2 concentration.

Another parameter that must be optimized is the concentration of solids in the scrubbing slurry. The recycle of previously precipitated calcium sulfite and sulfate is necessary to supply nucleation sites and thereby to minimize accumulation of crystals on scrubber surfaces as scale. A high concentration of solids in scrubbing slurry, however, not only minimizes scale formation, but also increases erosive wear in equipment and increases carryover of solids into the mist eliminators. The fresh or clarified water used to wash solids from the mist eliminators becomes mixed with the scrubbing slurry, and thereby reduces solids concentration. In each FGD system, therefore, an optimum balance must be made between the tendency to create scale and the tendency to plug the mist eliminators and exacerbate equipment erosion. This balance is expressed in a concentration of solids in the slurry and is measured by the density of the slurry. Optimum solids concentration is around 8% solids in lime-based FGD systems and 12–15% in limestone-based systems. Equivalent densities range between 1.06 and 1.12 g/cm^3.

Within limits, some FGD system operators also consider the cost of reagent in defining optimum conditions of operation. Especially in limestone-based systems, lower PH and higher solids content provide a greater percentage of utilization of the reagent added. However, the quality of the reagent in regard to the thoroughness of lime slaking or fineness of limestone grind probably has a great effect on reagent utilization.

The optimum pH and slurry density for a specific FGD system must be established, at least in part, by trial and error. Chemical analyses of a scrubbing slurry can establish such characteristics as scaling potential and degree of reagent utilization to speed up the stepwise procedure of defining optimum parameters. It is not yet possible to obtain a complete prediction of these values from theoretical calculations because data that define the precise effect of combinations of various operational and design options have not been fully developed.

20.4.1.1 System Health

The other facet of normal operation is the routine program of checks and examinations that are conducted to spot potential problems before they increase in complexity and result in costly correction. Following is a list of diagnostic checks that constitute a typical program of this type.

1) *Scrubber/absorber (slurry recycle).* Check the recirculation slurry header pressure at the branch take-offs to the scrubber. If the header pressure has increased at one or more of the branches, it is indicative of nozzle pluggage. If it has decreased, it indicates nozzle wear. The pressure should be checked and recorded once per shift by using handheld gauges (static pressure sensor taps frequently become plugged).
2) *Scrubber/absorber.* Check the differential pressure across the absorber section. Slow buildup can indicate scale formation, although scale formation can be quite rapid if control of the process chemistry is lost. Suggested frequency is two to four times per shift. If a low-pressure drop is detected, further checks should be made to determine whether there is low scrubbing liquor flow or low flue gas flow due to a partially closed damper, inoperable or damaged variable venturi throat, where applicable, or some similar constriction. A sudden high-pressure drop probably indicates a sudden increase in gas flow rate, possibly due to an improper fan damper setting. If the fan amperage is too high, the gas flow should be restricted immediately to avoid fan-motor overheating and serious damage.
3) *Scrubber/absorber.* Check the SO_2 removal efficiency by using data from the continuous SO_2 monitors. Decreased removal efficiency could indicate a chemical imbalance or poor gas flow distribution. The gas velocity could be exceeding design limits in one or more of the sections. The SO_2 removal efficiency information coupled with pressure drop and slurry chemical composition data can be used to determine whether poor gas flow distribution may be the problem. Suggested frequency of checking SO_2 removal efficiency is twice per shift.

4) *Scrubber/absorber*. Another indication of poor gas flow distribution is entrained solids or mist carryover. If inspection ports are available, the amount of carryover from the scrubber to the mist eliminator as well as downstream of the mist eliminator can be assessed. In severe cases, entrained solids and mist carryover can be observed. Another indication of carryover would be excessive fan vibration due to slurry deposits on the fan blades. Suggested frequency of inspection and notation in the operating log is once per day.
5) *Mist eliminator*. The differential pressure across the mist-eliminator section should be checked and visually inspected (through portholes) the appearance of blade surfaces. A pressure increase indicates scale buildup or pluggage, which can be verified by visual inspection. Excessive moisture in the scrubber outlet duct is another indication of mist-eliminator malfunction. Suggested frequency is once per shift.
6) *Mist eliminator*. The composition of the spray liquor should be checked, especially for suspended solids, alkalinity, and pH. Often, the spray liquor source is the overflow from the clarifier, although many systems use the mist-eliminator wash as the source of freshwater makeup; thus, checking the composition of the spray liquor often ties in with assessing the performance of the clarifier. High solids content in the spray liquor can lead to erosion; more significantly, however, high pH or dissolved alkalinity can lead to removal of additional SO_2 in the mist eliminators and scale formation. If the potential for scaling exists, more freshwater makeup should be used, consistent with overall water-balance limitations. The composition of the wash water should be checked and noted several times per week.
7) *Mist eliminator*. The check timing sequence and valve operation should be checked. Flow in deluge-type systems can be checked for proper timing by observing valve actuators and by feeling the headers for flow. A significant number of deluge-type systems have been found to have leaks in the automatic shut-off valves which lead to upsets in the overall water balance. Leaks can be detected by observing the pressure in the supply line with flow to all points from that line shut off and comparing that pressure with the pressure in the line with the supply open to the mist-eliminator header and flow from the mist eliminator curtailed only by the automatic shut-off valves. The ability to perform this check and the frequency of such checks depend upon the overall piping configuration and the degree of concern about the water balance.
8) *Slurry recirculation pumps*. A slow loss of discharge pressure indicates probable impeller erosion. High pressure can indicate restricted flow. Intake restrictions to the pumps can be caused by plugging of the intake lines or screens, which can lead to cavitation. Slurry flow can be checked by touching the pipe; a warm pipe (about 120°F) indicates flow. The discharge pressure should be noted at least once per shift.
9) *Slurry recirculation pumps*. Seal-water consumption should be individually monitored for each pump. Excessive water consumption indicates seal wear, which is significant because it may necessitate restricting the freshwater supply to the mist eliminators in order to maintain the overall water balance. Seal-water consumption should be noted once per shift.
10) *Scrubber slurry*. The density meter controlling the bleed rate to the thickener should be calibrated every shift by using a simple pulp tester or comparable device. This will permit control of the solids concentration in the spray liquor. The scrubber slurry also should be manually checked for pH once per shift to verify the in-place pH instruments. In addition, on limestone systems the available limestone in the slurry should be checked, because pH is a poor control indicator in such systems.
11) *Thickener underflow*. The solids concentration in the underflow should be checked. The sludge blanket depth should be maintained in the 3- to 7-ft range and blown down only at 4-ft intervals; continuous blowdown is to be avoided. The blanket depth can be checked from the top of the thickener with a dipstick. Too high or low a concentration can be controlled by adjusting blowdown frequency, the solids levels of incoming system bleed streams, the amount of flocculant addition, and rake operation. In severe instances, the sludge blanket will float so that it cannot be pumped out by the underflow pumps. This may be due to too much flocculant addition or too high a rake speed (both compounded by inattentive operation). The incoming bleed stream must be shut off, as well as the rake. Where possible, the rake should be moved up and operated in an on–off fashion to try to promote settling (e.g., 30 seconds on every 3 minutes). If it is possible to introduce the system bleed at another point, this may also help initiate settling. Proper thickener operation is critical for overall system operation and should be carefully monitored by a full-time operator.
12) *Thickener overflow*. The solids concentration in the thickener overflow should be measured periodically to determine system performance. Visual inspection of the overflow clarity is usually sufficient for assessing performance on a routine basis and should be done at least twice per shift.

20.5 Startup

The initial checkout and operation of a new FGD system is a critical phase in the life of the unit. An FGD system is a complex process unit, and no two are exactly alike. Unexpected problems develop during startup, and the available personnel are often unable to handle all the problems at once. Incidents of severe damage to FGD systems have occurred during the first days of operation.

Three considerations are of primary importance during the startup phase:

1) *Full and complete mechanical checkout of every component in the system.* There is usually strong economic pressure to get the unit on line, but extended delays and greatly increased costs can result from attempts to operate the plant before construction and checkout are completed.
2) *Prevention of system overheating.* Loss of recycle slurry flow will cause hot flue gas to pass through the scrubber/absorber, which can severely damage rubber linings, plastic mist eliminators, and metal components not designed for any temperature excursion. Loss of recycle flow can occur through pump or electrical failure, severe plugging from inadequately mixed slurry, pipeline failure, loss of liquid level in the recycle tank, and many other reasons. At all times during operation, but especially during startup, the flue gas must be immediately prevented from passing through the absorber vessel if the recycle slurry flow stops. Various systems permit this to be accomplished in a number of ways, ranging from opening a bypass damper to shutting down the boiler.
3) *Prevention of scaling.* When the process chemistry is badly out of balance, as it often is during startup, severe scaling can occur in a matter of a few hours. Speed in approaching normal pH and solids content during startup is of prime importance.

Actual startup is not difficult. It consists of first establishing slurry recycle, then starting the flow of flue gas. The slurry should be as close to optimum conditions as possible. If another FGD system is operating at the station, underflow from the second thickener can be used to prepare a slurry for the new unit. If a second unit is unavailable, best practice includes the preparation of a synthetic slurry using purchased gypsum. In either instance, a rapid adjustment of pH is important, and the pH controller should be placed in the automatic mode as soon as the proper pH value can be attained.

Many FGD systems include a startup line to permit the return of thickener underflow slurry to the recycle tank. The objective is to allow a more rapid approach to operating slurry density. If the necessary recycle line has not been provided, the use of a portable pump and a hose line may be indicated. If a sludge filter is used in the sludge disposal area, manual return of the filter cake to the recycle tank is another option. Still, another technique is the use of reagent slurry that is more concentrated than normal.

Especially for use prior to a planned shutdown in which the recycle tank will be drained, special sumps with the necessary piping are designed into some FGD systems to store slurry and thereby simplify restart. Accumulation of these solids in the thickener is also possible unless the thickener is common to several modules.

20.6 Maintenance

In a large, complex, and continuously operating process unit, programs of maintenance have evolved almost into a science, which cannot be discussed fully in this chapter. Two general areas of maintenance are recognized: emergency and preventive. Emergency maintenance follows an unexpected failure of a plant component, which often disrupts plant operation, sometimes causes a major shutdown, and almost always results in an abnormally high repair expense. Preventive maintenance is scheduled so as not to disrupt operations, and it does not involve overtime use of maintenance personnel and the high costs of emergency procurement of parts or tools. The purpose of preventive maintenance is to minimize emergency maintenance, and a judicious program of preventive maintenance is vital to the continued satisfactory performance of an FGD system.

The program is based on an inventory of every component of the plant, categorized with respect to its potential for failure and its availability for service. Some components, such as rotating machinery, slurry nozzles, and control valves, will wear more rapidly than major process vessels or utility piping. Preventive maintenance programs must assign a frequency of attention to all these items; although the components least likely to fail will be examined less frequently, they must not be ignored entirely.

Availability for service defines when each component will be examined. Components will be classified into four groups. A program of routine renovation may not be applied to a few noncritical or expendable items, which can be quickly repaired or replaced by shift personnel. Electrodes for pH measurement may be in this category. Three schedules of availability will be prepared, however, to cover all other equipment items.

A program of regular renovation is practiced while the system continues to operate on those equipment items where redundant equipment was included as part of the original design. In some FGD systems, this may be limited to pumps and fans; in other systems, it may include entire scrubbing modules. A second program of renovation covers items that, through minor modifications to operation, can be removed temporarily from service for cleaning, replacement of parts, or total replacement with a spare unit. This program includes such items as ball mills, sludge filters, and most instruments. Finally, a third program of renovation covers those equipment items that cannot be removed from service without shutting down the plant. This may be a major part of some systems, whereas in others it may only involve low-maintenance devices such as clarifier vessels and major sections of ductwork and piping. An effort must be made to refurbish this class of equipment during scheduled shutdowns, which also usually involve the boiler and the power-generation equipment.

A well-conceived maintenance program replete with appropriate spare equipment and trained personnel, coupled with a systematic approach to monitoring and control of process operations, will pay significant dividends.

20.7 Improving Operation and Performance

Optimum conditions are expressed primarily in the chemical and physical characteristics of the scrubbing slurry. Of all the measurable characteristics, slurry pH is the most important. If all other variables are constant, a consistent pH ensures that the amount of reagent fed to the scrubber/absorber is in the proper ratio to the amount of SO_2 being absorbed from the flue gas. If the pH varies, the variation will be reflected in other parameters throughout the system. If pH is too low, less SO_2 will be absorbed and more SO_2 will pass through into the stack. If pH is too high, an excess of reagent will be present in the slurry and formation of scale may result.

In general, most lime-based FGD systems operate best at a pH between 8.0 and 8.5, and limestone-based systems operate best at a pH between 5.5 and 6.0. The optimum pH will be different for different systems, however, depending on the location within the system where pH is measured, the reactivity of the reagent, and details of the unit design. The degree to which pH can be permitted to vary from optimum conditions also differs for different systems. Units designed for low- and medium sulfur coal can usually tolerate a rather wide variation in the percentage of total SO_2 that is absorbed from the flue gas and still remain within emission limits. Since high-sulfur coal produces a flue gas richer in SO_2, an equivalent percentage variation cannot usually be tolerated, and closer control of pH is necessary.

20.8 Conclusions

FGD continues to be a very important technology for the control of SO_2 emissions from combustion sources. The selection of an FGD process for a specific plant generally includes preliminary screening of applicable processes and final evaluation of the selected processes. The preliminary screening would consider performance requirements, site-specific constraints, raw material requirements, and desired end products. The final evaluation should consider operational experience with similar systems, control efficiency, energy needs, and costs.

As noted earlier, the most advanced FGD processes from a commercialization standpoint are wet lime and limestone scrubbing. Proper system design is crucial to the successful operation of any FGD system. Most of the operational problems of FGD systems are related to two primary design considerations: mechanical and chemical. The mechanical design considerations consist of materials of construction, process control, scrubber and recycle tank sizing and configuration, component redundancy, gas and liquid flow distribution, and accessibility for operation and maintenance. The chemical design considerations include recycle slurry pH, liquid/gas ratio, reagent stoichiometry, degree of oxidation, and homogeneous precipitation of solids in the recycle tank. Inadequate mechanical design will lead to plugging and erosion of mechanical components in the scrubber and auxiliary equipment. Inadequate chemical design, or loss of chemical control, will lead to scaling (deposition of calcium sulfite/sulfate/carbonate solids) and corrosion in the FGD system. Many of the problems are interrelated, and it is essential to recognize the key variables in the design phase.

Control of calcium sulfite scale formation is achieved by controlling the slurry pH and reagent stoichiometry below a certain level and operating at a high liquid/gas ratio and recycle tank residence time. Control of calcium sulfate formation is accomplished either by coprecipitation, when the oxidation level is below 16% and the scrubbing liquor is subsaturated with gypsum, or by circulating gypsum seed crystals for supersaturation, when the oxidation level is above 16%. Supersaturation can also be controlled by operating in a forced-oxidized mode. Magnesium additives (MgO or $MgSO_4$) or buffer additives (adipic or benzoic acid) have also been used to control scaling and improve SO_2 removal efficiency and reagent utilization.

Specification of proper materials of construction is another important consideration. The materials generally preferred for the scrubber vessel are 316 and 316L stainless steel, although rubber or resin lining is also widely used. The capital-cost advantage of lining should be weighed against the potential failure due to improper application or contact with hot flue gases during upset conditions. Higher-grade alloys are used for special areas in the scrubber. The use of natural rubber, PVC, and FRP predominates in the liquid circuits of the FGD system. The materials for stack lining when high-sulfur, high-chloride coal is burned require careful consideration, and no single material has appeared as the predominantly successful choice.

The process control systems using integration of automatic and manual control techniques have proven to be much more successful than fully automatic systems. The primary variables that require process control include gas flow, recycle slurry pH, reagent feed rate, recycle slurry density, and overall water balance.

An FGD system is essentially a chemical processing plant; thus, the installation philosophy used in the chemical industry should be followed. The rotating machinery, such as fans, pumps, and agitators, should be installed with sufficient free space for maintenance. Easy accessibility is also essential for high-maintenance instruments such as SO_2 analyzers and pH electrodes. The initial checkout and startup is a critical phase for an FGD system. The most important considerations in system startup are complete mechanical checkout of every component, prevention of hot gas flow through the scrubber, and rapid attainment of stable system chemistry.

During initial normal operation, the optimum values of process parameters should be identified and organized operating and maintenance procedures established to hold the system close to the optimum conditions. As noted, the optimum pH at the scrubber inlet varies between 8.0 and 8.5 for lime, and 5.5 and 6.0 for limestone systems. The optimum recycle slurry solids content is 8–10%, and 12–15% for the lime and limestone systems, respectively. Typical manual diagnostic checks include pressure in the recycle header, pressure drop across the absorber and the mist eliminator, SO_2 removal efficiency, pH and composition of recycle slurry and mist-eliminator wash water, and slurry solids content for the reagent feed, recycle, and thickener underflow streams.

It is essential that personnel be assigned exclusively to the FGD operation and maintenance team. These operators should be well trained in emergency and preventive maintenance programs. It is only through proper training of operator and maintenance personnel that careful attention is paid to design, construction, and initial operation phases will yield the long-term dividends of successful system operation.

References

1 **T. Devitt et al.**, *Flue Gas Desulfurization System Capabilities for Coal-Fired Steam Generators, 2 Vols.* EPA-600/7-78-032, Research Triangle Park, NC, 1978.

2 **Electric Power Research Institute**, *Lime FGD Systems Data Book*. Final Report, FP-1030, Research Project 982-1, Palo Alto, Calif, 1979.

3 **Electric Power Research Institute**, *Lime/Limestone Scrubber Operation and Control Study*. Final Report, RP-630-2, Palo Alto, Calif, 1978.

4 **H. Head and S. Wang**, *EPA Alkali Scrubbing Test Facility: Advanced Program – Fourth Progress Report, 2 Vols.* EPA-600/7-79-244, Research Triangle Park, NC, 1979.

5 **B. Laseke, and T. Devitt**, *Status of flue gas desulfurization, Chem. Eng. Prog.*, 75(2), 37–50, 1979.

6 *Proceedings of the Symposium on Flue Gas Desulfurization*, 2 Vols. EPA-600/7-79-167, Las Vegas, Nevada, 1979.

7 **A. Slack, and G. Holliden**, *Sulfur Dioxide Removal from Waste Gases*, Noyes Data Corp., Park Ridge, NJ, 1975.

8 **J. Smith et al.**, *EPA Utility FGD Survey: December 1978 – January 1979. Updated through July—September 1979.* EPA-600/7-79-022, Research Triangle Park, NC, 1979.

9 **L. Theodore**, *Chemical Reactor Analysis and Applications for the Practicing Engineer*, John Wiley & Sons, Hoboken, NJ, 2012.

10 **L. Theodore**, *Air Pollution Control Equipment Calculations*, John Wiley & Sons, Hoboken, NJ, 2008.

21

Biofiltration

While covering odorous objects with soil is a longstanding practice for odor control, the commercial use of biofiltration is a relatively recent development. Biofilters reproduce and improve the soil cover concept that is used in landfills and leach fields, e.g., to remove pollutants from air exhaust streams. In a biofilter, an off-gas containing biodegradable volatile organic compounds (VOCs) or volatile inorganic compounds is passed through a bed packed with damp, porous organic particles. The volatile organic and inorganic compounds are sorbed by the biologically active filter bed. Microorganisms attached to the wetted filter material aerobically degrade the sorbed chemical compounds.[1]

The biofiltration process is similar to the conventional activated sludge treatment for wastewater purification since in both instances microorganisms oxidize the organic compounds to carbon dioxide and water in a moist, oxygen-rich environment. The principal difference is that in a biofilter the microorganisms are immobilized on the solid filter material, while in wastewater treatment they are dispersed as a suspension in the liquid phase.

Biofiltration is often a highly efficient and low-cost alternative to other, more conventional air pollution control (APC) technologies discussed earlier in this book such as thermal oxidation, catalytic incineration, refrigerated condensation, carbon adsorption, and absorption by chemicals and water scrubbers. Until recently, the majority of US environmental professionals were unfamiliar with the technology and rarely considered biofiltration for appropriate APC applications. For example, the control of odorous gases in the United States is at present predominately accomplished by one of the conventional APC methods previously discussed.[2] While these traditional techniques can effectively reduce odors, they often have serious capital and operating cost drawbacks including large fuel, chemical, or water requirements. In Europe and Japan, where there is greater technology awareness, biofiltration is the most commonly employed odor control technique.[2] About 500 biofilters are currently operating in Europe[3] versus less than 50 in the US.[4] In Germany, biofilters represent approximately 60% of the APC devices utilized for the treatment of odors at wastewater treatment plants.[5]

While most initial applications addressed odor control, biofilters have been successfully employed in a wide range of applications for removal of VOCs and other air toxics. Based on experiences in Europe, biofiltration can provide economic and other advantages in comparison to alternate APC technologies when applied to air streams with low concentrations (typically less than 1,000 ppm methane) of readily biodegradable air pollutants.[3] Biofilters have consistently demonstrated high destruction efficiency (>90%) for many common air pollutants including organic compounds such as alcohols, aldehydes, amines, and inorganic compounds such as hydrogen sulfide and ammonia.[6] Successful commercial biofilter applications include the treatment of off-gas emissions from composting operations, rendering plants, food and tobacco processing, chemical manufacturing, foundries, and other industrial facilities.[6] Since biofiltration utilizes the same natural biological process occurring in soils and water, it is generally considered a safe, environmentally friendly technology. Biofiltration, like other biological processes, has the advantage that pollutants are converted to harmless, or much less harmful, oxidation products such as carbon dioxide and water. Also, biological processes typically have low energy requirements and do not generate additional environmental problems since the pollutant is not transferred to another medium, thereby creating a secondary discharge.[7]

Optimizing Air Pollution Control Equipment Performance: Operation & Maintenance, First Edition. Jay Richardson and Louis Theodore.

21.1 Description of Control Device

Biofiltration is an APC technology in which microorganisms immobilized on organic packing material treat an off-gas containing biodegradable volatile organic or inorganic compounds. In removing pollutants from off-gas streams, biofilters are fixed-film bioreactors that provide a large surface area for mass transfer between the gas and liquid phases. Biofilters utilize the processes of adsorption, absorption, and microbial degradation to sorb and eliminate the pollutant.

The process of biofiltration, illustrated in Figure 21.1, involves:

1) The collection of the raw waste gas from the emission sources with transport via a ductwork and blower system through the pretreatment equipment and biofilter reactor.
2) Pretreatment of the raw gas is typically required to remove particulates, adjust the temperature, and increase relative humidity to saturation.
3) Following pretreatment, the influent gas is uniformly dispersed throughout the biofilter reactor via an air distribution system.

In a biofilter, the contaminated off-gas stream typically flows upward through one or more packed layers of organic filter material. Soil or compost are the most common filter materials.

The filter material performs three functions[8]:

1) It provides a substrate for microbial attachment.
2) It is a source of elemental nutrients required by the microorganisms for growth.
3) It provides a porous structural matrix through which the influent gas can readily pass.

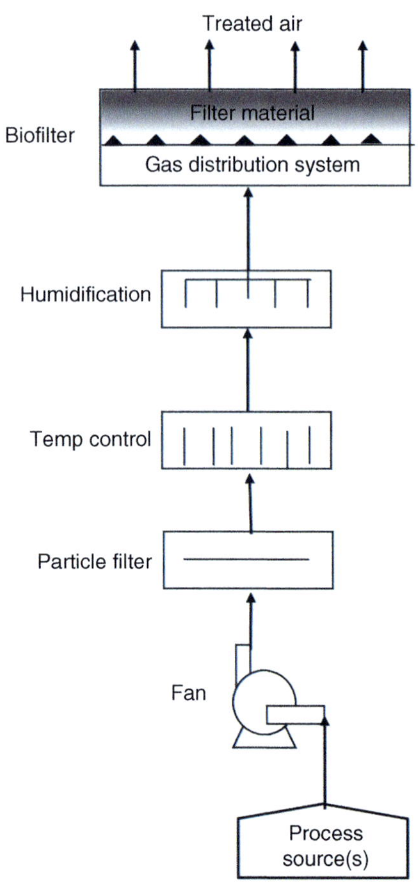

Figure 21.1 Schematic flowsheet illustrating the individual elements of an open, single-layer biofilter system.

As the off-gas flows through the organic filter material, the contaminant compounds are adsorbed onto the porous solid particles of the filter material and absorbed into the water film coating the particles. Microorganisms require moisture for survival. In a biofilter the microbial population abides in the water film that surrounds the wetted filter particles. The microorganisms are the active agents which utilize the chemicals in the waste gas as a food source to transform and destroy the pollutants to oxidation end products (usually CO_2, H_2O, and mineral salts).

Biofiltration blends the adsorption of an activated carbon filter with the washing effect of a water scrubber and with biological oxidation. Activated carbon and water scrubbers simply sorb the chemicals, and either regeneration or a water purge is required to maintain removal efficiency. The advantage of biofiltration is that the microorganisms transform the pollutants by oxidation, thereby continuously regenerating sorption capacity.

The waste gas must reside within the biofilter for a time sufficient to achieve a high sorption efficiency. Biofilters operate at temperature and moisture conditions which are favorable for microbial growth. These conditions, however, are not optimum for sorption so a biofilter requires a larger volume in comparison to a carbon adsorber in order to achieve equal removal efficiency. Biofilter residence times of at least 15 seconds are typically longer are required for high destruction efficiency, depending on the concentration and biodegradability of the pollutants. Residence time is expressed as the filter material volume divided by the volumetric gas flow rate. The biofilter volume required is specific to application and target removal efficiency and is a function of the raw gas loading, pollutant concentration, and biodegradation capacity of the filter.

Biofilters can be constructed in an open bed design with the filter material covering the air distribution system, or an enclosed design with the filter contained within a structure, protected from weather conditions.

The oldest and simplest biofilter design is a soil bed. A soil bed uses a network of perforated pipes buried about 2 ft below soil level to distribute the waste gas. Following the initial soil bed usage, biofiltration applications were relatively inactive in the United States Research, development, and installation

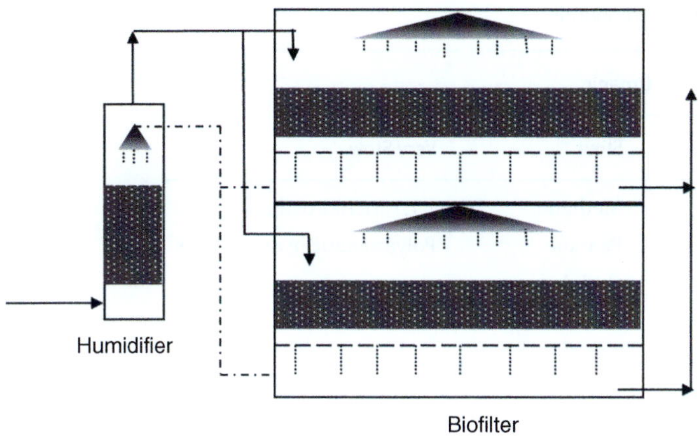

Figure 21.2 An enclosed two-level biofilter with a supplemental water spray system for moisture control and recirculation of the filter bed drainage in the humidifier.

of the technology were more actively pursued in Europe. In the 1980s more advanced biofilter designs were developed, most notably in The Netherlands and Germany. Open bed systems were designed that utilize concrete blocks or slabs with venting slots for influent waste gas distribution. Enclosed biofilters were also developed whereby the influent gas can be either upflow or downflow, and constructed as a single-layer bed or in multilevel design as illustrated in Figure 21.2. Enclosed multilevel biofilters with stacked beds can operate either in series or parallel.

21.2 Design Considerations

Systems can be used to control air emissions contaminated with low to moderate concentrations of biodegradable VOCs, hazardous air pollutants (HAPs), and odorous compounds. Biological control systems include three types of process designs:

1) Biofilters
2) Biotrickling filters
3) Bioscrubbers

In biofilters and biotrickling filters, moist biofilms are attached to a porous media in a packed-bed bioreactor. The contaminated gas stream is passed through this biologically active packed bed, where the pollutants are degraded by the microorganisms. The principal difference between biofilters and biotrickling filters is that in biotrickling filters, the liquid phase moves (trickles) through the porous media, whereas the liquid phase is stationary in biofilters. In bioscrubber systems, absorption of the pollutants takes place in one unit – a packed column, spray tower, or bubble column scrubber – while biodegradation primarily takes place in a liquid-phase bioreactor, similar to an activated sludge reactor.

Biofiltration is an appropriate technology to consider for treatment of air emissions where the waste gas contains a low concentration of biodegradable contaminants. Several attributes of the raw gas influence the feasibility of biofiltration and its removal efficiency.

Since pollutants in the raw gas are destroyed by microorganisms, the pollutant compounds in the gas must be amenable to biodegradation. The raw gas must contain an adequate oxygen content for aerobic degradation. The raw gas pollutants, both individually and as a mixture, must not be present in concentrations that are toxic to the microorganisms. Some compounds, for example, formaldehyde, are readily degraded at low concentrations, but are toxic to microorganisms at higher concentrations. Compounds that are most rapidly degradable in a biofilter are generally of low molecular weight and highly water soluble, and for VOCs, molecules with functional groups containing oxygen, nitrogen, or sulfur.[1] The biodegradability of individual raw gas components and classes of compounds are listed in Table 21.1.[1,6]

Biofilters were originally employed for odor control applications where the off-gas was generally very dilute and sulfur compounds were the primary components. In the 1980s, biofiltration technology began to be successfully employed for the treatment of industrial air emissions containing VOCs and other toxics in much higher concentrations than most odor control applications. Table 21.2 lists examples[9–11] of several full-scale biofilters which were installed for the control of industrial VOC emissions.

Table 21.1 Biodegradability of individual and classes of volatile compounds.

	Organic			
Inorganic Rapid	**Rapid**	**Good**	**Slow**	**Very Slow**
Hydrogen sulfide	Methanol	Ethyl acetate	Methane	1,1,1 trichloroethane
Ammonia	Butanol	Acetone	Pentane	Polyaromatic hydrocarbons
Sulfur dioxide	Formaldehyde	Benzene	Cyclohexane	
	Acetaldehyde	Styrene		
	Amines	Methyl mercaptan		
	Butyric acid			

Table 21.2 Examples of operating full-scale biofilters.

Industry Ref.	Location	Inlet Gas (Flow, m^3/hr)	Inlet Gas (Conc., g/m^3)	Compounds	Efficiency (Volatile Organic Compound [VOC] Removal)
Yeast production	East Brunswick, NJ	60,000	1.0	Ethanol aldehydes	85%
Casting foundry	Northridge, CA	17,000	1.9	Mainly ethanol	85%
Polymer	Trenton, MI	35,000	0.4–0.7	Alcohol aldehydes esters	90–95%
Plastics production	Bergen op Zoom, Netherlands	1,000	0.5	Toluene phenol acetone	80–95%

The five main components and design considerations of a biofiltration system typically include raw gas collection and transportation, preconditioning, humidification, influent gas distribution, and filter material. These are listed below:

1) Raw gas collection and transportation
2) Raw gas preconditioning
   a) Preconditioning particulates
   b) Preconditioning temperature
   c) Preconditioning other
3) Humidification
4) Influent gas distribution system
5) Filter material

Additional details are available in the literature[12].

The required biofilter gas residence time (filter material volume) for a specific application depends upon:

1) The influent gas:
   a) Volumetric flow rate
   b) Concentration of the pollutant compounds
   c) Biodegradability of the compounds
   d) Variability of the volumetric flow rate and concentration of the pollutant compounds
2) The required destruction efficiency
3) The biofilter's degradation capacity
4) The biofilter design, and the type and height of the filter bed packing material

The factors which influence capital cost of a biofilter project are sizing, type of design, space availability, and preconditioning requirements:

1) Biofilter size is influenced by influent gas characteristics, required destruction efficiency, and the biofilter degradation capacity.
2) Biofilter design (type of air distribution system, open or enclosed, multilevel) is influenced by the availability and cost of land. The large size of a biofilter often makes space constraints a major design concern.
3) Preconditioning requirements can add significantly to the cost if, for example, large amounts of particulate need to be removed or substantial heating or cooling is necessary.

21.3 Operation and Maintenance

Instrumentation should be installed to monitor the important factors which affect biofiltration performance. Instrumentation normally includes influent gas flow rate, temperature, and humidity before and after humidification, pressure drop through the filter material, for each of the biofilter cells: gas flow and filter material temperature.[13]

The filter material is the habitat for the microbial population. Maintaining a favorable environment is critical to sustain high destruction efficiency. The important filter material attributes to maintain are moisture, nutrient content, pH, and low gas flow resistance.

There are two sources of waste from the biofiltration process: water drainage from the filter bed and the disposal of spent filter material.

Drainage of excess water from the filter bed is a source of wastewater discharge. The leachate from the filter generally contains dissolved off-gas pollutant compounds as well as biomass and fine filter particles. Water drainage from the filter is often recirculated to the humidifier to minimize the quantity of wastewater. If filter drainage is recirculated, the humidification system should be designed to prevent fouling with solids. Periodic discharge of the leachate is usually required.[3] Since the leachate can contain contaminants, the discharge should be treated in compliance with local wastewater regulations.

The rate of decomposition of the filter material can vary depending on the influent gas, operating conditions, and type of bed material. Several European biofilters have operated successfully with the same filter material for over five years.[13,14] The selection of the filter material should consider its eventual disposal. The presence of hazardous components (such as heavy metals) should be avoided. In Europe, spent filter material has been used for gardening and landfill cover.[3] Prior to discarding, the used filter material should be tested and disposed of by appropriate methods to fulfill all applicable environmental regulations.

At startup, when the waste gas is first introduced into the filter, the microbial population begins to acclimate to metabolize the pollutant compounds. Destruction efficiency is typically lower at startup than the design target. Over time, efficiency increases as the microbial population distribution shifts toward cultures which degrade the pollutants in the off-gas. Acclimation time to achieve higher steady-state destruction efficiency can vary from a few days to as long as several months depending on the biodegradability of the pollutants and the initial population of the active microbial strains in the filter bed. For easily biodegraded compounds, the natural filter bed materials generally contain suitable cultures, making seeding with cultivated strains unnecessary. For many easily degraded air pollutants acclimation to steady-state typically takes about ten days.[7]

During periods of inactivity longer than a few hours, maintaining a minimum flow of air through the filter is required to sustain a sufficient level of dissolved oxygen in the biofilm for microbial growth and activity.[13] To maintain aerobic conditions during extended periods of waste gas flow interruptions, the filter is normally aerated by feeding fresh air, with the inlet fan operating either intermittently or at a reduced flow rate.

Also of concern is the potential for a reduction in biofiltration performance and a period of reacclimation following an extended interruption of waste gas loading. Different results have been reported in the literature[13] on the effect of extended off-gas flow interruptions. In some situations, an interruption of up to two weeks has almost no effect and, in others, a reduction in performance has been observed following a three-day shutdown.

21.4 Improving Operation and Performance

Biofilter efficiency is measured by the reduction between the raw and clean gas in specific and total chemical concentrations and/or odor intensity. Measurement methods are established based on performance efficiency of interest, odor, or chemical reduction. Measurement analysis is by standard detection methods such as chemical analysis by flame ionization and gas chromatography.

With enclosed biofilters, gas samples are taken from the exhaust stack by standard techniques. With open biofilter beds the exhaust is considered to be a short, wide stack and samples are taken by means of an inverted funnel (flux chamber). A standard procedure for sampling an open biofilter has been established by the Society of German Engineers (VDI) and is published in the *VDI Handbook of Air Regulations*.[6] The procedure describes taking the biofilter exhaust sample in an inverted funnel (hood) designed to capture the gas without the influence of wind. The hood should cover a minimum biofilter area of $1\,m^2$ with the outlet vent diameter sufficiently large for minimal impairment of flow.

Sampling sites are selected over the bed surface to be representative of the design and loading of the biofilter. The gas flow rate out of the inverted funnel is measured at the different sampling sites to determine flow uniformity of the gas distribution system. Samples for efficiency analysis are taken after allowing sufficient time for exhaust gas to displace the original air in the hood chamber. Heated tubing should be employed to avoid condensation in the sample line. Monitoring for total organic carbon is typically analyzed using a flame ionization detector or photoionization detector. Samples can be either continuously analyzed or collected for later analysis.

In situations where the concentration of the inlet compound varies, continuous analysis over a period of time is recommended to determine destruction efficiency.[3] Continuous monitoring is recommended since the biofilter's large volume reacts slowly to loading changes and inaccurate samples could result in a false efficiency conclusion. In the case of a multicomponent inlet gas, the detectors used for analysis will have a distinct response factor for each component. Gas chromatography or other chemical specific analysis on grab samples is recommended to determine the destruction efficiency of the individual components in the off-gas.

21.5 Conclusions

1) Biofiltration is an emerging APC technology where microorganisms immobilized on organic filter material treat gas emissions containing biodegradable volatile organic and inorganic compounds. The microorganisms utilize the pollutants as a food source to transform and destroy the compounds into harmless oxidation end products. Biofiltration is similar to activated carbon and water scrubbing, in regard to the sorption of the pollutants. However, it has the advantage of biodegrading the compounds to continuously regenerate sorption capacity.
2) Biofiltration has been fairly widely applied in Europe and Japan, mainly to control odors and more recently to treat industrial emission containing VOCs and other toxics. The advantage of biofiltration is that it is a relatively simple, environmentally benign APC technology which is generally lower in capital and operating cost when employed for appropriate applications.
3) Based on current experience, biofiltration has been applied most successfully in the treatment of waste gases with a dilute concentration of easily biodegraded pollutants. Destruction efficiency above 90% can be achieved for many common VOCs. Efficiency can exceed 99% for odors and increases with low concentrations of readily degraded compounds.
4) While the commercial use of biofilters is presently limited in the United States, regulatory and public demands for a cleaner environment are expected to stimulate greater interest in the technology.
5) In ozone nonattainment areas, local environmental regulators are increasingly targeting small VOC sources. For example, the Los Angeles and San Francisco Bay Air Quality Management Districts now require the control of ethanol emissions from large commercial bakeries. Many of these smaller VOC sources are low in VOC concentration, which tends to favor biofiltration.
6) The public's increased environmental awareness has stimulated demand for control of nuisance odors emitted from public health sector and industrial sources. While many objectional smells are not considered toxic or harmful to public health, odor is a very noticeable form of air pollution. The largest number of complaints to local environmental regulatory agencies is provoked by odors. Biofiltration was originally developed to control odors where for many applications it is the most cost-effective APC option.
7) While it is not used extensively in industry, for many ozone and odor control applications, biofiltration is a cost-effective, environmentally benign alternative option.

References

1 **H. Bohn**, *Consider Biofiltration for Decontaminating Gases*, pp. 34–40, Chemical Engineering Progress, New York, NY, 1992.

2 **E. Allen, and S. Phatak**, *Control of Organo-Sulfur Compound Emissions Using Biofiltration – Methyl Mercaptan*, Paper Presented at the Air & Waste Management Association, 86th Meeting and Exhibition, June 13–18, Denver, CO, 1993.

3 **G. Lesson, and A. Winer**, *Biofiltration: an innovative air pollution control technology for VOC emissions, J Air Waste Manag Assoc*, 41(8), 1045–1054, Pittsburgh, PA, 1991.

4 **IPS**, *Odor Control Completing The Composting Process*, International Process Systems, Inc., Glastonbury, CT, 1990.

5 **F. Frechen**, *Biological Deodorization of Odorous Air from Waste and Wastewater Treatment Plants*, Paper Presented at the Air & Waste Management Association 86th Meeting and Exhibition, Denver, CO, 1993.

6 **VDI Richtlinien 3477**, *Biological Waste Gas/Waste Air Purification, Biofilters*, VDI—Handbuch Reinhalten der Loft, Band 6, 1991.

7 **S. Ottengraf**, Exhaust gas purification, in biotechnology, in *Microbial Degradations*, Schonborn, W., Vol. Ed., Ch. 12 Rehm, H. J. and Reed, G. Eds, Vol. 8, pp. 427–452, VCH Publishers, Weinheim, Germany, 1986.

8 **Biofiltration Committee**, *Institute of Clean Air Companies*, Biofiltration White Paper, unpublished draft, Washington, DC 1993.

9 **D. Dragan**, *Criteria for Feasibility of Biofiltration Technology in Air Pollution Control*. Report for AAS/EPA Environmental Science and Engineering Fellowship Program, US Environmental Protection Agency, Washington, DC, 1993.

10 **J. Yavorsky**, *Biofiltration for Control of Gas Streams Containing Low Concentrations of Volatile Organic Compounds*, Paper Presented at the Air & Waste Management Association 86th Meeting and Exhibition, Denver, CO, 1993.

11 **R. Polmeroy**, *US Patent 2,793,096 De-odorizing of Gas Streams by the Use of Microbiological Growths* 1957.

12 **J. Mycoch, J. McKenna, and L. Theodore**, *Handbook of Air Pollution Control Equipment Technology*, CRC Press Taylor & Francis Group, Boca Raton, FL, 1995.

13 **D. Eitner**, *G+E Umwelt-Technologie GmbH*, Aachen, Germany, personal communication, 1990.

14 **F. Laulanie**, *SAPS*, Le Coudray en Thelle, France, personal communication, 1990.

22

Stacks

Note: Parts of this chapter are an edited and updated version of an earlier chapter written by Eldon Dille for a 1981 Prentice-Hall publication titled "Air Pollution Control Equipment: Selection, Design, Operation, and Maintenance" which was edited by L. Theodore and A.J. Buonicore. The rights of the book were later transferred to Theodore Tutorials during the 1985–1986 time period.

After a waste gas has been treated, its release to the atmosphere is usually by means of a stack. One should differentiate between the casual term "stack" and a stack or chimney which is actually used for dispersion. A stack has come to mean any vertical vent pipe that discharges gases into the atmosphere from any one of a thousand different processes. The gases may be warm or cold; they may have high exit velocities or low exit velocities, and the stack is merely a means of discharging them out of the operating area into the atmosphere where they can be diluted and dispersed.

Tall stacks or chimneys have both a historical use and a present value. Tall stacks were designed to provide a natural draft for combustion processes, but with no concern for the safe atmospheric dispersion of the various waste gases which pass through them. In fact, in most situations, the stack was designed to bring the discharge from a furnace out of the building instead of having it back up into the building. Today, however, one can see a different use for high stacks as an ultimate disposal method for certain wastes which would otherwise be difficult to treat. Stack heights throughout the world have been increasing in height for the purpose of providing better dispersion.

22.1 Description of Control Device

After the 1970 Clean Air Act passed the US Congress, the world of coal-fired power plant stack design and maintenance entered a period of dynamic change. Use of electrostatic precipitators and flue gas desulfurization (FGD) units drastically altered the environment inside stacks. Gas escape velocities were cut in half, stack temperatures decreased, and the incidence of acid condensation on the stack liner increased markedly. Stack maintainers found themselves with an increased rate of liner disintegration and failure, while stack designers found themselves specifying new liners that had no proven track record.

Part of the problem came from the scrubbers. During the 1970s, there were 15 major and 20 minor scrubber suppliers in the United States.[1] Each one produced several process design combinations. Engineers specified different processes from different manufacturers, so the power plant owner usually ended up with an assortment of components – one from column A, one from column B, and so on.

Another part of the problem came from the power production equipment. Each power plant had a different combination of boiler design, control equipment type, and associated pieces of equipment. The amounts of sulfur trioxide, sulfur dioxide, and nitrogen oxide compounds found in the flue gas varied with the new boiler designs. Also, the quantities of fly ash allowed to enter the FGD system varied, depending on the design of the control devices. The final result was a flue gas at each plant that was probably like no other in the world.

Each flue gas shared one common thread – the creation of a stack environment conducive to sulfuric acid condensation on the liner wall. To reduce this acid attack, many engineers recommended the installation of reheaters to get the stack temperatures back up above the sulfur trioxide dew point, but not everyone agreed to their use. Some engineers argued that

Optimizing Air Pollution Control Equipment Performance: Operation & Maintenance, First Edition. Jay Richardson and Louis Theodore.

gas- or oil-fired reheaters were too expensive to operate and too difficult to maintain; besides, there was another way to get stack temperatures up, divert heat directly from the boiler to the stack. This could be done economically and the final gases would still meet clean air requirements. Of course, this bypass approach would mean increasing the volume of corrosive acids in the stack, but "the stack could take it."

Perhaps the one thing that has not changed since the 1970 Clean Air Act is this common attitude that "the stack can take it." This attitude has been around a lot longer than this latest debate that generated it. Stack care and maintenance have always been a low priority for plant owners. The stack is a passive structure. It is easily forgotten until a crisis occurs. But the acidic nature of the environment found in modem stacks mandates a reexamination of this negligent attitude. Some stacks built before and since 1972 may not be able "to take it."

Advances in scrubber technology, which were once made at an exponential rate, may or may not solve many of the problems associated with wet, low-temperature flue gases. Even with the best control device and the most efficient FGD system, some percentage of sulfur dioxide, sulfur trioxide, hydrochloric acid, hydrofluoric acid, nitric acid, etc., that result from coal combustion will end up in the stack. What these acids do there will affect inspection and maintenance programs and the need for costly repairs.

A good portion of this chapter on maintenance concentrates on stack design and the corrosive effects of sulfuric acid on the stack and related equipment. Of course, one will find procedures for inspection, maintenance, and repair, but not find any operational procedures here. That is simply because there is nothing on a stack to operate.

22.2 Design Considerations

The function of a modem stack is to lift plant flue gases to a suitable elevation above the plant so that prevailing horizontal winds can disperse and help dilute the effluent contents.

Reinforced-concrete stack design is one of two types: boiler stacks designed to operate from 120°F to 600°F, and manufacturing-process stacks designed to operate up to 2,000°F. This chapter concentrates on the first type of stack and the changing nature of designs caused by air quality requirements.

To do its job right, the average utility boiler stack has to rise about 600 ft above the general plant elevation. Some stacks, depending on plant location and elevation, may go up only 350 ft, whereas others may rise to 1,000 ft. For smaller package boilers, stacks can be as low as 50 ft.

The stack height of conventional coal-fired power plants is determined from 44 FR 2608, dated January 12, 1979. This is a federal regulation that places the stack height at the building height plus one and a half times the lesser height of the building or width projection facing the prevailing wind. For example, a power plant building 250 ft high by 200 ft wide by 500 ft long would have a stack height of 250 ft plus 300 ft or a total height of 550 ft.

A typical stack consists of a concrete shell designed to withstand the stresses and loadings associated with a 500- to 600-ft vertical structure, plus a lining of independent brick, steel or concrete, corbel-supported brick, or gunned concrete. The type of lining used depends on the temperature tolerances and corrosive resistance required and the seismic zone in which the stack is located. On the stack top there is a cap made of metal, ceramic, or fiberglass.

Because of the stack environments caused by clean air regulations, most modem stacks were built in a period of design flux. Stacks built for power plants with FGD systems are all state-of-the-art designs. As such, some will perform better than others.

Stack designers use the American Concrete Institute's *Specification for Design and Construction of Reinforced Concrete Chimneys* (ACI 307-69) to plan for material, construction, and design requirements. As noted earlier, the words "chimney" and "stack" are used interchangeably in this chapter. Individuals involved in the construction of this unit usually refer to it as a chimney; individuals involved in its use, particularly those in air pollution, refer to the unit as a stack. This specification spells out the basic design standards needed to fit the right type of stack to a particular power plant. Some designers criticize the specification as being too conservative in some areas, but deviations from it are rare.

ACI 307-69 sets forth recommended loadings for the design of reinforced-concrete chimneys and recommended methods for determining the stresses in the concrete and reinforcement resulting from these loadings.

Although the specification is on firm ground about its structural standards, ACI 307-69 admits that the sections on lining materials "do not guarantee the adequacy of these materials to resist temperatures, abrasion or corrosion resulting from chimney gases." As one might expect, the area where engineering experience is least defined is where most maintenance problems occur – the liner is usually a maintainer's major concern.

When a maintenance problem occurs, it may be tempting for the maintainer to point the finger of blame at the designer or the contractor. One may be right in doing so, but in the vast majority of cases stack maintenance problems stem from the flue gases the stack must handle the natural forces that a 500- to 600-ft concrete spire must face. If the same maintenance problem is experienced repeatedly, however, it could be caused by a construction or design flaw and should be reported.

22.2.1 The Stack Foundation

Stacks are often supported on pile foundations that support a concrete mat footing, usually a circular slab. A lead baseplate is located between the footing and the concrete slab. In most stack foundations, the slab provides a wearing surface and a drain for water that might collect at the stack base.

ACI 370-69 (3.10) states:

1) *In chimneys conducting heated gases, consideration shall be given to the protection against the radiant heat of the gases for any part of the foundation which is exposed within the chimney and concrete floors either at the base of the chimneys or supported from the chimney column.*
2) *It has been demonstrated that radiated heat from gases causes high temperatures at top surface of foundation or floor if not insulated. Even in modern power plants with moderate flue gas temperatures, this radiant heat may be sufficient to crack the chimney foundation or floor although flue openings enter the chimney an appreciable distance above the foundation.*
3) *A satisfactory protection for the foundation floor is a ventilated insulation in the form of hollow tile covered with insulating brick or fire brick depending upon the temperature involved, with the voids of the hollow tile communicating directly to the atmosphere by means of air ducts when the chimney is lined or by openings in the shell when the chimney is unlined.*

Lead pans at the bottom of the flue(s) can be eaten away by acid attack. Since these pans are located below 1–1½ ft of concrete, an attack of the lead pan by the acid will occur after the concrete has been attached. When a flue begins to leak, a subverted lead pan beneath the concrete may be the cause.

In brick-lined stacks, some mortars in the lining are designed to contend with acid attack but not water. During the annual maintenance shutdown or whenever there is an extended boiler outage, the top of the stack should be covered if there is a possibility that rain may enter the stack. Water in the stack without flue gas present can result in a corrosive attack on the mortar. Water can change the acid-resistant mortar into a paste that has no structural or cohesive strength. This is especially true at the base of the stack if water is allowed to collect. This explanation of the way the mortar fails is continuously reevaluated by the industry.

22.2.2 The Shell

All concrete stack shells built with ACI 370-69 recommendations are reinforced with vertical and circumferential reinforcing steel. The size, spacing, and amount of reinforcing bars used in construction depends on loading factors and the wall thickness required. Shell walls that are 18 in. or thicker have inner-face reinforcement of No. 4 bars on 24-in. centers vertically, and No. 4 bars on 12-in. centers horizontally. Some older stacks may not have inner-face reinforcing, so they are more subject to thermal and wind-load cracking than reinforced inner faces.

ACI 370-69 states: "The minimum shell wall thickness for any chimney with an internal diameter of 20 ft or less shall be 7 in. When the internal diameter exceeds 20 ft, the minimum thickness shall be increased ¼ in. for each 2 ft increase in internal diameter."

Since the external shell is concrete, it is resistant to most forms of mechanical attack (for example: precipitation and wind). But eventually, every concrete shell will experience spalling. Thermal changes from normal day and night exposures can cause small cracks in the concrete. These will gradually increase in size until spalling occurs. In colder climates, water driven into these cracks can freeze and cause physical damage more rapidly. With the introduction of FGD systems, there can be acid downwash along the top of the shell. This can result in a chemical attack on the concrete. (One should consider coating the outside of the concrete shell with a coating that is resistant to sulfuric acid attack. A coating based on polyester or epoxy resins could prove adequate in this environment. Fluoroelastomeric coatings also have the necessary acid resistance plus the advantage of bridging small hairline cracks in the concrete.)

Sulfuric acid in hairline cracks from downwash or internal infiltration can form salts and accelerate spalling. Structural failure may occur if the acid and water are allowed to reach the reinforcing bars. If the attack is allowed to go this far without detection, a major sectional repair or total replacement may be required.

Wind can also cause a cracking problem, or make an existing problem worse, especially in unreinforced areas of concrete. A timely system of crack inspections and a policy of instant repair will help avoid the difficulty and expense of repairs or reconstruction.

Another form of mechanical attack is seismic disturbance. The shell design in earthquake-prone areas is usually based on the minimum lateral seismic force data given in ACI 370-69. A dynamic analysis based on the local seismic-response spectrum may also be a part of the design planning. A dynamic analysis gives insights into the local geology which may directly affect design decisions.[2]

22.2.3 The Liner

Unlike the concrete shell, the brick liner is not built to precise engineering specifications. The type of liner used in the stack was chosen according to seismic conditions, the type of service (constant or intermittent peaking), gas temperature and gas composition in the flue, and the degree of reliability required. Structural requirements were, of course, a part of the liner selection.

The function of a liner is to accept the thermal stress in the stack column and to help protect the shell from the corrosive acids of coal-fired combustion. There are six basic types of liners: independent brick, steel, gunned concrete (gunite), fiberglass-reinforced resin (more commonly known as fiberglass-reinforced plastic liner [FRP]), alloys, and cast-in-place concrete. The use of high-alloy metal for liners is nascent; cast-in-place concrete liners are all but extinct. A liner that was popular in the past – sectional brick liners supported by corbels in the concrete shell – is rarely found today because acid penetration can subvert the support corbels.

Each modern-era liner will be discussed separately, but first, consider the environments a stack may encounter.

The major concern in liner design is acid formation which leads to corrosion. Acids generated in the coal combustion process include sulfuric acid, sulfurous acid (an early stage of sulfuric acid), hydrofluoric acid, hydrochloric acid, and nitric acid. The dewpoints of sulfurous, hydrochloric, and hydrofluoric acids are usually well below the dew point of sulfuric acid. Prior to the introduction of FGD systems, these did not pose a threat to the stack. Today, temperatures in the stack can be within or below the dewpoints of sulfurous, hydrochloric, and hydrofluoric acids.

Hydrofluoric acid is the Achilles' heel of inorganic monolithic gunned linings, potassium silicate mortars, bricks, and fiberglass fillers. Concentrations of this acid have been detected in linings and are being monitored to determine if they are reaching problem concentration levels.

Nitric acid, an oxidizing acid, will readily attack most components of the stack. All the corrosion products of nitric acid are water-soluble, so corrosion will proceed until the nitric acid is used up. Only scattered reports have been made of nitric acid presence in power plant stacks. It is expected that new boiler designs that minimize the formation of nitrogen oxide compounds will result in minimizing the production of stack-bound nitric acid.

22.2.4 Brick Liners

Independent brick liners, the type most widely used, evolved from corbel brick liners, the standard design of the 1930s and 1940s. The corbel liner consisted of 30- to 40-ft lengths of brickwork, usually 4 in. thick, that were supported by corbels (ledges) built into the liner face of the concrete shell. A space called an annulus, 2 in. or more, separated the lining from the column. The annulus was often filled with insulation to offset thermal stresses within the shell.

The corbel design worked fine when stack temperatures were 400–600°F and the stack was self-drafting. But after World War II, power plant designs became more efficient, stack temperatures dropped to 300°F, and the positive pressure inside the stack forced acid through the brick to the corbels and to the shell. The high alumina-based mortar used in the corbel brick liners began to soften as the pH reached 4.5, then the mortar failed completely at 3.5 pH. Mortars containing sodium compounds mixed with the sulfuric acid to produce growth crystals that swelled to 300% of their original volume. Because of these failures, the days of the corbel brick liner were numbered.

Replacing the corbel liner was an independent brick liner that was actually a stack built within a stack. The minimum annular space between the liner and the shell was increased to about 2 ft and newly developed acid-proof potassium silicate mortars were used. All the inherent structural strength and high-temperature resistance of the independent brick design was retained while acid penetration of the shell was reduced by the larger annular space. And fewer mortar failures due to acid attack were reported.

Then came wet FGD systems. Large quantities of sulfuric acid at different concentrations formed on the liner face as flue temperatures dropped to 130°F and positive pressure in the stack increased. The acid worked its way through the brick and attacked the steel tension bands used to control thermal brick expansion. It also attacked the buckstays supporting these bands, the testing doors and ports, and the concrete shell itself. To solve this problem, the annular space was pressurized so that the flow of acid from the inside of the stack to the outside of the brickwork would be stopped. To get an effective pressure in the annular space, a cap had to be put on the stack top. After the cap was made as leak-tight as possible, all the doors and openings in the stack were sealed. This allowed the annular space to be pressurized with a relatively minimum-sized blower.

Not all modern-day independent brick-lined stacks are pressurized. Some power companies have chosen to closely monitor the shell conditions and pressurize only if there is evidence of acid attack. There is mounting evidence that the positive pressure in the stack from an FGD system causes acid migration through the brickwork and that a pressurized annular space may become a universal design requirement on future brick-lined stacks.

22.2.5 Independent Brick Liners

Brick liners in plants with FGD units should be built with ASTM Specifications C-279 acid-resistant brick. If a wet scrubber is used, the mortar should be sulfuric acid- and sulfonation-resistant. Expect stacks made of standard brick and nonacid resistant mortars to fail from acid attack of the mortar once an FGD system becomes operational.

Since a brick liner is subject to positive pressure from the FGD system, there has to be some type of protection for the metal portions of the stack shell. Annular space pressurization and metal coatings that are acid and sulfonation-resistant help prevent this acid attack. The stack cap and the exterior of the shell can also be damaged by corrosive acids. Condensed gases can be carried down the outside of the shell by a draft and attack the concrete shell, the protective system, the navigational lights, access ladders, and service platforms.

Variations on the general theme drawn so far may exist in an independent brick liner. There are no strict design standards here; most of the techniques used in the design and construction of independent brick liners are a function of experience. According to a paper prepared by Custodis Construction Company:

> *Excepting for the design procedures which we have applied for seismic conditions (which allow us some design control for areas of moderate seismic activity), and for reviewing stresses between openings, if two openings are built close together, there are no rational design procedures or codes available to enable us to design the work to the "limits" of safety. Therefore, a great deal of the "design" of brick linings is based upon our in-house standards and "rules of thumb," which have been proven over a long period of time to provide a very stable and conservative brick lining, often lasting the lifetime of the plant.*[3]

It is almost inevitable that cracks will appear in a brick lining. These cracks are controlled by steel tension bands spaced regularly up the liner height.

Here's how the bands work: consider a typical brick liner with a 12-in.-thick wall and an internal diameter of 20 ft. Assume that a temperature gradient of 150°F exists between the hot and cool faces. For these conditions, calculations show that the inner face of the lining tends to "grow" by only a fraction of an inch. Neglecting its own "growth," the steel band must resist this strain, or a portion of it, to control crack formation.

One can see that unless the bands are perfectly tight and in contact with the brick at all points around the circumference, they will not be effective in controlling cracks. Too often, bands are installed with gaps between steel and brick large enough to pass a finger. Such bands are of absolutely no value.[4]

Around openings in the lining, where the buckstay is supposed to accept the band loadings, this steel beam may or may not be stiff enough to do its job. A shear key (stepped bricks that tie the buckstay, to the chimney shell), helps relatively light structural sections to transfer uniform band reactions directly to the brickwork as compressive stresses.[5]

In a radial brick chimney, the vertical open space in the bricks is sometimes filled with mortar during construction. If this mortar is a conventional calcium-based mortar, salts will be formed when acid migrates into the brick. Since the brick is usually in compression, the inside brick face will fall off due to the internal pressure generated by the acid/mortar reaction.[6] This internal spalling is difficult to detect while the stack is in operation and may lead to liner failure if it goes undetected for an extended period of time.

Sodium silicate mortars are also susceptible to salt's growth. The stress caused by a water reaction with the sulfated mortar can cause failure of the brickwork. Evidence of this failure is bending brickwork and loss of strength in the mortar. Brickwork in this condition is subject to any stress caused by vibration, including earthquakes and pressure changes in the boiler.

Two types of mortar are acceptable for use in liner construction. These are Class A Type 4 potassium silicate mortars and Class A Type 5 silica mortars. Most of the brick stack liners built in the United States since 1955 use potassium silicate mortars. These mortars have to be exposed to an acidic environment of a pH below 5.5. If the pH of the liquid on the brickwork goes above 5.5, these mortars begin to lose their desirable properties. This is the mortar type that may suffer damage at the top of the flue when exposed to rainwater. One should remember that potassium silicate mortars were designed to withstand acid, not water.

Class A Type 5 silica mortars are not affected by rainwater or sulfuric acid. However, they are more expensive to install than potassium silicate mortars and all bricks must be wire-cut or scratched before laying. This helps compensate for a lower bonding strength.[7]

A third type of mortar is emerging from European designs. This is a sodium-free potassium silicate which decreases the effects of water on the still curing mortar.

22.2.6 Steel Liners

Steel liners were introduced around 1960 to help power plane owners cope with corrosion problems then experienced with corbel brick linings. The stated benefits of a steel liner were a lower first cost than brick liners, resistance to earthquakes, and high resistance to corrosive forces.[8] Prior to the passage of the 1972 Clean Air Act, the performance of steel liners met most expectations. There were some cases of liner buckling, but steel liner failures were very rare.

Steel liners were able to cope with acid attack because they were exposed to sulfuric acid in its concentrated form. Also, the temperature of the flue gases passing through the steel liners was above the acid dew point of the sulfuric acid in the flue gas. Some of the steel liners were given additional protection by coating the liner with inorganic materials that formed a monolithic, protective layer on the inner face of the liner. This liner within a liner was either formed, troweled, or gunited into place (guniting is a process that pressure sprays refractory concrete or calcium aluminate cement onto a surface).

With the introduction of FGD systems, the first protection given to steel liners was to coat them with organic resins, polyesters, and combinations thereof, many of which contained glass flakes and fibers. Rolling, brushing, and spraying were the techniques used to apply these coatings 60–70 ml thick.[9] As power plants began to bypass the scrubber units and stack temperatures increased beyond the organic coating tolerance, the incidence of coating failures increased.

The present use of organic coatings on steel linings (in stacks that handle scrubbed and bypass gases) is restricted to the use of an organic lining material equivalent to or surpassing the capabilities of a fluoroelastomeric coating. As of this writing, fluoroelastomeric coatings are the only organic coatings that can withstand both the relatively high temperatures and high concentrations of sulfuric acid found in bypass gases and the low-temperature, high-humidity, dilute acid conditions found in scrubbed gases. Conventional polyester and epoxy coatings can withstand the latter condition, but not the former. Fluoroelastomeric coatings are capable of resisting temperatures up to 400°F continuously and 85% sulfuric acid concentrations without any adverse damage to the coating. This coating is not expected to give indefinite service life, however. All organic coatings are sacrificial in nature and will have to be maintained at some interval. The precise interval required between repairs will depend on the service conditions experienced by the coating, quality of the application of the coating, and the compatibility of the coating with the actual chemical conditions inside the liner.

If a separate liner is used for the scrubbed gases, a conventional polyester resin-based coating can be used to protect the liner. This choice is based on the low-temperature, low-acid concentration found under this operating condition. The upper section of this liner can be made of a high-alloy metal to resist the corrosion caused by thermal upset conditions.

For a polyester resin to protect carbon steel in a power plant liner, it now appears that the moist flue gases should not go over about 180°F. This means that bypass unscrubbed gases cannot be passed through a liner coated with polyester resins. A design being used to minimize the attack of sulfuric acid on the stack is dual flue liners in a single concrete shell. The design has one steel flue lined with an organic coating optimized for the expected chemical environment for the FGD system. The other steel liner is for the bypass gases. Since this dual flue liner concept is relatively unproved, it will have to be followed closely to see how the design actually performs in service. The bypass liner and FGD liner are fabricated of high-alloy metal for the upper exit section of each liner. This was done to minimize the corrosive attack of the down gases into each liner at the top of the liner. Should failure occur in these liners, it would probably be in the uncoated liner. The mode of attack would come from sulfuric acid deposited on the steel that is then diluted with water absorbed from the operation of the FGD system. To minimize the chance of this failure, the high-alloy upper liner section was used. There is the possibility of operating the bypass liner at a low gas flow level to minimize the effect of the wet cool gases entering it from the FGD liner. A fluoroelastomeric coating over carbon steel was tested in the bypass flue. This coating could possibly be used to protect the carbon steel bypass liner should a protective coating become necessary.

Another technique for handling the corrosion problems is to keep both liners in service all the time. This could be considered if the total emissions for both liners would meet the EPA requirements. The use of the dual-liner concept has been tried and is closely monitored to determine if this is a viable alternative to an alloy liner.

Another liner uses dual materials of construction. The bottom portion of the liner is fabricated of brick with acid-proof mortar and the upper section is made of nickel-chromium high alloy. The concept behind this design is that the upper portion of the stack liner has the most severe chemical exposure and thus needs maximum protection.

There are three types of inorganic coatings used in steel liners: inorganic silicate, modified silicate base, and modified inorganic silicate base.

The first is an inorganic silicate composition which resists all acids (except HF), as well as water, oil, most solvents, and temperatures to 955°C. It is suitable for use over a pH range of 0.0–7.0. This acid-proof concrete may be applied by guniting, pouring, or forming. During the past years, it has been on the market, millions of pounds have been used successfully to withstand corrosion, abrasion, and elevated temperatures.

Now there is a new generation of corrosion-resistant inorganic monolithic linings available. The first is a modified silicate base which is a single-component system, a powder to be mixed with water. It hardens within 24 hours at 20°C to produce a high-strength, corrosion-resistant lining with extremely good adhesion to steel, concrete, and brick. It is not affected by exposure to acids (except HF), mild alkalis, water, oil, solvents, and temperatures to 955°C and is suitable for use over a pH range of 0.0–9.0.

The second product is a modified inorganic silicate base, which is thermally insulating and lightweight. It is also a single-component system, a powder to be mixed with water when used. This material develops a compressive strength of 3,900–4,200 psi and a tensile strength of 400–500 psi. It has a coefficient of thermal expansion of 7.1×10^{-6} which is very close to that of steel, thereby minimizing cracking due to temperature changes. It weighs approximately 98 lb/ft^3 and is considerably lighter in weight than most other corrosion-resistant linings, thereby reducing the dead load. It has a "k" factor of 2.25–2.50 and will provide a temperature drop of approximately 50%/in. of thickness. In other words, a 2-in. thickness with the hot face at 260°C will drop the temperature on the cold face to 130°C. It has excellent resistance to abrasion and a bond strength of 300 psi to steel. It withstands temperatures to 955°C and is unaffected by water, moisture, steam, acids, and alkalis in a pH range of 0.0–9.0.[10]

The acid-proof concrete lining, while not completely impervious to the sulfuric acid, has a low initial effective porosity and acts as an efficient filter. Due to this low effective porosity, the penetration rate of the corrodent to the surface of the base metal is greatly reduced. This slow penetration allows the acid reaction to proceed to completion with the alkaline concrete lining material to form sulfates or anodic inhibitors. The sulfates fill any available pores in the lining. At the same time, a cementitious form of silicic acid is also precipitated from the matrix and the reaction reduces the porosity of the lining and makes it a stronger and denser physical barrier to the acid. When the acid eventually reaches the base metal there is no fluid flow and so no kinetic motion by which deposits may be carried away. Since the initial acid reaching the metal surface will be weaker, it will react more slowly and thus form a protective film of insoluble iron salts on the face of the metal. These salts further inhibit the corrosive action and form an integral cathodic inhibitor barrier held in place by the concrete lining.[11]

As long as the monolithic acid-proof concrete lining maintains its physical integrity and remains intact, there is no corrosion at the interface of the base metal. To have corrosion it is necessary for oxygen, water, or an electrolyte to be present, so if the lining is broken or cracked by accident, oxygen can then reach the metal and cause localized cathodic corrosion and damage to the metal shell.[12]

Another design aspect of interest is the use of insulation on the steel liner exterior. This insulation is installed to prevent heat from escaping through the liner. If more heat is retained in the stack, the chances of condensation on the liner's interior are reduced. The insulation also helps keep thermal stresses on the concrete shell within tolerable levels.

Most insulation is attached to the steel liner sections on the ground before the liner sections are assembled. The spun-glass insulation board is installed by centering stick pins into the boards no more than 16 in. apart and no more than 4 in. from the insulation board edges. A stud welding gun secures the boards to the liner. Wire mesh is placed over the insulation board, then stud-welded into the liner. The mesh ends are laced together where they overlap.[13]

Sometimes, tape is used instead of mesh to secure the insulation boards and/or to seal joints. The selection of the tape or the wire mesh should be made after consulting with the plant engineer.

22.2.7 Fiberglass Liners

The lining materials of construction for power plant liners alternated between carbon steel and brick during the pre-FGD era. With the lowering of the flue gas to around 120°F without reheat and to 180°F with reheat, a new lining material was needed. In an attempt to find a new material in the 1960s, a coating of fiberglass-reinforced plastic coating was applied to steel liners. This composite material experienced some failures. However, the coating did withstand the corrosive liquids found in the liner.

Based on the chemical resistance of the plastic itself, a stack fabricator erected the first fiberglass-reinforced plastic stack liner in 1968. Installed at the Electric Reduction Company plant in Long Harbour, Newfoundland, the liner was designed to resist chemical attack from flue gases from an oil-fired boiler and hot gases from a phosphate production plant. The flue gases contained hydrochloric acid, sulfuric acid, hydrofluoric, and fluorosilicic acids. Numerous FRP liners were then installed in various industries.[14]

Finally, in 1974, the first FRP was installed in the power industry. This first liner was 805 ft high and 15 ft in diameter. It was installed at East Kentucky Rural Electric Company's Spurlock Station, Unit No. 1, Maysville, Kentucky. This liner was completed in 1976 and went into service in 1977.[15]

The latest generation of fiberglass liners have helically wound FRP linings that have an angle of helical windings that change with the elevation of the fiberglass in the stack. This angle change provides the axial and hoop strengths required.[16]

Fiberglass liners have several advantages over conventional brick liners. An outstanding feature is their chemical resistance to most liquids found in the flue gas. Another desirable feature is the resistance to high lateral forces from earthquakes.

FRP liners are not without problems, however. Usually, the FRP liner has a higher expansion due to heat than carbon steel. Since the FRP liner has a lower physical strength than steel, provisions must be made for additional structural support. Not to be overlooked are the sulfuric acid concentrations and temperatures expected on the surface of the liner.

Several layers are required in FRP liners to achieve the desired physical properties. The outer layer will contain an ultraviolet inhibitor and be essentially all resin. Next, the structural layer is fabricated of 60–65% fiberglass filament and 35–40% resin. The helix angle of the fiberglass filament will vary depending on the location of the section in the stack. (On one liner, this angle varies from 8° to 30°.) Next comes the interior layer. This layer consists of 20–30% glass and 70–80% resin. Finally, the innermost layer is the chemical barrier layer. This layer is 100% resin, usually, but may contain some glass veils. Since this layer is the one that conducts the static charge from the liner, it is filled with conductive carbon pigment. It is important that a continuous electrical path be made between the inner carbon-filled layers. This continuous electrical path can be made by physically connecting each section on the inner layer with woven grounding cables. These cables should be designed to withstand the corrosive liquids expected in the liner.

One of the weak points in the design of an FRP liner is the fiberglass filament. This filament is an excellent structural material unless the flue gas that passes through the liner contains a significant amount of hydrofluoric acid. Hydrofluoric acid in high concentrations will readily dissolve glass and thus cause failure of the lining. Should the hydrofluoric acid concentration be excessive, a synthetic fiber such as Dacron should be considered for the main structural filament. Since the fiberglass liners are usually designed to operate after FGD systems, there is a high probability that the hydrofluoric acid will be removed by the FGD system and will not appear in the fiberglass liner in significant quantities. This should be determined on each individual FGD system before a design contract is issued.

Since a tall fiberglass liner cannot support its own weight, special design considerations must be made. Several techniques have been developed for compensating for the weakness of FRP liners. One of them is a counterweight and pulley system. This patented system allows the liner to expand and contract without incurring structural damage to the liner. Another support technique uses intermediate supports with expansion joints between the individual sections. This keeps the compressive loads on the liner below the design limits.

One of the major considerations in the design of the FRP liner is the temperature limitations on the liner. These limitations are determined by the resin used in the liner fabrication. Failure to observe the temperature restrictions of the resin used in the fabrication of the liner results in structural failure of the liner. Excessive temperatures in the liner can cause softening of the resin and result in the bending of the liner.

A major concern in the fabrication of FRP liners is flammability. The addition of a fire-retardant chemical (antimony trioxide) to certain layers of the FRP liner will retard a fire initiation in the liner. And once the fire begins, the antimony trioxide continues to restrict the rate of burning. It is standard practice to install a water spray system in the ductwork ahead of the fiberglass liner. This emergency water spray system is activated by a present temperature excursion in the FRP liner. The precise temperature that activates the emergency water-quench system should be selected based on the physical properties of the resin used in the fabrication of the liner.

The precise design used for introducing the quenching water into the FRP liner is also important. Care should be taken in the design to ensure that the fire protection quenching water does not directly impinge onto the FRP liner. Direct impingement can result in liner damage. Whenever the emergency fire protection system has been used, an inspection of the FRP liner should be made for possible thermal damage.

Fiberglass liners with flue gas passing through them act as a static electricity generator. This makes the electrical continuity of the grounding system very important. The inspection and maintenance of this grounding system cannot be overstressed. It would be helpful to have a static electricity meter to use when working around the FRP liner. This instrument allows the inspector and workers to determine if the grounding system is operating before they approach the liner.

22.2.8 Alloy Liners

An increasing interest is being shown in the use of high-corrosion-resistant alloys for the fabrication of stack liners. Since each coal and FGD system results in differing corrosive liner environments, the precise alloy being used varies. The most often used materials are nickel and chromium alloys with a 6% or higher molybdenum content.

These materials are known for their resistance to sulfuric acid corrosion and a varying resistance to chloride pitting. The precise chloride resistance of each alloy is dependent on the pH together with the temperature and chloride concentrations the alloy will experience.

A test that can help determine the proper alloy for corrosion resistance is the coupon test. This test involves a set of various alloys assembled on a corrosion coupon rack as per ASTM G4-68. Installation of these racks in the liner at various places of concern allows an in-place test of the corrosion effect on a variety of alloys. By checking the corrosive attack of each alloy sample, the most economical alloy for the liner can be selected. It is a practice to select a slightly higher corrosion-resistant alloy than the coupon tests indicate. The ultimate evaluation of the alloys and materials of construction is to fabricate the test material into the wall of the liner. The resulting tests should be identical to the results obtained when the new construction material is put into service.

The corrosion coupons are usually round in shape and are electrically insulated from each other by Teflon spacers. The rack used to hold the coupons is fabricated from an alloy considered to be sufficiently resistant to the chemical environment of the liner to remain intact for the duration of the test. The coupons are measured for thickness and weighed before and after the exposure test. The corrosion rates can be determined by use of the formula:

$$p = \frac{W\left(365 \times 10^{3}\right)}{\left(2.54\right)^{3} Adt} \tag{22.1}$$

where p = average penetration, mpy (mils/yr)
d = density of metal, g/cm^3
t = time of exposure, days
W = weight loss, g
A = specimen surface area, in.^{2}

The corrosion coupons can be installed on the wall of the liner or on special racks installed in the liner ports. The precise location depends on the time of the test, probability of the unit being offline for removal of installed coupons, and the configuration of the lining. The tests can run from 1 months to 6 months (the normal test should be long enough for 10–15% weight loss of the coupons).

These test coupons will not corrode at the same rate as alloys installed in the lining. The coupon test results should be used for comparative purposes (for example, comparing one coupon with another). Since the results obtained from these coupons are not available quickly, it is a good policy to install the test coupons when the lining is first put into service. Then, if any liner problem should occur, there are some data indicating what a suitable alternative might be.

At least one power plant has an entire stack lining of ASTM B443-72 alloy. This alloy was chosen after extensive stack testing of various alloys. Sections of several high alloys were welded onto the base of the liner and exposed to the various operation conditions of the boiler and FGD system. With the particular coal and operating parameters found in this stack, ASTM B443-72 alloy was the most corrosion-resistant.

Other power plants are being fabricated with high-nickel/chromium/molybdenum alloys as just a part of the stack. Some stacks have high alloy where the metal parts contact the brick liners; others have a partially or completely fabricated liner of high alloy.

To assure that the drains and various metal parts of any stack with FGD systems are not corroded by the environment of the scrubbed gases, some engineers specify that all parts of the stack base be fabricated of high-nickel-chromium-molybdenum alloy. The baseplate of the stack has been fabricated of lead for some plants.

Inside many carbon steel liners coated with protective organic fluoroelastomeric coatings, the top section of the liner has been fabricated of high-alloy materials. The particular alloy used depends on the precise chemical conditions extant. Other carbon steel liners coated with the protective organic fluoroelastomeric coatings have the coating extended to the top of the liner with no alloy to protect the top section of the carbon steel metal liner. The fluoroelastomeric coating may also be used on the outside of the carbon steel liner above the cap.

The top platform of many new stacks uses high alloys, ranging from 316L, stainless to nickel/chromium/molybdenum alloys. One of the difficulties experienced by designers has been finding standard fabricated alloy parts. Some substitutions are being made with the free knowledge that they will have to be replaced at frequent intervals due to corrosive attack.

Conventional materials used on some of the new FGD stacks have a short life span. In less than 1 year of operation, conventional galvanized steel at the top of the FGD stack may have the galvanized coating completely removed by corrosion. This results in rapid attack of the remaining steel and failure of the entire structure (working platform and operating warning lights). Each FGD system and its respective coal should be carefully evaluated to determine the chemical environment that will be found inside the stack liner and at the top of the stack.

There will be acid penetration through the brick liner to some degree even with a pressurized annular space. This migration of acid can occur when the seals on the doors in the chimney are not kept airtight or properly sealed. Any migration of acid into the bands (conventional carbon steel) around the brick liner will result in a rapid attack of the carbon steel metal. Failure of the bands will allow the brick liner to expand when it cracks and can result in total failure of the brick liner. The use of the proper protective coating or lining in steel bands will assure continual retention of the brick liner.

One concept in alloy fabrication being examined for possible use in the stack is explosion-clad carbon steel. The explosion cladding process allows the primary structural strength of the liner to be supported by carbon steel. The carbon steel base metal in the cladding is on the nonwetted side of the liner. Thus, standard construction techniques can be used between the liner and the shell of the stack. However, on the wetted side of the cladding, high-nickel alloy is used. This high-nickel alloy is selected to have complete chemical resistance to the expected chemical environment in the liner. It is important that the cladding alloy have sufficient thickness to resist the corrosion of the chemical environment in the liner for the life of the power plant. Otherwise, corrosion will destroy the alloy and the carbon steel will rapidly disappear.

Special fabrication techniques are required when constructing a liner with explosion-clad alloys. A technique that has been tested for corrosion resistance consists of welding the alloy side of the cladding with a three-pass weld of high-alloy welding wire. The use of cladding allows a given amount of high alloy to be used in the fabrication of a number of liners.

Other types of cladding can be used for the fabrication of clad liners. These primarily consist of rolling a corrosion-resistant alloy and carbon steel through roller mills to form a single-bonded metal. The welding techniques required to join these sheets of clad metal would basically be similar to those required for joining explosion-clad metal sheets.

The fabrication of stainless steel and nickel-based alloys presents a completely new problem for the power industry. The fabrication techniques previously used with carbon steel are not necessarily applicable to alloys. As a matter of fact, it can be assumed that welders will have to be retrained to assure a satisfactory fabrication of the high alloys. Nickel alloys require the use of special welding techniques.

The remainder of this section is a discussion of some of the characteristics and techniques associated with corrosion-resistant alloys, as noted by the Technology Division of High Technology Corporation.[17]

22.2.9 Hot Cracking

Welding nickel-based alloys can be a major problem because of "hot cracking." Hot cracking is caused by stress and the presence of a "strain-intolerant microstructure." Since the stress of the metal is a natural occurrence during welding, care must be taken to minimize the stress by proper welding design and proper welding technique. The minimization of the strain-intolerant microstructure can be achieved by minimizing the heat generated during the welding process. This is accomplished by keeping the current required on the welding machine as low as possible and using a proper-sized electrode to maintain the minimum current load. Excessive heat input will increase the amount of metal exposed to the crack-sensitive temperature range and increase the duration of the elevated temperature exposure. A common source of excessive heat input is from "weaving" the electrode instead of using the preferred stringer bead welding technique. The weaving should not exceed three times the diameter of the electrode being used; however, the best precaution is to rely on sound welding practice and good common sense. A welder has described the proper welding current for alloys as the current required to weld so that it feels like you are taking the metal from the electrode and putting it on the metal.

22.2.10 Fusion-Zone Hot Cracking

Bend contour and shape are important factors in minimizing fusion-zone (weld metal) hot cracking, especially when using a butt welding plate of pipe which requires multiple passes. The individual stringer passes or beads should be deposited with a distinct crown. Weld beads that are deposited with a convex or flat surface are more susceptible to centerline cracking. If coated electrodes are used to make the welds, all the slag should be removed between passes so that there will not be any slag inclusion in the weldment. All flaws in the weld should be removed before another pass is placed in the welding area.

Cracks in the alloy can also occur at the edge of the heat-affected zone. These cracks are caused by excessive welding heat input coupled with stresses associated with welding.

Some manufacturers recommend that no welding of any kind occur if the metal temperature is below 32°F. Ideal welding temperature is 72°F. Interpass temperatures should not exceed 200°F. No cooling techniques should be used between passes that can introduce oil, grease, dirt, or contamination into the weld. This includes shop air, oil-soaked rags, hard-water deposits, etc. For best results, one should allow the metal to cool naturally.

The surfaces of all areas adjacent to the welding joint should be cleaned to bright metal prior to welding. This means that all grease, oil, crayon marks, dirt, magic marker marks, etc., should be removed by wiping the area with the recommended solvent and lint-free rags. Failure to do this can result in porosity and increased susceptibility to corrosion at the weldment.

Coated electrodes should be stored in sealed moisture-resistant containers until they are used. Electrodes that are removed from the original containers should be heated at least 15°F above the ambient temperature to assure that the electrodes remain dry. Wet electrodes should be dried as recommended by the manufacturer. One manufacturer recommends that the electrodes be dried at 250–300°F for at least 2 hours before they are used.

Porosity can occur in shielded metal arc (SMA) welding when welded vertically. This probably occurs because gravitational forces prevent the molten slag from properly protecting the weld pool. Another cause may be the breaking of the molten slag layer by improper rod movements. As much as possible, the SMA welding should be performed in the horizontal position. One should check with the supplier of the alloy and the welding rod to determine the proper rod for vertical welding if it is required.

Two other techniques are also used in welding alloys. These are gas tungsten arc (GTA) and gas metal arc (GMA). In both processes, the gas shielding the molten pool should precede the arc initiation by several seconds. Also, the gas flow should not be excessive. Excessive gas flow can result in porosity of the resulting weld. This excessive gas flow causes air to be brought into the weld puddle region. Should the gas flow be too low, there is insufficient gas flow to properly protect the molten pool, and air is brought into the molten pool. When welding with either GTA or GMA, care should be taken to keep the diffuser screen clean and free of splatter.

22.2.11 Other Shielding-related Problems

Another problem related to proper shielding is the inability to deposit a successful root pass, especially when butt welding pipe. The underside of a root pass should be adequately shielded with inert gas or else the weld bead will not "wet" and "flow" to produce the desired penetration effect. Proper shielding should also include the use of purge dams. When initiating the root pass, a temporary covering of tape over the remainder of the joint will minimize the loss of purge gas through the root opening.

The surface of weld beads deposited by the GTA process and GMA process will occasionally appear to have very thick oxide rafts, often referred to as "slag spots." Although these artifacts may be questionable from a cosmetic viewpoint, they very rarely interfere with the subsequent weld passes or with the overall integrity of the welded joint.

22.2.12 Repair of Welds

All rejected welds should be removed by a suitable grinding method. Remelting of the weld beads or depositing additional weld beads over the defect is definitely not recommended.

In joining a high alloy to carbon steel, an alloy welding filler material should be used that will give a final metal composition that is equal to or better than the parent metal.

One of the alloy manufacturers has provided a suggested list of fabricating tips when working with corrosion-resistant alloys. These tips are guides for maintaining the corrosion-resistant properties of the alloys.[18]

DO:

1) Follow acceptable welding practices as set forth in the welding section of the ASME Boiler and Pressure Vessel Code.
2) Use filler metal of the same composition as the alloy base metal.
3) Achieve complete penetration of the weld to avoid crevices which might lead to accelerated corrosion and stress raisers in the fabrication.
4) Remove all traces of foreign material from the alloy before heating to above the heat range when welding or for any other reason.
5) 100% dye-check all welds to assure crack-free joints for corrosion applications as required by ASME Boiler and Pressure Vessel Code.
6) 10% X-ray where required, to conform to code.
7) Solution-heat-treat after fabrication if the alloy and the corrosion application warrant it.
8) When heat treating, use fuels of low sulfur content to avoid sulfur penetration and subsequent detrimental effect on mechanical properties and corrosion resistance.

DO NOT:

1) Allow direct flame impingement on the alloys.
2) Overheat when solution-heat-treating or hot working. Overheating of some of the alloys by as little as 25–50°F above the recommended solution heat treatment temperature can cause incipient melting.
3) Age the alloys for prolonged periods of time in the 1,100–1,800°F range. Aging causes carbide precipitation in the grain boundaries, which leads to loss of ductility and greater susceptibility to corrosive attack. If aging does occur, solution heat treatment at the recommended temperature will restore much of the original ductility and corrosion resistance to the alloy.
4) Use oxyacetylene for welding the alloys unless otherwise recommended by the manufacturer.
5) Use brazing alloys that contain copper.
6) Make intermittent cuts when machining these alloys. They have a tendency to work-harden, which makes subsequent cuts much more difficult.
7) Use shot blasting to clean the alloys. Iron particles that are embedded in the surface of the alloys can cause serious product contamination and/or severe corrosion problems in some media.

22.3 Sulfuric Acid Attack

Sulfuric acid is primarily responsible for the corrosion attack on any material used in the stack liners. It is not the only acid you might find eating away at the liner, but it is the most plentiful and potentially the most destructive. This acid comes from the sulfur trioxide and sulfur dioxide produced in the boiler during fuel combustion. Other acids that can increase the corrosion rate of sulfuric acid are hydrochloric and hydrofluoric acids.

The Stationary Source Performance Standard of 1979 allows 10% of the sulfur dioxide emissions from solid and solid-derived fuel combustion to pass through the FGD system into the stack. Since this sulfur dioxide is saturated with water when it leaves the wet FGD system, there will be sufficient moisture to make an acidic solution in the stack.

There is also sulfur trioxide present in the flue gas when it comes from the boiler. Depending on the sulfur content of the coal, oil, or other fuels, the boiler design, vanadium content of the ignitor oil, vanadium pentoxide deposits on the boiler tubes, the pattern of varying loads on the boiler, and so on, various amounts of sulfur trioxide can be produced. The Shawnee Pilot Plant tests have shown that about 58% of the sulfur trioxide present in the unsaturated flue gas is removed by the FGD system. Some sulfur trioxide is also removed in the cold-side electrostatic precipitator. At some plants, if the fly ash being collected in the electrostatic precipitator has a very high electrical resistivity, it is sometimes the practice to add sulfur trioxide before the precipitator to improve the removal efficiency of the unit.

Sulfur trioxide that passes through the precipitator and FGD system goes through the demisters and into the stack, where it is available to make sulfuric acid. The quantity of sulfuric acid will depend on the precise amount of sulfur trioxide present in the treated flue gas.

Sulfur trioxide has a high affinity for water. When the sulfur trioxide concentration goes below its dew point, sulfur trioxide reacts with water vapor in less than one second to form minute droplets of sulfuric acid. These droplets will be very concentrated, approximately 98% sulfuric acid, but they will continue to react with water until the concentration on the liner is lowered to its boiling point equilibrium concentration. This medium-concentrated acid will either react with components of the flue lining or continue to absorb water and form more dilute sulfuric acid. The pH value of dilute sulfuric acid that finally forms after a wet FGD system is 0.5–2.2.

Sulfuric acid can be found whenever the flue gas temperature drops below 400°F. In the past, flue gases from the preheater were around 350°F. Even though this temperature is below the maximum acid condensation temperature, there were only minimal corrosion problems. In the contemporary stack environment of about 130°F, copious amounts of sulfuric acid are formed when sulfur trioxide gas drops below its dew point.

There has been considerable discussion about what the dew point of sulfuric acid is. It can be defined as the temperature at which liquid sulfuric acid forms in equilibrium with the sulfuric acid in the vapor state. Research to develop some instrument to accurately determine the sulfuric acid dew point has produced several mechanical techniques. These instruments generally determine the sulfuric acid dew point by the use of electrical conductivity probes. These probes are put into the flue gas stream and then slowly cooled. At the point that sulfuric acid and water condense on the probe, a sudden change will be seen on the meter, indicating the conductivity change across the probe. The temperature at which this sudden change in conductivity occurs is the dew point of the flue gas. Since sulfuric acid is an unusual compound, this dew-point temperature does not mean that liquid sulfuric acid will be seen on the walls of the stack.

Care must be taken when using these instruments to assure that a value is obtained indicative of what is happening in the liner. Gibson and Kufta suggest that the corrosion rate with sulfuric acid increases from the theoretical dew point of the acid to a temperature about 85°F below the sulfuric acid dew point.[18] From this temperature to the dew point of water, the corrosion rate decreases. Below the water dew point, the corrosion rate rapidly increases. This lower condition is the primary area of interest for scrubbed flue gases.

To illustrate the effect of trace amounts of sulfur trioxide on the dew point of sulfuric acid, here is an example taken from a paper by Robert R. Pierce:

> *Thus, if a combustion gas containing as much as 10% water vapor by volume also contains as little as 40 parts per million sulfur trioxide by volume (0.004%), it has a weight percentage of sulfuric acid (based on sulfur trioxide and water alone) of only 0.22%. The vapor phase in equilibrium with the liquid phase has a concentration of 82.5% by weight sulfuric acid at its boiling point at 0.1 atmosphere pressure which is 148°C (298°F). Thus, as soon as the temperature of such a mixture of 10% water vapor and 40 parts per million sulfur trioxide by volume drops below 148°C, liquid must form because the gas becomes supersaturated with 82.5% sulfuric acid. This powerfully acidic liquid will condense on any solid surface which is cool enough, which is why it is called the dew point or sometimes called the acid dew point. If the gas is cooled below the dew point temperature by radiation, etc., a mist of acid droplets will be formed. It may seem strange that a gas containing a mixture of sulfur trioxide and water corresponding to only 0.22% by weight of sulfuric acid will form a droplet or a film on a surface containing 82.5% by weight of sulfuric acid but such is the nature of this acid. Of course, only a very small amount of the water in the gas is condensed out with the acid.*[19]

As was mentioned earlier, considerable work has been done to develop some type of instrument to accurately determine the sulfuric acid dew point. But in the literature reviewed, there seems to be no agreement about the accuracy of the available mechanical techniques. There is, however, a method for determining the sulfur trioxide concentration in the flue gas. Method 8 from the Federal Register, Appendix A.32, can be used to determine the sulfur trioxide concentration (C). The concentration determined by the EPA Method 8 is in grams per dry standard cubic meter or pounds per dry standard cubic foot. This value can be converted to parts per million (ppm) by use of the equation:

$$SO_3(ppm) = \left(4.82\times10^6\right) C_{SO_3}\left(1b/dscf\right) \tag{22.2}$$

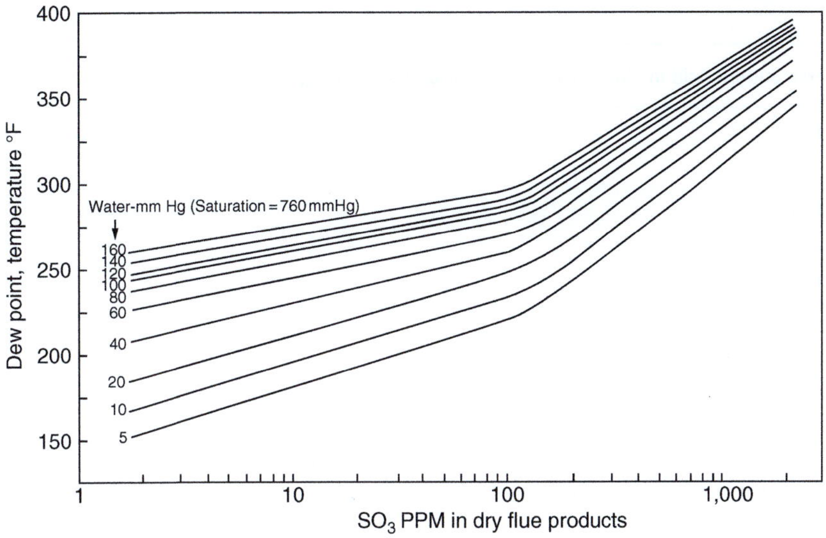

Figure 22.1 Determination of SO_3 concentration in flue gas.

To use the data presented in Figure 22.1, a moisture determination needs to be made. This can be done from EPA Method 8, paragraph 4.2. This moisture determination is used to determine the mm Hg water for use in Figure 22.1. The conversion of the volume of water collected in Method 8 converts to mm Hg by the equation:

$$\text{water}\left(\text{mm Hg}\right) = 760\frac{V_{\text{w std}}}{V_{\text{m std}} + V_{\text{w std}}} \tag{22.3}$$

where $V_{\text{w std}} = 0.0474\,\text{ft}^3/\text{ml}\ (V_{\text{lo}})$
V_{lo} = ml of condensate collected
$V_{\text{w Std}}$ = equivalent cubic feet of moisture in vapor state at STP
$V_{\text{m Std}}$ = dry gas measured in cubic feet at STP

Not only is sulfuric acid present in a dilute concentration in the lining, but at the low temperatures after the scrubber, hydrochloric acid can also be present. If there is some bypass of the unscrubbed flue gas and mixing of this unscrubbed flue gas with scrubbed flue gas, hydrochloric acid will be present in the liner of the stack. A set of conditions found in one stack that burned pulverized coal with 2.67% sulfur and 0.066% chlorides with no scrubber were discussed by Pierce and Kossler. Their data are presented in Table 22.1.[20]

From this data one can determine that hydrochloric acid can exist after a wet scrubber. The precise concentration of the hydrochloric acid will have to be determined and will probably not reach the concentrations shown in this table.

In 1962, Pennsalt Chemical Corporation began a program in 11 stacks to determine the sulfate content, free sulfuric acid, and pH. Their data are shown in Table 22.2.[20]

Table 22.1 Conditions in stack burning pulverized coal.

Temperature	Sulfuric Acid %	Hydrochloric Acid %
242	64.6	—
233	48.8	—
244	43.8	—
124	38.8	8.8
106	12.8	17.2
90	0.6	0.4

Table 22.2 Location of acidity in stack linings.

Height	Sulfate in Mortar (%)	Free Sulfuric Acid in Mortar (%)	Acidity of Mortar (pH)
Top Level	31.8	10.0	2.3
75% Level	31.6	9.7	2.0
50% Level	28.9	14.7	1.8
25% Level	32.1	12.9	1.9
Bottom	21.3	10.1	2.0

The data from Table 22.2 shows that the same acidic condition exists throughout the entire height of the lining. In the case of brick liners, it seems that the damage from the acid attack is more evident at the top of the lining. This is probably due to the fact that there is less weight on the bricks at the top of the lining. Also, there is more bending stress at the top of the lining. Another consideration is that the top of the lining is more likely to be exposed to rain from the environment, with resulting sulfation and hydration of the mortar.

One of the operating conditions that has become more apparent with the introduction of FGD systems is the formation of acid smut in the stacks. Gills and Lees reported[21]:

Based on many years of study of this problem, including examinations of the inside of chimneys, microscopic examination and analysis of the deposits, and examination of the smuts, it has been concluded that acidic smuts consist of large agglomerates of black spheres bonded together with sulfuric acid or acidic salts.

Smuts form where chimney surfaces are below 250°F and occur chiefly with metal stacks, but they have also been observed with concrete and brick stacks. It is possible that they can also form in ducts at the base of stacks.

It is believed that the four stages in the formation of smuts are:

1) Sulfuric acid solution is first deposited from the flue gas stream onto the cool surface where it reacts with the metal and forms an acidic salt.
2) This is a sticky layer and consequently soot particles adhere to it. Careful control of combustion conditions and use of good burners will reduce the concentration of these carbonaceous particles, but they will not be entirely eliminated. Hence smut emission will be reduced, but not eliminated by these measures.
3) Part of the acidic layer containing carbonaceous matter then flakes off due to a change in load or temperature.
4) The detached flakes break up into agglomerates or smuts. Some fall to the bottom of the chimney (where they form a thick acidic sooty layer) and some are carried out by the gases into the atmosphere. Owing to their large size and weight they fall near to the stack. These appear as large black smuts, often shaped like a cone up to one-half-inch diameter. Because of their acidic nature, they will:
   a) Damage the cellulose finish on cars.
   b) Damage fabrics, particularly synthetic fibers, such as nylon.
   c) Cause a brown stain on fabrics and asbestos roofs due to the acidic iron salts present in the smuts.
   d) Damage leaves of plants.
   e) Appear as sticky black smudges on surfaces on which they fall. This is often mistakenly described as an oil smudge. It is not oily, but acidic.[21]

Up until the introduction of FGD systems, the basic technique to minimize the formation of acid smut was to operate the stack above the sulfuric acid dew point. Now that the flue gas is cooled by the FGD system, there is a good possibility that acid smut will be formed in the stacks and ducts. Fortunately, as previously mentioned, an FGD system does remove part of the sulfur trioxide from the flue gas.

Another technique for the control of the acid smut mentioned by Gills and Lees[20] is the neutralization of the sulfuric acid by an additive. The injection of finely divided magnesium hydroxide gives effective results in some cases. In this country, a number of companies offer additives for reduction of sulfur trioxide. Most help decrease sulfuric acid formation in the stack.

Another cause of acid smut is air infiltration at the top of the stack. This effect can occur with any stack liner and should be considered in the initial design. Because of this effect, some liners are now designed with a high-corrosion-resistant upper liner section. Highly corrosion-resistant material should be considered inside the existing liner or in place of the existing liner (depending on the recommendations of the engineer), if an accelerated corrosion rate is discovered at the top of the stack.

To get an indication of the corrosion rate in the stack, one should conduct a coupon corrosion test. Discussed earlier in the chapter, these tests are described in detail by ASTM Standards G4 and D2688.

A device to determine the actual corrosion rate experienced by the metal liner is an ultrasonic thickness gauge. With a gauge capable of reading 0.001-in. differences in metal thickness, initial readings can be taken at various locations on the liner. Then, at selected intervals of plant operation, additional readings can be taken at the same locations and the corrosion rate of the liner determined. This measurement technique has the advantage of allowing to make measurements from the dry side of the liner while the liner is in operation. The main limitation is that the liner must be fabricated of solid metal.

There are a number of methods in addition to corrosion coupon tests and sulfur trioxide EPA Method 8 sampling for determining the corrosive conditions in the liner. These include an air-cooled constant temperature corrosion probe, the deposition rate probe, and the electrical conductivity dew-point meter. One of the problems with sulfuric acid is that the maximum rate of deposition of the acid occurs 20–90°F below the dew point of the acid. It also appears to be a function of the flue gas sulfur trioxide concentration. In a paper by Gibson and Kufta[22] a comparison is made between the various cold-end corrosion monitoring methods. If any of these methods are being considered for determining the corrosive environment in your liner, it is suggested that this reference be reviewed.

22.4 Inspection and Repair of Liners

With the introduction of FGD systems, the inspection of the stack liner has taken on increased importance. It is suggested that one contact a stack inspection firm to make a formal inspection of the stack liner after its first year of operation. The frequency of additional inspections should be determined after consultations with the stack inspectors and the engineer.

Inspection and repair information in this section will concentrate on five liner types:

1) Independent brick
2) Steel
3) Gunned concrete
4) Fiberglass
5) Alloy

22.4.1 Independent Brick Liners

Some of the inspection criteria for brick liners have already been discussed in the section on the design of brick liners.

One of the more frequent defects in the brick liner is cracking. There are basically three types of cracks. The first is the crack caused by a settling foundation. These cracks generally run upwards or zigzag through mortar joints. Next is the crack caused by excessive wind loads. These cracks usually go through the mortar joints at 45° angles. The last is the thermal crack that extends upward through brick and mortar. If any of these cracks are observed, special attention should be given to the metal tension bands holding the bricks together. It is possible that acid has migrated through the cracks in the bricks and attacked the metal bands. One should repair 311 cracks based on the recommendations of the stack designer and replace all damaged bands.

Should brick liners be out of service for an extended period of time, place a cap over the top of the liner. If rain or dew is allowed to wet acid-saturated brickwork, sulfation of the mortar joints can occur. This will lead to cracking and spoiling when the stack is put back into service.

Inspection of the liner should occur during the annual outage. Failure to make adequate preventative inspections can result in major failure of the liner. During the inspection look for:

1) Stack – Loose, missing, or spoiled brick, weathered joints, and cracked or weathered corbels caused by water or acid draining from the exterior stack wall projections.
2) Cap – Tightness of the seal to the shell, loose or missing lightning rod points, loose or broken cable point connections, and corrosion in the entire system.
3) Ladders and platforms – Corrosion of the structure, proper attachment to the shell.
4) Lining – The metal bands – Are they structurally sound and not being corroded? Is the packing intact at the breeching entry?

As part of the annual outage, soot should be removed from the lining. On old pre-FGD stacks, one should use a long-handled hoe to remove the soot through the cleanout door. Do not use water for soot removal water can cause rapid chilling of the hot masonry walls, which can lead to excessive thermal stress and possible wall cracking. On new FGD system linings, water on the lining can result in mortar attack and failure of the structure.

Stacks that were previously used for non-scrubbed gases are sometimes subjected to scrubbed flue gas due to retrofit FGD systems. These linings will probably have to be inspected more frequently for corrosion attack. This is true whether there is reheat or not. The addition of the FGD system will add moisture to the stack and lower the temperature of the gases in the stack. One firm has suggested that a solution to the acid attack of these retrofitted flues is to shotcrete the existing lining with an acid-resistant material. Another solution is to line the existing lining with a fiberglass liner. This liner will restrict the maximum operating temperature of the flue gases to 350°F or the upper operating limit of the resin used in the fiberglass liner. Lining the brickwork with an alloy liner is an alternative to totally replacing the brick liner. Installing any type of liner inside a brick liner should only be considered if the existing brickwork is in excellent structural condition or can be restored to excellent structural condition.

It is best to consult the engineer and stack contractor before any repairs are made to the stack. By doing this one can ensure that all design considerations are taken into account before any changes are made to the stack.

22.4.2 Steel Liners

Steel liners are found in two basic forms in the power industry. The first is the bare carbon steel liner that is used on a conventional power plant that does not have an FGD system. The second is the carbon steel liner coated with some organic or inorganic material used after an FGD system.

Inspection of the bare carbon steel liner has been a standard procedure for a number of years. The main areas of concern are the places where the temperature is allowed to drop below the acid dew point for sulfuric acid. At these locations, there will be corrosion of the steel. Areas susceptible to this form of corrosion are the entry doors into the liner, sample ports, inlet breechings, and the liner that extends beyond the cap. The basic repair procedure is to remove the corroded steel and replace it with new metal.

An inspection should be made of the insulation around the steel liner. Any missing insulation should be replaced to maintain the steel temperature above the acid dew point.

Failure of the gunited inorganic coating in a steel liner will be either mechanical, chemical, or both. Mechanical failure produces extensive cracking, spalling, and chunks falling from the coating. Chemical failure appears as a soft and mushy texture of the lining.

The first repair step is to contact the lining manufacturer. Depending on the particular coating used, it may or may not be possible to join new lining patches to the existing lining. In general, defective portions of the existing inorganic coating can be removed from the steel liner with water lances. Repairs should be made to the remaining coating according to the manufacturer's recommendations.

Since repairs of organically coated steel liners require the services of the coating applicator, the recognition of coating failures are discussed here.

NACE Publication 6D170, The Causes and Prevention of Coating Failures[23], describes some types of coating failures:

1) Chalking
2) Color fading or color change
3) Cracking, checking, and alligatoring
4) Peeling, flaking, and delamination
5) Rusting
6) Blistering
7) Lifting and wrinkling
8) Failures due to chemical attack or solvent attack
9) Failures around weld areas
10) Edge failures

This list does not include mechanical damage, erosion, temperature excursions, etc.

A detailed explanation of each of these modes of failure is beyond the scope of this chapter, but here is a brief description of each mode of failure:

1) *Chalking*. Term used to describe the formation of a powdery layer on a coating surface exposed to the weather. This powdery layer is usually white.
2) *Color facing or color change*. Is usually caused by the reaction of the pigment and/or binder with sunlight or the chemical environment.
3) *Cracking, checking, and alligatoring*. Cracking and checking are caused by shrinkage within the film during aging. Alligatoring is a film rupture which is usually caused by the application of a hard, brittle film over a more flexible film.
4) *Peeling, flaking, and delamination*. These are caused by poor adhesion. Flaking is a failure where the coating comes off in small flakes. Peeling is a failure where the coating comes off in large sections or sheets. When flaking or peeling occurs between coats, this failure is called delamination.
5) *Rusting*. Usually appears as (1) spot rusting in damaged areas, (2) pinhole rusting at minute areas, (3) rust nodules breaking through the coating, and (4) under film rusting, which eventually causes peeling or flaking of the coating. When rusting reaches 25% of the total coated surface area, some companies remove the entire coating and replace it.
6) *Blistering*. This describes small or large round projections or pimples that appear on the coating surface.
7) *Lifting and wrinkling*. Lifting describes a phenomenon that occurs when the solvent of a coat of paint softens the previous coat of paint too readily. Wrinkling describes the appearance of a paint surface that has not dried to a smooth layer, but rather into a roughly ridged surface.
8) Failure due to chemical attack or solvent attack. This will be evident by the apparent disintegration or dissolution of the film.
9) *Failures around weld areas*. Weld defects result in under-film rusting, formation of rust nodules, and film peeling.
10) *Edge failures*. Usually appear in the form of rusting through the film and eventually rust creepage under the film.

22.4.3 Gunned Concrete

Problems encountered with gunned concrete liners are the same as those experienced with gunned coatings on steel liners. The repair and inspection procedures are also the same.

22.4.4 FRP Liners

Specifications used as part of the specifications for fiberglass liners include:

1) ASTM Standard D3299
2) NBS Voluntary Product Standard PS15-69
3) ASTM Standard D2393-1972
4) ASTM Standard D2471-1971
5) ASTM Standard D2583-1975
6) ASTM Standard D2584-1972
7) ASTM Standard B84-1975

All resins used in fiberglass liners should contain a fire suppressant (antimony trioxide – less than 5% concentration) and an ultraviolet absorber to increase resistance to the destructive effect of the sun on the resin. Where a coating that is resistant to the chemical environment in the absorber is applied over the fiberglass liner, the life of the fiberglass is increased. (These coatings should contain a high ultraviolet-reflecting pigment such as titanium dioxide.)

If FRP liner experiences a thermal excursion or fire, it should be inspected. One should look for delamination of the resin-rich layers from the fiberglass-rich layers, and blistering, charring, and fracturing of the resin-rich layer.

22.4.5 Alloy Liners

Alloy liners are constructed of various metals. The precise alloy selected for a liner depends on the chemical environment it will face. To get maximum corrosion resistance from alloys, it is important that they be kept clean. That is true for all types of alloys. There is a saying in the alloy field that "stainless steel to be stainless must be clean." To a certain extent, this applies to nickel alloys as well as to stainless steels.

To examine the alloys, it is necessary to remove the dust and scale deposits from the metal surface. The engineer should recommend the type of tools and cleaning devices needed for the alloy type.

Because there are a number of alloys being used in the fabrication of liners, it is hard to generalize about the precise modes of failure that might occur in the alloy. There are several generalizations that can be made, however. For example, the use of steel shovels to clean an austenitic stainless steel can leave iron on the surface of the stainless steel. These iron scrapings will rust when the liner is put back into service and possibly leave pits in the stainless steel. These pits could then become points where chloride pitting might occur. Therefore, it is standard practice to use stainless steel tools when working with stainless steel.

In an austenitic stainless steel, such as 316, 326L, 317, 317L, 317LM, and 904L, a loss of passivity of the metal will appear as a light layer of rust on the alloy surface. Removal of this rust will probably reveal pitting corrosion underneath. These alloys will first show corrosion in areas of the liner that do not have rapid gas flow, e.g., where sample pipes go through the wall of the liner, at the surface of the wall of the liner, and recessed flanges out of the main flow of gas through the vessel.

In nickel-based alloys, there should not be any iron-colored corrosion products on the metal surface. Should a rust-like corrosion be found on the surface, it probably came from damage caused by iron tool scraping of the alloy.

Should scale be allowed to form and remain on the surface of the alloy, a form of crevice corrosion can begin. The concentration of corrosive salts under the scale can exceed the concentration found in the main liquid flow in the liners. Some recent corrosion studies by one of the major alloy suppliers have verified this increased corrosion effect in FGD systems and thus in liners of stacks. Scale that forms on the alloys should not be removed with carbon steel equipment – stainless tools must be used.

Handling of the nickel alloys requires special techniques in order to maintain the corrosion resistance for which they are noted. The engineer should be involved in the inspection and advise the owner about repairs. Each alloy supplier has an excellent selection of instructions for the welding and repair of their alloys. These should be followed in detail.

Alloy welding rods of metal inert gas (MIG) or tungsten inert gas (TIG) welding techniques are needed for all alloy liner repairs. This may require the purchase of new pulse arc welding machines and the retraining of plant personnel. Corrosion-resistant alloys cannot be treated as carbon steel and retain their corrosion-resistant properties.

Some FGD system owners have already found that they must clean the stainless steel in their FGD systems to keep the corrosion to a minimum. This is also true of the stainless steels in the stack area since it is not as feasible to clean the stack liner as it is to clean the FGD system, it seems logical to construct liners of more corrosion-resistant metals than are used in the FGD system.

When welding and repairing the alloy metals, one should be sure that the correct welding rods or wires are used. If there is any chance that carbon steel can be brought into the molten weld pool, the alloy rod or wire must have sufficient metallurgical composition not to lose the desired corrosion resistance when the iron is brought into the weld. It is usually estimated that about 25% of the molten weld pool, when welding an alloy to carbon steel, can come from the carbon steel. The alloy manufacturer can recommend the proper welding rod or wire for its alloy to minimize the effect of this iron dilution.

It is important that good fabrication techniques be used in working with the alloys in the stack. For example, the alloy should be wiped on both sides of the area to be welded with a solvent-soaked rag. This is to remove any organic material that might be present. During the welding process, this organic material can be burnt and brought into the molten weld metal. This will increase the carbon content of the final weld and lower the corrosion resistance of the weld.

22.5 Recent Advances

The basic design, construction, and operation of stacks has remained relatively unchanged over their first use. However, there are new designs and construction materials used in state-of-the-art stacks. In theory, each design will give satisfactory service for its life. However, until they actually have been in service for 30 years, the state-of-the-art designs will have to be considered "new" designs. Three relatively "new" design techniques used in power industry stacks are described below:

1) One method that has proven its ability to handle all types of acid condensation while permitting complete flexibility of operation with any type of fuel is a double-column construction using a concrete outer column and a completely separate inner brickwork column. Proper brick and the use of sodium-free potassium silicate mortars can make the inner column resistant to acid condensation. The space between the two columns can be narrow or, preferably, sufficiently wide to permit inspection and repair. In this design, the acid-resisting requirements of the inner column and

the physical requirements of the outer column can be separated. In addition, several methods can be used to prevent acid condensation from reaching and damaging the foundation.

2) Another variation of the double-column design is a concrete outer column and a steel inner column. The carbon steel column is insulated on its outer surface (facing the concrete column) to keep the inside surface of the steel above the acid condensation temperature. When this method is used, the temperatures at the inside steel surface must not be permitted to drop below the acid condensation point. Should the steel surface temperature drop below its acid dew point, a corrosive attack of the carbon steel will occur at a rapid rate. This corrosive attack could occur at a rate of 1/4 in. of steel per month of operation.
3) Still another variation is the fiberglass liner inside a concrete outer column. These fiberglass liners have to be joined to the concrete outer column with metals that are resistant to the corrosive liquids inside the liner. This assures the structural integrity of the liner if the acid should permeate the liner. Special precautions must also be taken to allow for the high thermal expansion of the fiberglass liner. Should these expansion characteristics not be taken into account, there will probably be structural damage to the liner. Also, since the fiberglass is flammable, special fire protection systems must be installed and maintained to assure fire protection. In addition, the sulfuric acid concentrations and flue gas temperatures that will be on the surface of the FRP liner must be considered.

22.6 Conclusions

From the day a stack is built, both external and internal forces are at work to destroy it. External villains are much easier to design for than internal ones, so exterior inspection and maintenance procedures can usually be based on known design parameters and realistic life expectations of construction materials. Despite this relative stability, however, internal corrosion may affect external performance. It cannot be overstressed that programmed inspections, timely crack repairs, and scheduled maintenance procedures are the keys to continuing stack shell integrity.

Inside the stack, knowing the strengths and weaknesses of a particular type of liner is just as important as knowing the kind of environment a liner must face. Independent brick liners and the emerging family of alloy liners have the best record and highest potential for success in the battle against low-temperature, wet stack environments, but even with these design Goliaths, there may be a David hiding in the post-FGD or bypass effluent. One must determine effluent temperature at various stack heights, dew-point condensation, and a myriad of other stack environment data. Only then can one tell what a liner can tolerate and for how long.

Continuing research and development portend optimism for increased stack longevity and performance reliability. Someday there may be a stack that "can take it," but that day is not here yet.

References

1 **R. Snook**. *The Impact of Flue Gas Desulfurization on Chimney Design in the United States*, Third International Chimney Design Symposium, Munich, Germany, 1978.
2 **B. Cooley**, *Consider dynamic wind and earthquake loads when designing reinforced-concrete stacks*, *Power*, 24, 1977.
3 **Custodis Construction Company**, *Design and Construction of Independent Brick Chimney Linings*, Revised Paper, Chicago, Ill 1979.
4 **B. Cooley**, *Consider Dynamic Wind*, p. 27.
5 Ibid.
6 **D. Kossler, and R. Pierce**, *Power Industry Chimneys: Brickwork Lining Performance*, Canadian Electrical Association Thermal Power Section Eastern Zone Meeting, Mont Gabriel, Quebec, 1964.
7 **Kossler, and Pierce**, Ibid.
8 **R. Snook**. *Interim Report of the ASCE Task Committee on Steel Chimney Liners*, American Power Conference, p. 1, Chicago, Ill, 1972.
9 **G. Read Jr.**, *Corrosion-Resistant Linings for Chimneys, Scrubbers and Ducts of Flue Gas Systems*, American Power Conference, 40th Annual Meeting, p. 2, Chicago, Ill, 1978.
10 Ibid., p. 4.

11 **R. Snook**, *Interim Report of the ASCE Task Committee*, p. 8.
12 **G. Read Jr.**, *Acidproof Concrete Lining for Steel Stacks of Power Plants*, Sauereisen Cements Company, p. 4, Pittsburgh, PA.
13 *Detroit Edison First to Insulate Stack Liner with New High Temperature Fiber Glass Board*, p. 45, Combustion, 1977.
14 **Custodis Construction Company**, *Sample Specification for Fiberglass Reinforced Plastic Chimney Liners*, p. 2, Chicago, Ill.
15 Ibid., p. 3.
16 Ibid., p. 5.
17 *Weldability Characteristics of Stellite Division High Performance Alloys*, Technology Division, Cabot Corporation Stellite R&D Department, 1977.
18 **G. Gibson, and R. Kufta**, *Cold-end corrosion monitoring methods*, *Power Eng.*, 71, 1977.
19 **R. Pierce**, *Acid Dew Point Estimating Method*, Chimney Design Symposium and Exhibition, TH 2281.C55, 1976, Quarto Vol. 3, p. 143, 1976.
20 **Kossler and Pierce**, *Power Industry Chimneys.*
21 **B. Gills and B. Less**, *Insulated Steel Chimneys Avoid Corrosion and Smut Emission*, American Society of Mechanical Engineers Winter Annual Meeting, p. 2, Chicago, 1965.
22 **Gibson, and Kufta**, *Cold-End Corrosion*, p. 73.
23 *The Causes and Prevention of Coating Failures*, NACE Publication 6D170 prepared by Task Group T-6D-22 (A. K. Long, Chairman) and Unit Committee T-6D (R. R. Rosenthal, Jr.), pp. 33–36.

23

Ventilation

Dilution ventilation refers to the technique of exchanging outside for inside air for the purposes of comfort or contaminant control. For comfort control, this technique can use cooled air in the summer and warmed air in the winter. For contaminant control purposes, however, dilution ventilation has limitations. If the quantity of contaminant released is sufficiently large it would be impractical to dilute it adequately. Also, individuals who are exposed to the contaminant must be far enough away from the source so that they will only breathe the lowest concentrations produced when this technique is employed. The dilution ventilation procedure works best where the contaminants have low toxicity and the generation of contaminant is uniform in known locations. An example of where it has been used to advantage is in the control of organic vapors. It is seldom applied successfully in the control of dusts and fumes.

The advantages of local exhaust ventilation (LEV) are essentially complete capture of the contaminant, regardless of its evolution rate, toxicity, or physical state. Relatively low air volumes are required compared to dilution ventilation. Disadvantages are the complexity of some designs and the fact that once installed they are not readily moved to other locations. These are minor aspects in most industrial situations.

23.1 Introduction to Industrial Ventilation Systems

When designing manufacturing sites or other industrial construction, ventilation must be considered an integral part of the overall industrial process. This approach is essential to ensure a safe and efficient design of the ventilation system within the overall process. Naturally, much of this applies in other applications, including hospitals and other medical facilities. The application areas for ventilation systems are as follows:

1) Control of contaminants to acceptable levels
2) Control of heat and humidity for comfort
3) Prevention of fires and explosions

All of these factors are important considerations at the design stage. In general, the controlling design criteria will be to achieve acceptable contaminant levels in the location in question.

The following are four steps to be considered in solving industrial ventilation problems:

1) *Process modifications*. This should always be the first approach to establish whether the process can be modified, and the contaminant can be controlled to an acceptable level. The end result of this process modification approach is that there is no need for a ventilation system.
2) *LEV*. With the exception of process modifications, LEV systems will almost always be the most cost-effective technology for controlling contaminant levels in the workplace. Contaminants are controlled at the source with low volumes of gas.
3) *General work area* (i.e. general or dilution ventilation). The concept of general or dilution ventilation is based on supplying or exhausting large volumes of air to an occupied space or the complete processing/manufacturing complex. The ventilation air can be supplied and exhausted by natural ventilation or mechanical ventilation. A subsequent section will describe the features of general ventilation for various facilities.

Optimizing Air Pollution Control Equipment Performance: Operation & Maintenance, First Edition. Jay Richardson and Louis Theodore.

4) *Personal protective equipment (PPE).* For some specific applications, it may be necessary to use PPE to reduce the exposure for workers. In general, this is only applicable if process modification or ventilation is not feasible. PPE can be considered for nonroutine maintenance work or for processes or systems that do not have feasible engineering control solutions.

23.2 Dilution Ventilation

Air circulation within a confined area or space may be classified as natural ventilation or mechanical ventilation.

Natural ventilation is the displacement and infiltration of air through windows, doors, walls, floors, ventilators, and other openings by natural forces such as pressure (wind) and temperature (air density) differences between the building and its surroundings. The air movement may be caused by one or the other, or a combination of the two. It must be remembered that air movement by infiltration is not only dependent on differences in temperature and pressure but also on the physical characteristics of the building.

Mechanical ventilation is the creation of air movement by some mechanical or artificial method, such as by the use of wall, window, portable fans, roof ventilators, or the operation of industrial equipment. This is usually required when contaminant dilution is necessary, as infiltration or other natural ventilation normally cannot provide the desired volume of air. In comparatively small rooms or work areas, window or wall fans are generally used for exhaust. In larger single-level building's roof ventilators, which take advantage of the air temperature differences that exist, may be used as exhausters. Many other types of commercial roof ventilators are available, from simple propeller to centrifugal types which are capable of producing high static pressures and efficiencies, and thus exhaust a greater volume of air. Where roof ventilators are not practical, air may be exhausted through grills to ductwork and fans. General supply or makeup air is usually required to prevent the creation of negative pressures in the building when sufficient dilution ventilation is provided by local exhaust hoods within the building. It should also be noted that while supply air blower and distributor systems are needed with this type of system they may not always be provided.

If the material of concern escapes into the air at a comparatively low rate and more or less uniformly throughout the work area – not at one or more isolated locations – dilution ventilation may be the most practical means of controlling the contaminant at the breathing zone of the worker. However, if the process generates especially toxic or irritating substances, local exhaust should be used. The successful application of dilution ventilation depends upon the following conditions:

1) The material or contaminant generated should not be in excess of the amount that can be diluted with a reasonable volume of air.
2) The distance between the individuals and the point of generation of the contaminant should be sufficient to assure that the individuals will not be exposed to an average concentration in excess of the current established threshold limit values (TLVs).[1]
3) The toxicity of the contaminant must be low (i.e., must have a high TLV).
4) The material or contaminant should be generated at a reasonably uniform rate.

Dilution ventilation is used with the greatest success in the control of contaminants when the rate of consumption can be determined with some degree of accuracy, by a materials balance or some other method, and the volume of dilution air required can be calculated.

It may well be that a problem does not exist due to dilution ventilation created by the construction and other features of the work area because practically all rooms or areas have, to a larger or smaller extent, some amount of natural or mechanical ventilation. On the other hand, only a small amount of material is required to produce a problem condition in an air-tight room.

Dilution ventilation is used successfully to control the concentrations in the workroom air consisting of gases and organic vapors with comparatively low toxic properties, as well as heat and water vapor. Several reasons preclude the successful application of dilution ventilation for controlling the concentration of fumes and dusts. The amount of material generated is usually too great, and reliable data on the rate of generation of fumes and dusts are very difficult to obtain and quantify. In addition, the material may be quite toxic, which would require an excessive amount of dilution air.

Dilution ventilation rates are often expressed as "air changes per hour" or "minutes per air change." These units express the ratio of the volume of air supplied or exhausted to the volume of the room. They are useful when applied to hospitals, meeting rooms, offices, schools, warehouses, retail stores, laundries, and similar spaces where the purpose of ventilation is simply odor, temperature, and humidity control, and the only contamination of the air is from the activity of people. For purposes of control by dilution, however, such terms are of little value. Air changes per hour are dependent upon the size of the room, while the quantity of air needed to dilute a given contaminant, generated at a constant rate, is not affected by the size of the room.

Dilution ventilation rates should be expressed in cubic feet per minute or some other absolute unit of quantity flow. The room volume will determine the rate at which a concentration of contaminant will build up, and also the time required to purge the room after the generation of contaminant is stopped. A large room will act as a surge tank, so that for an intermittent generation of contaminant it may be safe to supply air to balance the average rate of generation rather than the maximum rate.

The air exhausted must be replaced by an equal amount of uncontaminated fresh air, either from a mechanical system or by entry through openings in the building. This will assure that the room is not under excessive negative pressure, which will reduce the possibility of individuals being exposed to dangerous concentrations of other gases or vapors, principally carbon monoxide. Provision must be made to heat the makeup air during cold weather. Care must be exercised to prevent drafts around windows, doors, and other sources of makeup air.

23.3 Local Exhaust Systems

The control of atmospheric contamination is often accomplished by means of LEV. By "local exhaust" is meant the capture and removal of the contaminant close to the point of origin, before the contaminant enters the breathing zone or is disseminated into the general environment. This type of ventilation is preferred over general or dilution ventilation in some cases for three reasons. Less air needs to be exhausted to control the contaminant. This means that a smaller fan will be used, and the power consumption will probably be lower. Also, the makeup required will be less, and in cold climates or air-conditioned areas, this may be an important economic factor. The contaminant is captured before it enters the breathing zone and not after it has been disseminated. With some highly toxic and radioactive substances, this is the only method of exhaust possible. The contaminant is more highly concentrated in the conveying air stream. This is an important factor where the air will be cleaned either to salvage some valuable product or to prevent air pollution.

The exhaust hood is the first element in a local exhaust system; its design will determine the amount of airflow. In theory, a minimum velocity is created in or near the hood, which captures the contaminant and draws it into the hood and connecting ductwork. For gases and vapors lower capture velocities are needed when compared with particulates. The size (mass) of the particles will determine their inertia and ease of capture. Large dust particles released at high velocity (e.g., from grinding) are difficult to capture unless directed into the hood. Fine dust below a few microns in size is typically captured without difficulty, in similar fashion to gases and vapors.

The exhaust hood must be designed so that it can function without interfering with the worker or the operation. All hood openings must be kept to a practical minimum to keep airflow requirements as low as possible. In some cases, the entire operation can be enclosed so that only minimal openings are provided for entry of air to remove contaminants in the enclosure. Inspection and maintenance openings should be covered with doors whenever practical.

Canopy hoods are large hoods located over an operation. With either natural or forced exhaust, they are most effective for hot processes where the rising thermal currents bring contaminants into the hood. The control of contaminants with this hood is lost when there are cross-drafts or process upsets. The canopy hood is not recommended where men must work over an operation, such as a plating tank or cementing table, because the contaminated air would pass through the worker's breathing zone.

It is only after the hood design has been determined that exhaust volume requirements can be calculated by multiplying the known open area by the desired face velocity of the hood. Established hood designs for various processes are available in the literature.[2] Where the hood system has been fully described, the required air volume is usually given as well.

If air temperature and corrosion problems are not severe, hoods are usually constructed of galvanized sheet metal. As with elbows and transitions, the metal should be at least two gauges heavier than the connecting duct. Reinforcement with angle iron and other devices is required except for very small hoods. For elevated temperatures up to approximately 900°F, black iron may be employed, the thickness of the metal being increased in proportion to the temperature. For temperatures in the range of 400–500°F, 10-gauge metal is most commonly employed. When the temperature of the hood is as high as 900°F, the thickness of the metal may be increased up to ¼ in. Over 900°F, up to about 1,600–1,800°F, stainless steel must be employed. If the hood temperature excursions periodically exceed 1,800°F or are in excess of 1,600°F for a substantial amount of the time, refractory materials are required. A variety of materials are available for corrosive conditions. Plywood is sometimes employed for relatively light duty or for temporary installations. A rubber or plastic coating may sometimes be applied on steel. Some of these coatings can be applied like ordinary paint. If severe corrosion problems exist, hoods must be constructed of sheets of PVC (polyvinyl chloride), fiberglass, or transite.

23.4 Selecting Ventilation Systems

Exposure to viruses in a home, hospital, or workplace can be reduced by proper ventilation. As noted above, ventilation can be provided either by dilution ventilation or by a local exhaust system. In dilution ventilation, air is brought into the work area to dilute the contaminant sufficiently to minimize its concentration and subsequently reduce worker exposure. In a local exhaust system, the contaminant itself is removed from the source through hoods.

It should be noted that a local exhaust is generally preferred over a dilution system for health/hazard control. In determining the quantity of dilution air required, one must also consider mixing characteristics of the area in question, in addition to the quantity (mass or volume) of contaminant to be diluted. Thus, the amount of air required in a dilution ventilation system is much higher than the amount required in a local exhaust system. In addition, if the replacement air requires heating or cooling to maintain an acceptable workplace temperature, then the operating cost of a dilution ventilation system may further exceed the cost of a local exhaust system.

As mentioned earlier the major components of any industrial ventilation system generally include the following:

1) Exhaust hoods
2) Ductwork
3) Contaminant control devices
4) Exhaust fans
5) Exhaust vents or stacks

Several types of hoods are available. One must select the appropriate hood for a specific operation to effectively remove viruses or contaminants from an area and transport them into the ductwork. The ductwork must be sized such that the contaminant is transported without being deposited within the duct adequate velocity must be maintained in the duct to accomplish this. Selecting a control device that is optimal for the removal of viruses or contaminants is important to ensure meeting any relevant control removal efficiency requirements.

The exhaust fan is the workhorse of the ventilation system. The fan must provide the necessary volumetric flow at the required static pressure and must be capable of handling contaminated air characteristics such as dustiness, corrosiveness, and moisture in the air streams.[3,4]

Properly venting the exhaust out of a building is often necessary to avoid viruses or contaminant recirculation into the air intake or into a building through other openings. As noted earlier, such problems can be minimized by properly locating the vent pipe in relation to the aerodynamic characteristics of the building. In addition, all or a portion of the cleaned air may be recirculated. Once again, primary (outside) air may be added, and is referred to as makeup air; the temperature and humidity of the makeup air may have to be controlled. It also may be necessary to exhaust a portion of the room air. Finally, when selecting a ventilation system, one should never overlook the most obvious corrective step – modify the process to generate smaller amounts of contaminant and take steps to prevent contaminants from entering the workplace.

23.5 Ventilation Models

Indoor air pollution is rapidly becoming a major health issue in the United States. Indoor pollutant levels are quite often higher than outdoors, particularly where buildings are tightly constructed to save energy. Since most people spend nearly 90% of their time indoors, exposure to unhealthy concentrations of indoor air pollutants is often inevitable. The degree of risk associated with exposure to indoor pollutants depends on how well buildings are ventilated and the type, mixture, and amounts of pollutants in the building.

As noted earlier, ventilation is the field of applied science concerned with controlling airborne contaminants and viruses to produce healthy conditions for workers, doctors, patients, etc., and a clean environment. However, ventilation cannot completely prevent contaminants from entering the workplace. More to the point, and within the realm of achievement, is the goal of controlling contaminant exposure within prescribed limits. To accomplish this goal, one must be able to predict – hopefully accurately – the concentration of a species in a control volume. There are a number of equations that can be found in the literature to estimate the concentration of species in the control volume. These equations can be used, provided they are dimensionally consistent. The choice is generally one of convenience.[5]

23.6 Model Limitations

There are numerous deficiencies or inadequacies in the understanding of how to not only design a ventilation system to satisfy standards of performance but also to accurately predict concentration variation in a ventilated control volume. Some of the later problems include:

1) Inability to predict the concentration
2) Effects of drafts and wakes in the control volume
3) New or unusual contaminant sources
4) Capture velocity
5) Source term(s)

All of these concerns are deficiencies that arise during the motion of contaminants from their source of origin to the outlet from the ventilation system.

Regarding the effects of drafts and wakes in the control volume, theoretical consideration of ventilation systems is often based on the assumption that surrounding the source is a quiescent body of air. Nothing could be further from the truth. Every workplace contains equipment and workers performing tasks that produce room air currents of unique and unpredictable character, i.e., drafts. Spurious room air currents can also be produced.

Other factors affecting the motion of contaminants are walls, partitions, equipment surfaces, and other fixtures. In some cases, impediments may be eliminated, but for the most part, they may have to be incorporated into the design of the ventilation system from the beginning. An even more vexing circumstance is trying to correct unwanted room air currents produced by equipment or new procedures put in place after the ventilation system has been installed.

Theodore has provided several nonideal approaches that can be divided into three main categories.[5]

1) Imperfect mixing
2) Bypassing/short-circuiting
3) Both imperfect mixing and bypassing/short-circuiting

Pockets of stagnant fluid or dead spots can exist. Here, the discharge stream is lower because the remainder of the fluid resides in the control volume for a shorter period of time. The insertion of baffles in the control volume can enhance the mixing process. Fogler[6] has classified these as:

1) Segregated flow in which the elements of fluid do not mix but follow separate paths through the system so that they have different residence times, and
2) Micromixing in which the adjacent elements of fluid are partially mixed.

Unfortunately, much of this information is not available for a real system, and other approaches must be employed. The important fact for the practicing engineer or applied scientist is to remember that deviations from ideal behavior negatively impact the performance of a ventilation system. Every attempt should be made to minimize or eliminate this effect.[5–7]

Finally, Theodore has also noted that if the enthalpy of reaction for any of the species, including viruses, is not negligible, or, if it is required to supply or remove thermal energy in the form of heat from the control volume, an energy balance around the system must be obtained in order to determine the temperature and its variation with control volume.[8,9]

References

1 **L. Theodore, and R.R. Dupont**, *Environmental Health Risk and Hazard Risk Assessment: Principles and Calculations*, CRC Press/Taylor & Francis Group, Boca Raton, FL, 2013.
2 **R. Heinsohn**, *Industrial Ventilation: Engineering Principles*, John Wiley & Sons, Hoboken, NJ, 1991.
3 **P. Abulencia, and L. Theodore**, *Fluid Flow for the Practicing Chemical Engineer*, John Wiley & Sons, Hoboken, NJ, 2009.
4 **I. Farag**, *Fluid Flow, A Theodore Tutorial*, Theodore Tutorials (originally published by USEPA/APTI, RTP, NC), East Williston, NY, 1995.
5 **L. Theodore**, *Chemical Reactor Analysis and Applications for the Practicing Engineer*, John Wiley & Sons, Hoboken, NJ, 2012.
6 **S. Fogler**, *Elements of Chemical Reaction Engineering*, 4th Ed., Prentice-Hall, Upper Saddle River, NJ, 2006.
7 **L. Theodore**, *Chemical Engineering: The Essential Reference*, McGraw Hill, New York, NY, 2012.
8 **L. Theodore, F. Ricci, and T. VanVleet**, *Thermodynamics for the Practicing Engineer*, John Wiley & Sons, Hoboken, NJ, 2007.
9 **L. Theodore**, *Heat Transfer Applications for the Practicing Engineer*, John Wiley & Sons, Hoboken, NJ, 2011.

Part III

Epilogue

Webster defines epilogue as "an appending section, as of a novel, sewing to complete the plan or scope of work." And indeed, that is exactly what these final 6 chapters constitute. In some respects, the attempt here is to review air pollution control equipment-related issues and those not addressed earlier.

Optimizing Air Pollution Control Equipment Performance: Operation & Maintenance, First Edition. Jay Richardson and Louis Theodore.

24

Atmospheric Dispersion

Sarah Forster

Department of Biological and Environmental Sciences, Long Island University Post, Brookville, NY, USA

Stacks discharging to the atmosphere have long been one of the methods available to industry for disposing waste gases. The concentration to which humans, plants, animals, and structures are exposed at ground level can be reduced significantly by emitting the waste gases from a process at great heights. This permits the gaseous pollutants to be dispersed over a much larger area and will be referred to as control by dilution.

Atmospheric contamination arises primarily from the exhaust generated by industrial plants, power plants, refuse disposal plants, domestic activities, commercial heating, and transportation. These pollutants – which are in the form of particulates, smog, odors, and others – arise mostly from combustion processes and also contain varying amounts of undesirable gases such as oxides of sulfur, oxides of nitrogen, hydrocarbons, and carbon monoxide. The expanding needs of society for more energy and advanced transportation technology, coupled with the rapid growth of urban areas, have led to ever-increasing amounts and concentrations of pollutants in the atmosphere.

This chapter focuses on some of the practical considerations of the dispersion of pollutants in the atmosphere. Both continuous and instantaneous discharges are of concern to individuals involved with environmental management. However, the bulk of the material here has been presented for continuous emissions from point sources – for example, a stack. This has traditionally been an area of much concern in the air pollution field because stacks have long been one of the more common industrial methods for disposing of waste gases. The concentrations of pollutants that humans, plants, animals, and structures are exposed to at ground level can be reduced significantly by emitting the waste gases from a process at great heights. This control by dilution permits the pollutants to be dispersed over a much larger area. Although tall stacks may be effective in lowering the ground-level concentration of pollutants, they still do not, in themselves, reduce the amount of pollutants released into the atmosphere. However, in certain situations, it can be the most practical and economical way of dealing with an air pollution problem. However, as in the case of any sink whose capacity is exceeded, the atmosphere, too, has its limits. Poor dispersion of pollutants, resulting from localized overloading of the atmosphere and the fluctuations of meteorological phenomena, allows the buildup of pollutant levels that may degrade the quality of the environment and endanger public health. The bulk of the material in this chapter has been drawn from the earlier work of Turner[1] and Theodore.[2]

24.1 The Nature of Dispersion

The release of pollutants into the atmosphere is a traditional technique for disposing of them. Although gaseous emissions may be controlled by various sorption processes (or by combustion) and particulates (either solid or aerosol) by mechanical collection, filtration, electrostatic precipitators, or wet scrubbers, the effluent from the control device must still be dispersed into the atmosphere. Fortunately, one of the important properties of the atmosphere is its ability to disperse highly concentrated streams of gaseous pollutants. Of course, the atmosphere's ability to disperse such streams is not infinite and varies from quite good to quite poor, depending on local meteorological and geographical conditions. Therefore, the ability to model atmospheric dispersion and to predict pollutant concentrations from a source are important parts of air pollution engineering.

A continuous stream of pollutants released into a steady wind in the open atmosphere will first rise (usually), then bend over and travel with the wind, which will dilute the pollutants and carry them away from the source. This plume of

Optimizing Air Pollution Control Equipment Performance: Operation & Maintenance, First Edition. Jay Richardson and Louis Theodore.

pollutants will also spread out or disperse both in the horizontal and vertical directions from its center line.[1] In doing so, the concentration of the gaseous pollutant is decreased because the same amount of pollutant is now contained within a larger volume. This natural process of a high concentration spreading out to lower concentrations is the process of dispersion. Atmospheric dispersion is primarily accomplished by the wind movement of pollutants, but the character of the source of pollution requires that this action of the wind be taken into account in different ways.

The dilution of air contaminants is also a direct result of atmospheric turbulence and molecular diffusion.[3] However, the rate of turbulent diffusion[3] is so many thousand times greater than the rate of molecular diffusion that the latter effect can be neglected in atmospheric diffusion analysis. Atmospheric turbulence and, hence, atmospheric diffusion vary widely with weather conditions and topography. Taking all these factors into account, a four-step procedure is recommended in performing dispersion health effect studies:

1) Estimate the rate, duration, and location of the release into the environment.
2) Select the best available model to perform the calculations.
3) Perform the calculations and generate downstream concentrations resulting from the source emission(s).
4) Determine what effect, if any, the resulting discharge has on the environment, including humans, animals, vegetation, and materials of construction.

24.2 Meteorological Concerns

The atmosphere has been labeled the aerial dump for air pollution. Industrial society can be thankful that the atmosphere cleanses itself (up to a point) by natural phenomena. Atmospheric dilution occurs when the wind moves because of wind circulation or atmospheric turbulence caused by local sun intensity. As described earlier, pollution is "removed" from the atmosphere by precipitation and by other reactions (both physical and chemical) as well as by gravitational fallout.

The atmosphere is the medium in which air pollution is carried away from its source and diffuses. Meteorological factors have a considerable influence over the frequency, length of time, and concentrations of effluents to which the general public may be exposed. The variables that affect the severity of an air pollution problem at a given time and location are wind speed and direction, insolation (amount of sunlight), lapse rate (temperature variation with height), mixing depth, and precipitation. Unceasing change is the predominant characteristic of the atmosphere; for example, temperature and winds vary widely with latitude, season, and surrounding topography.[3,4]

Atmospheric dispersion depends primarily on horizontal and vertical transport. Horizontal transport depends on the turbulent structure of the wind field. As the wind velocity increases, the degree of dispersion increases with a corresponding decrease in the ground-level concentration of the contaminant at the receptor site. This is a result of the emissions being mixed into a larger volume of air. The diluted effluent may, depending on the wind direction, be carried out into essentially unoccupied terrain away from any receptors. Under different atmospheric conditions, the wind may funnel the diluted effluent down a river valley or between mountain ranges. If an inversion (temperature increases with height) is present aloft, which would prevent vertical transport, the pollutant concentration may build up continuously.

One can define atmospheric turbulence as those vertical and horizontal convection currents or eddies that mix process effluents with the surrounding air. Several generalizations can be made regarding the effect of atmospheric turbulence on effluent dispersion. It should be obvious that turbulence increases with increasing wind speed and a corresponding increase in horizontal dispersion. Mechanical turbulence is caused by changes in wind speed and wind shear at different altitudes. Either of these two conditions can lead to significant changes in concentration of the effluent at different elevations.

A major factor affecting atmospheric dispersion is vertical transport, which depends on the rate of change of ambient temperature with altitude. The dry adiabatic lapse rate is defined as a decrease in air temperature of 10°C/km or 5.4°F/1,000 ft. This is the rate at which a rising parcel of unpolluted air under natural conditions will decrease in temperature with elevation into the troposphere up to approximately 30,000 ft. An isothermal condition is one in which the temperature does not change with elevation. Stable atmospheric conditions occur for nonsaturated air when the actual environmental lapse rate is less than the dry adiabatic lapse rate. Such conditions hinder vertical transport. A negative lapse rate is a condition where there is an increase in air temperature with altitude. This is an inversion that effectively prevents vertical mixing. The opposite condition, of course, is more desirable. If super adiabatic conditions exist, i.e., the temperature decreases with altitude faster than the dry adiabatic lapse rate, the result is substantial vertical mixing and rapid dispersion of effluents because of vertical transport.[5] For all intents and purposes, rapid atmospheric diffusion in the vertical is always bounded. Utilization of this approximately 8- to 10-km vertical dimension can take place at very different rates, depending on the thermally driven vertical wind.

Methods of estimating atmospheric dispersion have undergone considerable revision, primarily due to the results of experimental measurements. In most dispersion problems the relevant atmospheric layer is that nearest the ground, varying in thickness from several hundred to a few thousand meters. Variations in both thermal and mechanical turbulence and in wind velocity are greatest in the layer in contact with the surface. Turbulence induced by buoyancy forces in the atmosphere is closely related to the vertical temperature structure.[6]

Topography can also have a considerable influence on horizontal transport and thus on pollutant dispersion. The degree of horizontal mixing can be influenced by sea and land breezes. It can also be influenced by man-made and natural terrain features such as mountains, valleys, or even a small ridge or row of hills. Low spots in the terrain or natural bowls can act as sites where pollutants tend to settle and accumulate because of a lack of horizontal transport in the land depressions. Other topographical features which can affect horizontal transport are city canyons and isolated buildings. City canyons occur when the buildings on both sides of a street are fairly close together and relatively tall. Such situations can cause a funneling of emissions from one location to another. Isolated buildings, or the presence of a high-rise building in a relatively low area, can cause redirection of dispersion patterns and route the emissions into an area in which many receptors live.

Dispersion of air contaminants is strongly dependent on the local meteorology of the atmosphere into which the pollutants are emitted. The mathematical formulation for the design of pollutant dispersal is associated with open-ground terrain free of obstructions. Obstructions of either the natural or man-made type alter the atmospheric circulation and with it the dispersion of pollutants. Particularly well-cited are the effects of mountain valley terrain, hills, lakes, shorelines, and buildings.

Valley effects occur due to the channeling of winds. Valleys tend to make the wind flow in the general direction of the valley axis. Slope winds occur in valleys in the evenings when the air near the ground is cooled by the ground. Any air pollution released on the slopes or in the bottom of a valley will stay in the valley if an inversion situation also exists. The onset of the valley wind usually lags several hours after the onset of the slope winds. In reverse, on a clear morning with a light wind, heating can cause up-slope and up-valley winds.

Shoreline winds are created because of the differences in heating rates of the earth and the water. Under the same amount of sunshine, land absorbs heat faster than water. Therefore, on a sunny day, sea breezes are created which come from the water to the land. Land also cools faster than water, creating land breezes on cloudy days or at nighttime. Land breezes are usually lower in velocity and shallower than sea breezes. These can create undesirable downwash from a stack located near the shoreline. Hills can cause a varying degree of influence on air pollution removal. A smooth hill causes the least effect on the flow of air. Rough hills, conversely, caused turbulent eddies and good mixing, which promote improved air pollution removal. If a mountain is sufficiently high, it will cause the wind to deflect and flow parallel to the mountain and not over the mountain.[7]

24.3 Plume Rise

A plume of hot gases emitted vertically has both a momentum and a buoyancy. As the plume moves away from the stack, it quickly loses its vertical momentum (owing to drag by and entrainment of the surrounding air). As the vertical momentum declines, the plume bends over in the direction of the mean wind. However, quite often the effect of buoyancy is still significant, and the plume continues to rise for a long time after bending over. The buoyancy term is due to the less-than-atmospheric density of the stack gases and may be temperature or composition induced. In either case, as the plume spreads out in the air (all the time mixing with the surrounding air), it becomes diluted by the air.

Modeling the rise of a plume of gases emitted from a stack into a horizontal wind is a complex mathematical problem. Plume rise depends not only on such stack gas parameters as temperature, molecular weight, and exit velocity, but also on such atmosphere parameters as wind speed, ambient temperature, and stability conditions.[1,2]

24.3.1 Plume Behaviors

The behavior of plumes emitted from any stack depends on localized air stability.[1,2] Typical situations are generally encountered in the lower thousand feet of the atmosphere. Effluents from tall stacks are often injected at an effective height of several hundreds of feet to a couple of thousand feet above ground because of the added effects of buoyancy and velocity on plume rise. Other factors affecting plume behavior are the diurnal variations in atmospheric stability and the long-term variations that occur with changing seasons.

Looping is a diurnal phenomenon and is the result of solar heating under super adiabatic lapse rate conditions. This causes rapid dilution of the plume with only occasional ground contact. Looping is typical plume behavior if large-scale thermal eddies are present. **Coning** occurs when the actual lapse rate is less than the dry adiabatic lapse rate. If the air is saturated, the result is a slight instability of the atmosphere with some horizontal and vertical mixing. In this situation, pollutants emitted at considerable elevations can be returned to ground level. Coning is often associated with moderate to high speeds and overcast conditions.

Fanning occurs in the presence of very light winds as a result of an inversion aloft. There is no vertical mixing. Rising effluents essentially travel along the bottom of the inversion layer and are slowly dispersed due to horizontal transport. **Lofting** is the situation in which a super adiabatic region exists above a low-level or surface inversion. This condition effectively traps all pollutants emitted from stacks with low effective heights while allowing vertical mixing above the inversion layer. The plume prevents all gaseous and small-particulate emissions from reaching the ground, and thus is one of the major goals of tall stack operation of electric utility and industrial plants. **Fumigation** can arise in two ways. One type is a result of a radiation inversion which causes the plume to fan out from the stack during the hours of darkness. After sunrise, solar heating causes an unstable layer to form and grow upwind from the fanning plume, bringing it suddenly to the ground. Fortunately, this condition is usually of very short duration. The second type of fumigation is more serious. It is a result of low-altitude heating of air from man-made sources. Cities cause unstable conditions to exist at elevations up to two to three times their average height. This convection disrupts any stability the air may have and brings the contaminants to the ground.

Trapping occurs when the plume effluent is caught between the earth's surface and an inversion aloft. Trapping is similar to fumigation with one significant difference: the contaminant is not at ground level for the entire plume length. Although good vertical mixing may exist below the inversion, the net effect of trapping is an increase in surface concentrations. Additional details on plume rise are available in the literature.[1,2]

24.4 Effective Stack Height

Reliance on atmospheric dispersion as a means of reducing ground-level concentrations is not foolproof. Inversions can occur with a rapid increase in ground-level pollutant concentrations. One solution to such situations is the tall stack concept. The goal is quite simple, inject the effluent above any normally expected inversion layer. This approach is used for exceptionally difficult or expensive treatment situations because tall stacks are quite expensive. To be effective, they must reach above the inversion layer so as to avoid local plume fallout. The stack itself does not have to penetrate the inversion layer if the emissions have adequate buoyancy velocity. In such cases, the effective stack height will be considerably greater than the actual stack height.[3] The effective stack height (equivalent to the effective height of the emission) is usually considered as the sum of the actual stack height, the plume rise due to the velocity (momentum) of the issuing gases, and the buoyancy rise, which is a function of the temperature of the gases being emitted and the atmospheric conditions.

This effective stack height depends on a number of factors. The emission factors include the gas flow rate, the temperature of the effluent at the top of the stack, and the diameter of the stack opening. The meteorological factors influencing plume rise are wind speed, air temperature, shear of the wind speed with height, and atmospheric stability. No theory on plume rise presently takes into account all of these variables, and it appears that the number of equations for calculating plume rise varies inversely with one's understanding of the process involved. Even if such a theory were available, measurements of all of the parameters would seldom be available. Most of the equations that have been formulated for computing the effective height of an emission are semiempirical in nature. When considering any of these plume rise equations, it is important to evaluate each in terms of the assumptions made and these circumstances existing at the time the particular correlation was formulated. Depending on the circumstances, some equations may definitely be more applicable than others.

The effective height of an emission rarely corresponds to the physical height of the source or the stack. If the plume is caught in the turbulent wake of the stack or of buildings in the vicinity of the source or stack, the effluent will be mixed rapidly downward toward the ground. If the plume is emitted free of these turbulent zones, a number of emission factors and meteorological factors will influence the rise of the plume. The influence of mechanical turbulence around a building or stack can significantly alter the effective stack height. This is especially true with high winds when the beneficial effect of high stack gas velocity is at a minimum and the plume is emitted nearly horizontally. Ensuring that the exit velocity of the stack gas is more than 1.5 times the wind speed will usually prevent downwash in the wake of the stack.

Several plume rise equations are presented in the literature. The Davidson-Bryant method is empirical because it is based on Bryant's wind tunnel experiments. It is restricted to gases below 125°F and to stacks of moderate height and larger. It does not give maximum plume rise because it is only a function of momentum and not buoyancy. The method applies when the atmosphere neither resists nor assists the vertical motion of the plume and when the wind velocities are 20 mph or higher, because above this velocity, the vertical motion is insignificant compared with the horizontal motion.[8] Briggs developed a series of equations that are currently used by the United States Environmental Protection Agency (US EPA).[9] The Tennessee Valley Authority (TVA) plume rise equations resulted from an extensive investigation on stack emissions from large coal-fired power plants that indicated that for particular ranges of atmospheric stability, expressions may be determined for estimating the effective plume rise at various distances from the stack. The stack heights ranged from 152 to 244 m, with inside stack diameters ranging from 8 to 9 m. The temperature of the flue gases leaving the stacks was approximately 140°C.[10]

Among those worth noting in equation form is a simplified Holland equation (which is valid for effluent gases hotter than 125°F and for neutral conditions), the Davidson-Bryant equation[8] described above, and in the Briggs equation[9] (which EPA is mandated to use). Holland's equation was developed with experimental data, but it frequently underestimates the effective height of the emission. Therefore, its use often requires the inclusion of safety factors.

If the effective heights of emission were the same under all atmospheric conditions, the highest ground-level concentrations from a given source would occur with the lightest winds. Generally, however, emission conditions are such that the effective stack height is an inverse function of wind speed. The maximum ground-level concentration occurs at some intermediate wind speed, at which a balance is reached between the dilution due to wind speed and the effective height of emission.

At whatever height smoke and flue gases are discharged, gravity will sooner or later bring the larger particles of grit, dust, and soot to the ground. In addition, a proportion of the gases and freely suspended fine particles unaffected by gravity will reach the ground because of the natural turbulence of the atmosphere, which mixes the flue gases with the main body of air passing over the earth's surface. Broadly speaking, the higher the point of discharge and the greater the total heat content of the effluent, the more widespread will be the dilution of the gases and fine particles by the time diffusion brings any to ground level.[4]

24.5 The Pasquill-Gifford Model

The Pasquill-Gifford model has been the basis of most models developed and accepted today.[11–13] This model has served as an atmospheric dispersion formula from which the path downwind of emissions can be estimated after obtaining the effective stack height. There are many other dispersion equations (models) presently available, most of them semiempirical in nature. However, this section will concentrate on the Pasquill-Gifford model. The assumptions made in the development of this equation are:

1) The plume spread has a Gaussian distribution in both the horizontal and vertical planes, with standard deviations of plume concentration distribution in the horizontal and vertical directions.
2) There is a uniform emission rate of pollutants.
3) Total reflection of the plume takes place at the earth's surface.
4) The plume moves downwind with mean wind speed.

Although any consistent set of units may be used, the cgs system is preferred in these equations, Cota[14] has produced a program that has computerized many of the above calculations for a point source. Baasel[15] has developed a simple technique for determining the maximum ground level concentration (GLC) resulting from discharges from elevated point sources. Other approaches and equations[16,17] are available for performing dispersion calculations. Additional details on dispersion models are available in the literature.[1,18]

24.6 Types of Emission Sources

The Pasquill-Gifford model serves as a rough estimate of the behavior of the dispersion of a pollutant in the atmosphere. The reliability of a model is enhanced as its equations are fine-tuned to certain specific conditions. One parameter that must be considered in the estimation of dispersion behavior is the type of emitting source of the pollutant that is being examined. The aforementioned Pasquill-Gifford model is based on the assumption that the emitting source has point source properties. Point source, which has been mentioned briefly, is the most common type of emission source that is used in estimation. Two other types of sources most commonly used are area and line. Point, area, and line sources can be further considered as either instantaneous or continuous.

Point sources are characterized as having a finite quantity of mass at a given point and time in an infinite medium. For a sample case, consider an instantaneous point source. The medium is initially assumed stationary; that is, the bulk velocity is zero everywhere. At the instant of discharge, the mass of the pollutant is considered to be contained in a "small" volume. As time goes on, this volume grows and the concentration decreases. This change is brought about by diffusive effects. The main advantage of using the instantaneous point source approach is that it is based on a very simple physical concept, and this enables the solution of a large number of important problems to be written down immediately from first principles. The solution for the instantaneous point source is fundamental and is considered to be the starting point for all subsequent analyses.[19]

Engineers and applied scientists have paid more attention to the continuous point source than to the instantaneous. If the discharge rate is constant and has continued over an extended period of time, the model becomes (in the limit as time goes to infinity) that of a steady continuous point source. This model solution corresponds to the well-known classical source solution in hydrodynamics and heat conduction. Taking the instantaneous point source solution and integrating with respect to a surface, will provide an area source solution. It may also be integrated with respect to three spatial coordinates, to give a solution for a volume source. These may in turn be integrated with time to yield a continuous source solution.[18,20,21]

Basically, the entire development on atmospheric dispersion has been limited to emissions from a "point" (e.g., stack source). Although most dispersion applications involve point sources, there are instances where the location of the emission can be more accurately described physically and mathematically by either a line source or an area source, details of which are available in the literature.[18,20–22]

24.7 Choosing A Model

In choosing the proper model, many of which are Pasquill-Gifford based, various factors have to be analyzed and good personal judgment is important.[23] A simple principle of the model structure can be seen as:

$$X = \mathrm{DF} \times Q \tag{24.1}$$

where X is the concentration of emissions, DF is the dispersion factor which is a factor of wind speed, temperature, pressure, and many other conditions that can affect a flow, and Q is the mass emission rate at the source. Four simple questions the engineer or applied scientist needs to answer in order to narrow the search are the following:

1) What type of source?
   The most commonly used are continuous point, instantaneous, area, volume, line, and dense gas. (A combination of these sources has also been practiced.)
2) What terrain?
   Whether the land around the source is simple or complex.
3) What kind of receptor?
   The type of receptor depends on the characteristics of the site of concern.
4) Time frame?
   Some models work for large time spans, and some do not.

Models that do not fit these constraints are usually abandoned. The selection of models becomes more specific depending on what is needed.

Evaluation of the models themselves should begin now. Three factors should enter into this selection step: (i) the description of the models, (ii) a scientific evaluation of models, and (iii) a performance evaluation based on simulation and comparison of observed data.[24,25] Simply put, the models must be well understood in terms of what parameters are allowable and what needs to be predicted. Parameters can include distinguishing the Pasquill-Gifford stability class of the atmospheric air, allowable values of the dispersion coefficients in which directions, buoyancy and momentum flux, condition of the terrain, and the accountability of the presence of downwash, among many other things.

The problem of having various possible models is that various different predictions can be obtained. In order to establish some common reference, a standard was sought by the government. The *Guideline on Air Quality Models*[26,27] is used by EPA, by the states, and by private industry in the review and preparation of Prevention of Significant Deterioration (PSD) permits and State Implementation Plan (SIP) revisions. The guideline serves as a means by which consistency is maintained in air quality analyses. On September 9, 1986 (51 FR 32180), EPA proposed to include four different changes to this guideline: (i) addition of a specific version of the Rough Terrain Diffusion Model (RTDM) as a screening model, (ii) modification of the downwash algorithm in the Industrial Source Complex (ISC) model, (iii) addition of the Offshore and Coastal Dispersion (OCD) model to EPA's list of preferred models, and (iv) addition of the AVACTA II model as an alternative model in the guideline. In industry today, the ISC models are preferred models for permitting and therefore are used in many applications involving normal or "after the fact" releases. Models for emergency and/or accidental releases depend on which regulatory agency must be answered to.

The ISC model is available as part of UNAMAP (Version six). The computer code is available on magnetic tape from the National Technical Information Service or via modem through their Bulletin Board Services (BBS). It can account for the following: settling and dry deposition of particulates; downwash; area, line, and volume sources; plume rise as a function of downwind distance; separation of point sources; and limited terrain adjustment.

In order to prepare for the worst, screen models are applied in order to simulate the worst-case scenario. One difference between screening models and the refined models mentioned earlier is that certain variables are set which are estimated to be values to give the worst conditions. In order to use these screen models, the parameters of the model must be fully grasped.

In short, these models are necessary to somewhat predict the behavior of atmospheric dispersions. These predictions may not necessarily be correct, in fact, they are rarely completely accurate. In order to choose the most effective model for the behavior of an emission, the source and the models have to be well understood.

24.8 Conclusions

1) Both continuous and instantaneous discharges are of concern to individuals involved with environmental management. This has traditionally been an area of much concern in the air pollution field because stacks have long been one of the more common industrial methods of disposing waste gases. The concentrations that humans, plants, animals, and structures are exposed to at ground level can be reduced significantly by emitting the waste gases from a process at great heights.
2) Perhaps the most important consideration in dispersion applications is to first determine the acceptable ground-level concentration of the waste pollutants. The topography of the area must also be considered. Awareness of the meteorological conditions prevalent in the area, such as the prevailing wind, humidity, and rainfall, is also essential. Finally, an accurate knowledge of the constituents of the waste gas and its physical and chemical properties is paramount.
3) The effective stack height is usually considered as the sum of the actual stack height, the plume rise due to the velocity of the issuing gases, and the buoyancy rise, which is a function of the temperature of the gases being emitted and the atmospheric conditions.
4) Having determined the effective height of an emission, the next step is to study its path downwind using the appropriate atmospheric dispersion equation. There are many dispersion equations presently available, most of them semiempirical in nature. The Pasquill-Gifford model, or some modification of it, is usually employed in performing downwind concentration calculations.
5) Although most dispersion applications involve point sources, there are instances where the location of the emission can be more accurately described physically and mathematically by either a line source or an area source.
6) Cases of instantaneous releases (as from an explosion) or short-term releases (in the order of seconds) are often of practical concern. To determine concentrations at any position downwind, one must consider the time interval after the time of release and diffusion in the downwind direction as well as lateral and vertical diffusion.

References

1 **D. B. Turner**, *Workbook of Atmospheric Dispersion Estimates*. Environmental Protection Agency, Publication No. AP-26 (revised), RTP, NC, 1970.

2 **L. Theodore, and R. Allen**, *Air Pollution Control Equipment, A Theodore Tutorial*, Theodore Tutorial, East Williston, NY, 1993. (originally published by the APTI/USEOA. RTP, NC). 2009.

3 **L. Theodore**, *Transport Phenomena for Engineers*, International Textbook Company, Scranton, PA, 1971.

4 **H.E. Hesketh**, *Understanding and Controlling Air Pollution*, Ann Arbor Science, Ann Arbor, MI, 1972.

5 **R.M. Bethea**, *Air Pollution Control Technology*, pp. 39–59, Van Nostrand Reinhold, New York, 1978.

6 **M. Fogiel**, *Modern Pollution Control Technology*, pp. 2.8–2.9, Research and Education Association, 1978.

7 **W. Strauss**, *Air Pollution Control*, pp. 24–30, Wiley-Interscience, New York, 1971.

8 **W. F. Davidson**, *Transactions of the Conference on Industrial Wastes*, 14th Annual Meeting of the Industrial Hygienics Foundation of America, 38, Pittsburgh, PA, 1949.

9 **G. A. Briggs**, *Plume Rise, AEC Critical Review Series*. USAEC, Division of Technical Information, U.S. Government Printing Office, Washington, D.C.

10 **T.L. Montgomery et al.**, Results of recent TVA investigations of plume rise, *J. Air Pollut. Control Assoc.*, 22(10), 779, 1972.

11 **F. Pasquill**, The estimation of the dispersion of windborne material, *Meteorol. Mag.*, 90(33), 1063, 1961.

12 **F.A. Gifford**, Use of routine meteorological observations for estimating atmospheric dispersion, *Nucl. Saf.*, 2(4), 47, 1961.

13 **C.D. Cooper, and F.C. Alley**, *Air Pollution Control: A Design Approach*, pp. 519–552, Waveland Press, Prospect Heights, IL, 493–515, 1986.

14 **H. Cota**, *J. Air Pollut. Control Assoc.*, 31(8), 253, 1984.

15 **W. Baasel**, *J. Air Pollut. Control Assoc.*, 38(8), 866, 1981.

16 **O.G. Sutton**, The problem of diffusion in the lower atmosphere, *Q.J.R. Meteorol. Soc.*, 73, 257, 1947.

17 **C.H. Bosanquet, and J.L. Pearson**, The spread of smoke and gases from chimneys, *Trans. Faraday Soc.*, 32, 1249, 1936.

18 **H.S. Carslaw, and J.C. Jaeger**, *Conduction of Heat in Solids*, 2nd Ed., Oxford University Press, London, 1959.

19 **L. Theodore, J. Reynolds, and F. Taylor**, *Accident and Emergency Management*, John Wiley & Sons, New York, NY, 1989.

20 **US EPA**, APTI Course 503, in *Accident and Emergency Management Student Manual* Theodore, L., *et al.* Eds, U.S. Environmental Protection Agency, Washington, D. C., 1989.

21 *Practical Guide to Atmospheric Dispersion Modeling*. Trinity Consultants, Dallas, TX, 99–109, 1988.

22 **L. Theodore**, *Personal Notes*, East Willington, NY, 2021.

23 **P. Diosey**, *Malcolm Pirnie*, 10602, White Plains, NY 1993.

24 *Evaluation and Assessment of Models for Emergency Response Planning*, Environmental Consultants, Clarksville, IN, 1986

25 **A.C. Stern et al.**, *Fundamental of Air Pollution*, pp. 277–280, Academic Press, New York, 1973.

26 **US EPA**, *Guideline on Air Quality Models*. Publication No. EPA-450/2-78-027 (OAQPS No. 1.2–08), Research Triangle Park, NC, 1978.

27 **US EPA**, *Industrial Source Complex (ISC) Dispersion Model User's Guide, Second Edition, Vols. 1 and 2*. Publication Nos. EPA-450/4-86-005a and EPA-450/4-86-005b (NRTIS PB86 234259 and 23467), Research Triangle Park, NC, 1986.

25

Control Equipment Cost Considerations

Although the technical parameters influencing the selection and design of a given type of air pollution control system may be unique, cost is usually the only parameter relevant to all systems. This is especially true where different types of systems can control a source to achieve an emission limit. Cost is often the parameter used to select the optimum system from the alternatives available.

Economics also plays an important role in setting many state and federal air pollution control regulations. The extent of this role varies with the type of regulation. For some types of regulations, cost is explicitly used in determining their stringency. This use may involve a balancing of costs and environmental impacts, costs and dollar valuation of benefits, or environmental impacts and economic consequences of control costs. For other types of regulations, cost analysis is used to choose among alternative regulations with the same level of stringency. For these regulations, the environmental goal is determined by some set of criteria that do not include costs. However, cost-effectiveness analysis is employed to determine the minimum economic way of achieving the goal. For some regulations, cost influences enforcement procedures or requirements for demonstration of progress toward compliance with an environmental quality standard. For example, the size of any monetary penalty assessed for noncompliance as part of an enforcement action needs to be set with awareness of the magnitude of the control costs being postponed or bypassed by the noncomplying facility. For regulations without a fixed compliance schedule, demonstration of reasonable progress toward the goal is sometimes tied to the cost of attaining the goal on a different schedule.

An understanding of the economics involved in air pollution control is important in making decisions at both the engineering and management levels. Every engineer or scientist should be able to execute an economic evaluation of a proposed project. If the project is not profitable, it should obviously not be pursued and the earlier such a project can be identified, the fewer resources that will be wasted.

Before the cost of an air pollution project or study can be evaluated, the factors contributing to the cost must be recognized. There are two major contributing factors: capital costs and operating costs. Once the total cost of the project has been estimated, the engineer must determine whether or not the process (change) will be profitable. This involves converting all cost contributions to an annualized basis. If more than one project proposal is under study, this method provides a basis for comparing alternate proposals and for choosing the best proposal.

Detailed cost estimates are beyond the scope of not only this chapter, but also this book. Such procedures are capable of accuracies around $\pm 10\%$; however, such estimates generally require many months of engineering work. This chapter is designed to give the reader a basis for preliminary cost analysis only.

Note: The material presented in this chapter was adapted primarily from Shen et al.,[1] Theodore and Neuser,[2] Theodore and Moy,[3] and Santoleri et al.[4] The reader should also note that some of the materials have appeared elsewhere in the book.

25.1 Capital Costs

Equipment cost is a function of many variables, one of the most significant of which is capacity. Other important variables vary with the cost of equipment or process. Preliminary estimates are often based on simple cost-capacity relationships that are valid when the other variables are confined to narrow ranges of values; these relationships can be represented by an approximate linear (on log–log coordinates) cost equipment relationship of the form[5]:

Optimizing Air Pollution Control Equipment Performance: Operation & Maintenance, First Edition. Jay Richardson and Louis Theodore.

$$C = \alpha Q^{\beta} \tag{25.1}$$

where C = cost
Q = some measure of equipment or process capacity
α, β = empirical "constants" that depend mainly on the equipment type

It should be emphasized that this procedure is suitable for rough estimation only; actual estimates from vendors are preferable. Only major pieces of equipment are included in this analysis; smaller peripheral equipment such as pumps and compressors may not be included. Similar methods for estimating costs are available in the literature.[5] If greater accuracy is needed, however, actual quotes from vendors should be used.

Again, the equipment cost estimation model just described is useful for a very preliminary estimation. If more accurate values are needed and old price data are available, the use of an indexing method may be preferable, although a bit more time-consuming. The method consists of adjusting the earlier cost data to present values using factors that correct for inflation.

The usual technique for determining the capital costs (i.e., total capital costs [TCCs], which include equipment design, purchase, and installation) for a project and/or process can be based on the factored method of establishing direct and indirect installation cost as a function of the known equipment costs This is basically a modified Lang method, whereby cost factors are applied to known equipment costs.[4,5]

The first step is to obtain from vendors (or, if less accuracy is acceptable, from one of the estimation techniques previously discussed) the purchase prices of primary and auxiliary equipment. The total base price, designated by X – which should include instrumentation, control, taxes, freight costs, and so on – serves as the basis for estimating the direct and indirect installation costs. The installation costs are obtained by multiplying X by the cost factors, which are available in literature.[5–9] For more refined estimates, the cost factors can be adjusted to more closely model the proposed system, by using adjustment factors that take into account the complexity and sensitivity of the system.[4,5]

The second step is to estimate the direct installation costs by summing up all the cost factors involved in the direct installation costs, which include piping, insulation, foundation and supports, and so on. The sum of these factors is designated as the direct installation cost factor (DCF). The direct installation costs are the product of the DCF and X.

The third step consists of estimating the indirect installation costs. The procedure here is the same as that for the direct installation cost – that is, all the cost factors for the indirect installation costs (engineering and supervision, startup, construction fees, etc.) are added. The sum is designated as the indirect installation cost factor (ICF). The indirect installation costs are then the product of ICF and X. Once the direct and indirect installation costs have been calculated, the TCC[3] may be evaluated as:

$$\text{TCC} = X + (\text{DCF})(X) + (\text{ICF})(X) \tag{25.2}$$

This cost is converted to annualized capital costs (ACCs) with the use of the capital recovery factor (CRF), which is described in the literature.[4,6] The ACC is the product of the CRF and the TCC and represents the total installed equipment cost distributed over the lifetime of the facility.

25.2 Operating Costs

Operating costs can vary from site to site because the costs partly reflect local conditions – for example, staffing practices, labor, and utility costs. Operating costs, like capital costs, may be separated into two categories: direct and indirect costs are those that cover material and labor and are directly involved in operating equipment at the facility. These include labor, materials, maintenance, and maintenance supplies, replacement parts, waste (e.g., residues after incineration) disposal fees, and utilities and laboratory costs. Indirect costs are those operating costs associated with, but not directly involved in operating equipment at the facility; costs such as overhead (e.g., building land leasing and office supplies), administrative fees, local property taxes, and insurance fees fall into this category.

The major direct operating costs are usually associated with the labor and materials costs for the project, which often involve the cost of the chemicals needed for the operation of the process. Labor costs differ greatly but are a strong function of the degree of controls and/or instrumentation. Typically, there are three working shifts per day with one supervisor per shift. On the other hand, the plants may be manned by a single operator for only one-third or one-half of each shift; i.e., usually only the operator, supervisor, and site manager are necessary to run the facility. Salary costs vary from state to state and depend significantly on the location of the facility. The cost of utilities generally consists of that of electricity, water, fuel, and steam. The annual costs are estimated with the use of material and energy balances. Cost for waste disposal can be estimated on a per-ton-capital basis. The cost of landfilling ash can run significantly upward of $100/ton if the material is hazardous and can be as high as $10/ton if it is nonhazardous. For example, the cost of handling a scrubber effluent can vary depending on the method of disposal. If conventional sewer disposal is used, the effluent probably has to be cooled and neutralized before disposal; the cost for this depends on the solids' concentration. Annual maintenance costs can be estimated as a percentage of the capital equipment cost. The annual cost of replacement parts can be computed by dividing the cost of the individual part by its expected lifetime. The life expectancies can be found in the literature.[5] Laboratory costs depend on the number of samples tested and the extent of these tests; these costs can be estimated as 10–20% of the operating labor costs.

The indirect operating costs consist of overhead, local property tax, insurance, and administration, less any credits. The overhead comprises payroll, fringe benefits, social security, unemployment insurance, and other compensation that is indirectly paid to plant personnel. This cost can be estimated as 50–80% of the operating labor, supervision, and maintenance costs.[8,9] Local property taxes and insurance can be estimated as 1–2% of the TCC, while administration costs can be estimated as 2% of the TCC.

The total operating cost is the sum of the direct operating cost and the indirect operating costs, less any credits that may be recovered (e.g., the value of recovered steam). Unlike capital costs, operating costs are calculated on an annual basis.

25.3 Hidden Economic Factors

The main problem with the traditional type of economic analysis, as discussed above, is that it is difficult – nay, in some cases impossible – to quantify some of the not-so-obvious economic merits of a business and/or environmental program.[10]

Several considerations have just recently surfaced as factors that need to be taken into account in any meaningful economic analysis of a project effort. What follows is a summary of most of these considerations:

1) Long-term liabilities
2) Regulatory compliance
3) Regulatory recordkeeping
4) Dealings with the Environmental Protection Agency (EPA)
5) Dealings with the state and local regulatory bodies
6) Fines and penalties
7) Potential tax benefits
8) Customer relations
9) Stockholder support (corporate image)
10) Improved public image
11) Insurance costs and claims
12) Effect on borrowing power
13) Improved mental and physical well-being of employees
14) Reduced health maintenance costs
15) Employee morale
16) Worker safety
17) Rising costs of waste treatment and/or disposal
18) Training costs
19) Emergency response planning
20) Any combination of the above

Many programs have been quenched in their early stages because a comprehensive economic analysis was not performed. Until the effects described above are included, the true merits of a project may be clouded by incorrect and/or incomplete economic data.

Can something be done by industry to remedy the above problem? One approach is to use a modified version of the standard Delphi Panel. In order to estimate these "other" economic benefits, several knowledgeable individuals within and perhaps outside the organization are asked to independently provide estimates, with explanatory details, on these economic benefits. Each individual on the panel is then allowed to independently review all responses. The cycle is then repeated until the groups' responses approach convergence.[11]

25.4 Project Evaluation

In comparing alternate processes or different options of a particular air pollution control process from an economic point of view, it is recommended that the total capital cost be converted to an annual basis by distributing it over the projected lifetime of the facility. The sum of both the ACCs and the annual operating costs (AOC) is known as the total annualized cost (TAC). The economic merit of the proposed facility, process, or scheme can be examined once the total annual cost is available. Alternate facilities or options (e.g., a baghouse versus an electrostatic precipitator for particulate control, or two different processes for accomplishing the same degree of waste destruction) may also be compared. A small flaw in this procedure is the assumption that the operating costs remain constant throughout the lifetime of the facility. Since the analysis is geared for comparing different air pollution control alternatives, the changes with time should be somewhat uniform among the various alternatives, resulting in little loss of accuracy.

The conversion of the total capital cost to an annualized basis involves an economic parameter known as the CRF. These factors can be found in any standard economics text[12,13] or can be calculated directly from Equation 25.3:

$$\text{CRF} = \frac{(i)(1+i)^n}{(1+i)^n - 1} \tag{25.3}$$

where n = projected lifetime of the system (years)
i = annual interest rate (expressed as a fraction)

The CRF is a positive, fractional number. The ACC is computed by multiplying the TCC by the CRF. The ACC reflects the cost associated with recovering the initial capital outlay over the depreciable life of the system. Investment and operating costs can be accounted for in other ways, such as a present worth analysis. This is especially true when comparing environmental control systems having different depreciable lives. There are usually other considerations in such decisions besides the economics, but if all the other factors are equal, the proposal with the lowest TAC should be the most viable.

Once a particular air pollution control scheme has been selected, it is common practice to optimize the process from a capital cost, and operation and maintenance (O&M) standpoint. There are many optimization procedures available, most of them too detailed for meaningful application in this chapter. These sophisticated optimization techniques, some of which are routinely used in the design of conventional chemical and petrochemical plants, invariably involve computer calculations. However, the use of these techniques in environmental management analysis is usually not warranted.

One simple optimization procedure that is recommended is the perturbation study. This involves a systematic change (or perturbation) of variables, one by one, in an attempt to locate the optimum design from a cost and operation viewpoint. To be practical, this often means that the engineer must limit the number of variables by assigning constant values to those process variables that are known beforehand to play an insignificant role. Reasonable guesses and simple or short-cut mathematical methods can further simplify the procedure. Much information can be gathered from this type of study since it usually identifies those variables that significantly impact the overall performance of the process and also helps identify the major contributors to the TAC.

25.5 Future Trends

Economic analysis will continue to incorporate the costs and benefits environment. This trend is continuing to expand with the inclusion of both sustainability effects chemistry/engineering, and green chemistry/engineering.[6,9,10] The wide adoption of design specifications for operations would ensure that the economic evaluations are prepared to be more inclusive of these emerging concepts. This will hopefully result in economic assessments that are comprehensive, easily understood, and readily put to use.

Finally, it should be clear that existing or proposed processes require a standard procedure of economic evaluation to:

1) Closely predict costs – capital and operating
2) Test feasibility
3) Make a fair and balanced evaluation among alternative schemes

Even within the federal government, several methods of cost estimation have been developed over the years for a wide range of technologies. Often, such methods only deal with cost estimates and perhaps feasibility. More is needed for an economic evaluation – for example, an assessment of the process reliability, an appraisal of the extent of process development, and a definition of the project scope.[2]

References

1 **T. Shen, Y. Choi, and L. Theodore**, *EPA Manual Hazardous Waste Incineration*. (USEPA Manual) USEPA, Research Triangle Park, NC, 1985.

2 **L. Theodore, and K. Neuser**, *Engineering Economics and Finance, A Theodore Tutorial*, Theodore Tutorials, East Williston, NY, originally published by USEPA/APTI, Research Triangle Park, NC, 1996.

3 **L. Theodore and E. Moy**, *Hazardous Waste Incineration*. USEPA, Research Triangle Park, NC 1992.

4 **J. Santoleri, J. Reynolds, and L. Theodore**, *Introduction to Hazardous Waste Incineration*, 2nd Ed., John Wiley & Sons, Hoboken, NY, 2009.

5 **W. Vatavuk, and R. Neveril**, *Factors for Estimating Capital and Operating Costs*, pp. 157–162, Chemical Engineering, New York, 1980.

6 **L. Theodore**, *Chemical Engineering: The Essential Reference*, McGraw-Hill, New York, 2014.

7 **G. Ulrich**, *A Guide to Chemical Engineering Process Design and Economics*, John Wiley & Sons, Hoboken, NJ, 1984.

8 **G. Vogel, and E. Martin**, *Hazardous Waste Incineration, Part 3 – Estimating Operating Costs*, pp. 97–100, Chemical Engineering, New York, 1984.

9 **R. Dupont, L. Theodore, and K. Ganesan**, *Pollution Prevention*, CRC Press, Boca Raton, FL, 2000.

10 **R. Dupont, K. Ganesan, and L. Theodore**, *Pollution Prevention – Sustainability, Industrial Ecology, and Green Engineering*, 2nd Ed., CRC Press/Taylor & Francis Group, Boca Raton, FL, 2017.

11 **L. Theodore**, *Personal Notes*, East Williston, NY, 1989.

12 **E. DeGarmo, J. Canada, and W. Sullivan**, *Engineering Economy*, 6th Ed., Macmillan, New York, 1979.

13 **C. Hodgman, S. Selby, and R. Weast Eds**, *CRC Standard Mathematical Tables*, 12th Ed., CRC Press, Boca Raton, FL, 1961.

26

Measurement Methods

Vincenza Imperiale

Manhattan College, Bronx, NY, USA

Measurements of particle and gaseous air pollutants have been used for exposure assessment and epidemiological studies. Most methods for regulated gas pollutants (e.g., CO, NO_2, SO_2, and ozone) use in situ continuous monitors for hourly averaged concentrations.[1] The accurate quantitative analysis of the discharge of pollutants from a process must be determined prior to the design and/or selection of control equipment. If the unit is properly engineered, utilizing the emission data as input to the control device and the code requirements as maximum effluent limitations, most pollutants can be successfully controlled by one or a combination of the methods discussed earlier.

The objective of source testing is to obtain data representative of the process being sampled. The steps followed in obtaining representative data from a large source are:

1) Obtaining a measurement that reflects the time magnitude of the characteristics being measured at the location where the measurement is made.
2) Taking a number of measurements in such a manner that the data obtained from these measurements are representative of the source.

Sampling is the keystone of source analysis. Sampling methods and tools vary in complexity according to the specific task; therefore, a degree of both technical knowledge and sampling complexity analysis is needed to design a sampling activity. Sampling is performed to measure quantities or concentrations of pollutants in effluent streams, to measure the efficiency of a pollution abatement device, to guide the designer of pollution control equipment and facilities, and/or to appraise contamination from a process or source. A complete measurement requires the determination of the concentration and contaminant characteristics, as well as the associated flow. Most statutory limitations require mass rates of emission; both concentration and volumetric flow rate data are, therefore, required.

Methods include passive sampling (e.g., diffusion-based methods on an absorbent) and active sampling (e.g., pump or blower). Passive sampling uses entrapment or sorption in a diffusion tube, badge, detector tube, or similar device. Passive sampling is often more appropriate for measuring long-term exposure rather than acute exposure since the level of detection may be higher for a passive sampling device than an active one. For all passive sampling methods, the samples are analyzed using gas chromatography, spectroscopy, high-performance liquid chromatography, or a similar method, depending on the chemicals of concern.[2]

Active sampling uses a small air pump (or blower) to draw (or push) air through a filter, packed tube, or similar device. Unlike passive samplers, active sampling requires moving parts and electricity. Active monitors are available to measure particulate matter 2.5 μm and 10 μm in size ($PM_{2.5}$ and PM_{10}) using filters. The filter can be placed anywhere on the individual and the individual carries the pump and battery pack in a bag. Cyclone samplers are accessible to measure particulates. The sampler spins the particles in an air stream, forcing them to the edges of the device for collection.[2]

The method used to sample a given media will directly affect the accuracy and representativeness of the analysis. It is imperative that an appropriate methodology be selected to obtain the most reliable results possible; in effect, to be representative of the population from which the sample is drawn.[3,4] Several factors should be considered when selecting a sampling method, including:

1) The sampling program objectives (documenting exposures, determining regulatory compliance, locating a source)
2) The media to be sampled (air, water, soil, vegetation, sludge, etc.)

Optimizing Air Pollution Control Equipment Performance: Operation & Maintenance, First Edition. Jay Richardson and Louis Theodore.

3) The physical and chemical properties of the contaminant(s) of interest
4) Potential matrix effects as a result of characteristics of other contaminants in the sample
5) Regulatory requirements
6) Safety requirements
7) Difficulty of using a specific method
8) Cost
9) Scale of sample area (small-scale site related to individual persons versus a large-scale site)
10) Accessibility to the sample site
11) Short- versus long-term sampling requirements

In addition to these, basic professional judgment also plays a role in selecting a sampling method.

26.1 Source Sampling

Although details on actual measurement methods for both gaseous and particulate pollutants are beyond the scope of this book, the presentation of some general information on sampling appears warranted. An accurate quantitative analysis of the discharge of pollutants from a process must be determined prior to the design and/or selection of control equipment. If the unit is properly engineered utilizing the emission data as input to the control device and the code requirements as maximum effluent limitations, most gaseous and particulate pollutants can be successfully controlled by one or a combination of the methods. For a rectangular cross-section, the equivalent diameter is determined from

$$\text{Equivalent Diameter} = \left[\frac{2(\text{length})(\text{width})}{\text{length} + \text{width}}\right] \tag{26.1}$$

Once these traverse points have been determined, velocity measurements are made to determine gas flow. The stack-gas velocity is usually determined by means of a pitot tube and differential pressure gauge. Once a flow profile has been established, a sampling strategy can be considered. Since sampling collection can be simplified and greatly reduced, depending on flow characteristics, it is best to complete the flow-profile measurement before sampling or measuring pollutant concentrations. Specific details on source sampling procedures for both gaseous and particulate pollutants may be readily found in the air pollution literature.[5,6]

An ongoing air pollution sampling program also provides the basis for process control and produces a record of how the control system is operating. This information keeps the operating personnel informed of plant efficiencies and helps in predicting problems that are developing in the system. Because sampling results are a record of plant performance, they are often also evaluated by governing and/or regulatory bodies. For these reasons, it is essential that a control system's testing program produces complete and accurate results. A suggested schedule for performing the various sampling tests should be outlined. A list of reference materials for use should also be given. All references considered essential for proper laboratory operation should be provided. Samples of the laboratory worksheets recommended for use should be available.[5]

26.2 Sampling Guidelines

Once the sampling location is decided on, the flue cross section is laid out in a number of equal areas, the center of each being the point where the measurement is to be taken. For a rectangular stack, the cross-section is divided into equal areas of the same shape, and the traverse points are located at the center of each equal area, as shown in Figure 26.1a. The ratio of the length to the width of each elemental area should be selected. For circular stacks, the cross-section is divided into equal annular areas, and the traverse points are located at the centroids of each area, as shown in Figure 26.1b. The location of the traverse points as a percentage of stack diameter from the inside wall to the traverse point for circular stack sampling is given in Table 26.1. The number of traverse points necessary on each of the two perpendiculars for a particular stack may be estimated from Table 26.2.[6]

Once these traverse points have been determined, velocity measurements are made to determine gas flow. The stack-gas velocity is usually determined by means of a pitot tube and differential pressure gauge.[3,4] Where velocities are very low (less than 10 ft/sec) and where great accuracy is not required, an anemometer may be used. For gases moving in small pipes at

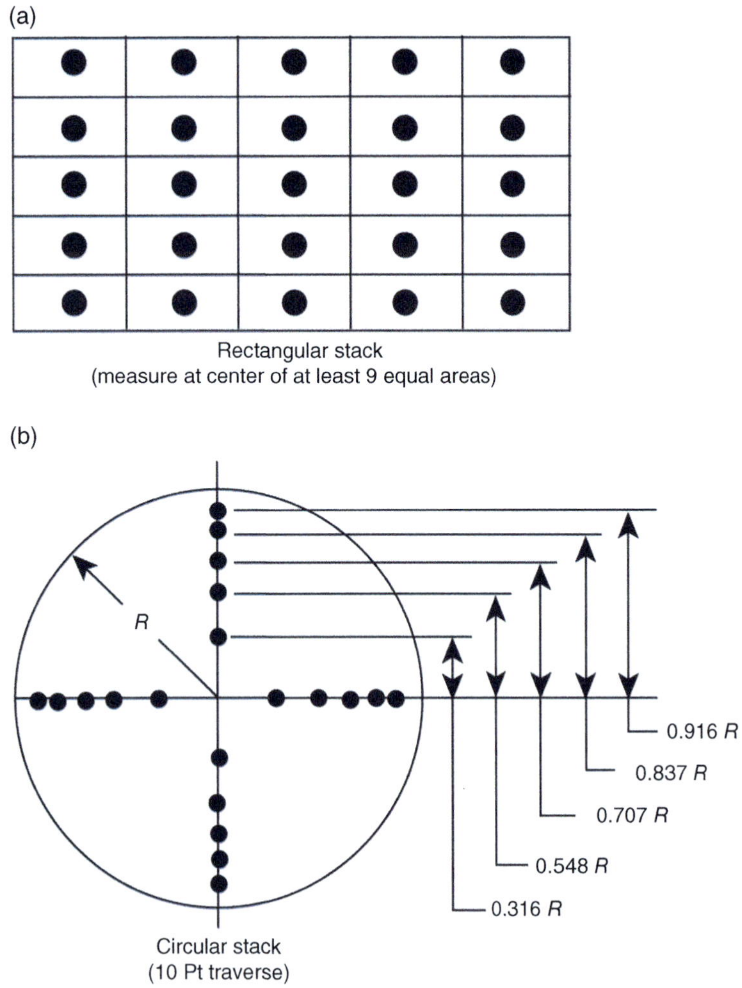

Figure 26.1 Traverse point locations for velocity measurement or for multicomponent sampling in (a) rectangular ducts or (b) circular stacks.

relatively high velocities or pressures, orifice disk meters or venture meters may be used.[5,6] These are valuable as continuous or permanent measuring devices.

Once a flow profile has been established, a sampling strategy can be considered. Since sampling collection can be simplified and greatly reduced, depending on flow characteristics, it is best to complete the flow-profile measurement before sampling of measuring pollutant concentrations. Source characteristics may be placed in one of four categories:

1) No time variation, with the emission relatively uniform across the stack cross-section. In this case, only one concentration measurement is necessary for accurate results.
2) Steady generation of contaminants, but nonuniform flow across the sampling location. In this case, a traverse is necessary to measure the average concentration. Typically, this is performed at the points selected for the velocity traverse. The time of sampling at each point should be the same to get a representative, composite sample.
3) Cyclical operation in which a satisfactory sampling location is available, and the variation across the stack is relatively uniform when the operation is running. Because the process involves time variation only, the sampling is conducted at one point for extended periods, usually related to one or more operational cycles.
4) Nonuniform source and flow conditions. This case requires the most complicated procedure. If there is some measurable cycle related to the process, the sampling can be conducted over this period using simultaneously collected samples. One sample is collected at a reference point and the others at selected traverse points. This is repeated until a complete traverse is made. Results are corrected by using the reference point data as a measure of the time variation.

When sampling for gases, it is also usually necessary to analyze the temperature variation across the stack. This is performed as part of the velocity traverse and indicates variation in gas distribution. If the temperatures are relatively

Table 26.1 Traverse point locations in circular stacks.

Traverse Point Number on a Diameter	Number of Traverse Points on a Diameter									
	6	8	10	12	14	16	18	20	22	24
1	4.4	3.3	2.5	2.1	1.8	1.6	1.4	1.3	1.1	1.1
2	14.7	10.5	8.2	6.7	5.7	4.9	4.4	3.9	3.5	3.2
3	29.5	19.4	14.6	11.8	9.9	8.5	7.5	6.7	6.0	5.5
4	70.5	32.3	22.6	17.7	14.6	12.5	10.9	9.7	8.7	7.9
5	85.3	67.7	34.2	25.0	20.1	16.9	14.6	12.9	11.6	10.5
6	95.6	80.6	65.8	35.5	26.9	22.0	18.8	16.5	14.6	13.2
7		89.5	77.4	64.5	36.6	28.3	23.6	20.4	18.0	16.1
8		96.7	85.4	75.0	63.4	37.5	29.6	25.0	21.8	19.4
9			91.8	82.3	73.1	62.5	38.2	30.6	26.1	23.0
10			97.5	88.2	79.9	71.7	61.8	38.8	31.5	27.2
11				93.3	85.4	78.0	70.4	61.2	39.3	32.3
12				97.9	90.1	83.1	76.4	69.4	60.7	39.8
13					94.3	87.5	81.2	75.0	68.5	60.2
14					98.2	91.5	85.4	79.6	73.9	67.7
15						95.1	89.1	83.5	78.2	72.8
16						98.4	92.5	87.1	82.0	77.0
17							95.6	90.3	85.4	80.6
18							98.6	93.3	88.4	83.9
19								96.1	91.3	86.8
20								98.7	94.0	89.5
21									96.5	92.1
22									98.9	94.5
23										96.8
24										98.9

Note 1: Figures in body of table are percent of stack diameter from inside wall to traverse point.

Table 26.2 Estimating the number of traverse points.

Number of Stack Diameters Upstream and Downstream of Flow Disturbance		
Upstream	Downstream	Number of Traverse Points on Each Diameter
8+	2+	6
7.3	1.8	8
6.7	1.7	10
6.0	1.5	12
5.3	1.3	14
4.7	1.2	16
4.0	1.0	18
3.3	0.8	20
2.6	0.6	22
2.0	0.5	24

Note 1: If a different number of traverse points is required by disturbances upstream and downstream, choose whichever number is greater.

constant, then a single sample point may suffice. If temperature variation is considered to be large (greater than 5%), then traversing is necessary and is done at the same points that the velocity traverse is made. The sampling rate and time of sampling at each point are often kept the same to simplify calculations.

In the gas sampling, a straight probe fitted with an integral filter is placed in the stack. The filter removes particulates at stack temperature, thereby preventing downstream fouling, as well as minimizing losses of gaseous pollutants due to reaction with the particulates on cooling. Suction on the nozzle draws the sample into a collection device. The volume of gas remaining after the collected constituent has been removed is measured with a wet or dry test-meter downstream of the collection device. Sampling is completed when either a cycle in the process has been ended or when sufficient sample has been obtained for analysis. Minimum probe lengths should be utilized, and extended flushing with stack gas should precede sample collection. This flushing is very important if losses to the walls of the probe and sample collector are to be avoided. Often the entire sampling system must be heated to prevent condensation of water, heavy hydrocarbons, or sulfuric acid mist.

Also important is knowledge of the average molecular weight of the flowing gases to be able to determine the actual velocity or volumetric flow rate in the stack. The more common measurement devices include:

1) Orsat Analysis – While Orsat Analysis is not extremely accurate, it is sufficient for the purpose of determining average molecular weight and many other unit operating parameters. In portable Orsat Analyzers, a measured gas sample is passed through several reagents and the decrease in gas volume after passage through each reagent is determined. Reagents that are reasonably selective absorbers include 20– 40% aqueous KOH for CO_2 (and other acid gases), alkaline pyrogallol for O_2, and acidic cuprous chloride for CO. Nitrogen can usually be determined by difference.
2) Gas Chromatographic Analysis – A sample is usually caught in a glass bomb or plastic bag and returned to the laboratory for analysis. This technique has the advantage of measuring minor constituents accurately, as well as the major ones.

26.3 Continuous Emission Monitoring

In most air pollution control/prevention programs, monitoring the emissions to the atmosphere is essential to the success of the program. Continuous emission monitoring systems (CEMS), as they are commonly called, are employed to monitor the emissions of most point or stack-type sources of air pollution and their associated effluent streams. According to the United States Environmental Protection Agency, CEMS is the total equipment necessary for the determination of a gas or particulate matter concentration or emission rate using pollutant analyzer measurements and a conversion equation, graph, or computer program to produce results in units of the applicable emission limitation or standard.[7] As with other pollution monitoring, selection of the appropriate methodology is critical to the overall accuracy and reliability of the program.[8]

For CEMS in particular, seven factors should be included as fundamental parts of any monitoring program. These factors are:

1) Knowledge of the process to be monitored.
2) Knowledge of pertinent regulatory requirements.
3) Procurement of correct monitors.
4) Institution of rigorous and repetitive calibration, quality control, and quality assurance procedures with skilled personnel.
5) Use of formal and lasting data collection, review, and reporting procedures by trained staff.
6) Reliance on a solid maintenance program with suitably trained personnel.
7) Comprehensive cost tracking, cost control, and budgeting.

Each of these are detailed below.

A thorough understanding of the process to which the CEMS will be added is necessary to obtain the most efficient and appropriate use of the CEMS. Regulatory requirements are process specific not only from the standpoint of type (e.g., refinery versus coal-fired boiler), but also of size (e.g., coal-fired boiler over 100 MMBtu/hr heat input versus coal-fired boiler under 100 MMBtu/hr heat input). In addition, certain process conditions such as startups, shutdowns, and malfunctions are treated, in some instances, differently from normal plant operation. Variances from certain facets of the requirements may be obtained if the meeting of these requirements is proven technically infeasible due to operational constraints. Consequently, a good understanding of the operation to be equipped with the CEMS is invaluable in ensuring the most compatible regulatory treatment.

Armed with the basic process understanding, the potential CEMS user can then begin to interpret the regulatory requirements. A complete and detailed understanding of these regulations is required. The regulations provide dictums on all facets of CEMS operation: from what parameters are to be monitored, to what performance criteria the individual monitors are required to meet, and even what calculations the monitoring system must use. Parts of many regulatory sections are subject to different interpretations. Review of these sections with the appropriate agency leads to a common interpretation between regulator and regulatee and helps to avoid potential future problems.

Procuring the correct monitors and other components of the CEMS, while easier to do than during the initial years of CEMS development, is still a significant effort. Many manufacturers make a CEMS for the same task but offer different features. The potential customer should try to find the system which is most easily integrated into his or her overall operational organization. Decisions about such things as whether to purchase an extractive or in situ measurement technique should be made. Considerations such as future available operational funds or level of maintenance personnel should be factored into the buying decision. Any potential CEMS should be reviewed with the personnel expected to operate and maintain the system. The more the procurement process is thorough and well thought out, the better the success of the future CEMS operation.

Scheduled and frequent calibrations, quality control, and quality assurance aid immensely in ensuring the success of the CEMS. These systems are renowned for needing frequent and extensive maintenance to ensure accuracy and reliability. In most cases, regulators require documentation of system accuracy and integrity. Calibrations, coupled with a quality control program, aid maintenance personnel by providing indicators of system accuracy as well as performance guidelines in the form of quality control limits. Quality assurance data help assure regulators that the data from the CEMS are precise and accurate, in other words, that the quality control program is successful. Personnel training in all aspects of these activities aids in obtaining meaningful data and facilitating the process as a whole.

Specific reporting requirements are found in the regulations. These range from minimum data capture rates to required accuracy levels and report contents. Individual regulatory agencies may also add their own reporting requirements. Deviations from these requirements will, at the least, cause confusion, and at most, result in enforcement action comparable to violation of a pollutant standard. Most data reporting requirements are met in some fashion with computerized capture and reporting networks. These networks, coupled with the level of detail required in regulations, make it necessary to dedicate trained personnel to these functions and develop standard operating procedures. These commitments further help assure regulators about the overall integrity of the CEMS.

The CEMS will not operate accurately and reliably without well-planned and frequent maintenance. Many failures of CEMS occurred in the early years of pollutant monitoring because of the misconception that the CEMS could be started up and then ignored. The CEMS, like any other mechanical/electrical system, requires maintenance. Unlike many other systems, however, the CEMS usually requires more frequent maintenance due, in large part, to its complex nature and poor area of application. Usually, a vibrating stack or duct, often with limited access and frequently with a dirty or corrosive ambient environment, is chosen as the location of the CEMS. This situation lends itself to high maintenance requirements. These requirements, coupled with the complex electronic nature of a CEMS, make it necessary to dedicate trained personnel to the maintenance of the CEMS. The utilization of these personnel also ensures that the proper priority is placed on CEMS maintenance, since in many cases the CEMS is demonstrating regulatory compliance.

As with other integrated systems at commensurate levels of complexity, the CEMS will be operated less expensively if costs are tracked and controlled. Budgetary goals should be set yearly. Major deviations from these goals should be explored to determine whether operational or system changes are needed. A review of the economic aspects of the CEMS performance can often point out areas where improvements are needed or possible. The operation of the CEMS provides no potential for a profit, but a company's financial situation can be improved through efficient operation of the CEMS.

26.4 Opacity Measurements

Many earlier pollution codes limit emissions of black smoke by setting Ringelmann number limits on the density or opacity of the plume. Smoke density is defined as the degree of blackness (opacity) of a plume expressed in terms of some arbitrary scale. The apparent blackness or stack plume opacity depends on the concentration of the particulate matter in the effluent, the size of the particulate, the depth of the smoke column being viewed, natural lighting conditions such as the direction of the sun relative to the observer, and the color of the particles. One of the oldest and most common methods of describing the black smoke or opacity from a source is to compare it to the shades of gray on the Ringelmann chart. The Ringelmann system is a scheme whereby graduated shades of gray, varying by five equal steps between white and black, may be accurately reproduced on a white background by means of a rectangular grid of black lines of definite width and spacing. To use a chart, it is supported on a level with the eye, at a distance from the observer where the lines of the chart merge into shades of gray and, as nearly as possible, are in line with the stack. The observer glances from the smoke (as it issues from the stack) to the chart and notes the number of the chart most nearly corresponding with the shade of the smoke; then this number is recorded with the time of observation. A clear stack is recorded as No.0; 100% black smoke is recorded as No. 5. In theory, such a number will correspond to the light transmittance of the plume. While the chart serves a useful purpose, it should be noted that the data obtained are empirical in nature and have definite limitations. Despite these limitations, the chart still provides the well-trained observer with good practical results. In fact, the Ringelmann number criteria have been replaced today by opacity measurement readings almost everywhere in the United States. Although the Ringelmann chart is only applicable to black smoke, "equivalent opacity" measurements (e.g., Ringelmann No.1 corresponds to 20% opacity) are also applied to smoke plumes of other colors such as white, brown, and blue.[9]

Many stacks require virtually continuous monitoring to control smoke emissions; consequently, in-stack opacity measurements are desirable. These measurements are not useful when there is a detached plume that is essentially invisible in the stack but becomes visible some distance beyond the stack. Also, when stack gases are opaque due to condensed water vapor, this method does not serve a useful purpose. Steam plumes, although aesthetically unattractive, present no air pollution problem unless the steam causes an adverse effect due to decreased visibility. The in-stack measurement of opacity, however, can still be very valuable in those situations where suspended particulates are the major contributor to opacity. To provide such continuous opacity measurements, a large number of stacks have been equipped with in-stack transmissometers (transmission photometers). Usually, a light or radiant-heat beam is projected across the stack. The intensity of radiation reaching the other side is proportional to the number and size of particles obstructing the beam at any time. Once the unit is installed, it can be calibrated in place in terms of stack opacity with visual observations of the stack. The major problems with these systems are their propensity for becoming fouled and the difficulties associated with in situ calibration. Attempts to maintain clean and dust-free optics have been directed toward recessing the components away from the stack and continually sweeping clean air over the exposed areas. Often, however, these actions are inadequate and measurement accuracy suffers. After the instrumentation has been in use for some time, the optics can become dirty and scratched, and the detector and light-source characteristics can change. Therefore, regular recalibration is required (often on a weekly basis). The problem is that there is no really acceptable means of accomplishing this recalibration, except for 0% transmission, unless the stack is taken out of use.

There are certain inherent characteristics, common to all opacity instrumentation, which all but preclude their valid use as an indication of stack-gas mass particulate loading, the major one arising because of varying optical effects produced by particulates of different sizes and nature. With present knowledge, it is all but impossible to relate the concentration of particulates in a gas stream with optical opacity or other measurable optical phenomena. This was mentioned previously and warrants emphasis. A given weight concentration of particulate matter in a gas stream will absorb and reflect light to varying degrees, depending on the size and nature of the particles involved. For example, a given concentration of particulate matter of typical magnitude will attenuate a light beam very slightly if the particles are large in size (10 μm or larger). Smaller particles (1–10 μm) will have a greater attenuation effect. Attenuation will be very great for those particulates having a size approaching the wavelength of the light source (0.1–1.0 μm). Hence, the optical effect measured may vary by several orders of magnitude with no change in particle concentration, only a change in particle-size distribution. A second limitation of this type of instrument is that it "sees" only a limited portion of the total flow area. These instruments should be considered on the basis of what they really are and what they are capable of doing, i.e., indicating major deviations from normal conditions resulting from poor combustion, failure of particulate collection systems, or soot blowing.

26.5 Sampling Statistical Analysis

The words *sample* and *sampling* have appeared liberally throughout this chapter. Both play an important role in monitoring and analysis. It is therefore important that the reader understand both the meaning and statistical implications of these two words. Shaefer and Theodore provide most of the following material.[3]

When there is a set of n measurements, one may wish to use the values of these measurements in many instances to estimate certain characteristics of a larger number of measurements or observations that have not been made. In other words, the interest is not directly on a particular set of measurements but, rather, in using them to estimate something about a potentially larger set. One refers to the measurements as a *sample*. The larger "supply" of measurement, which one may have, is called a *population*. From the characteristics of a sample of measurements, one can estimate similar characteristics for the population.

When one calculates some measured value of a sample, it is referred to as a statistic. A statistic is any characteristic set of measurements calculated from a sample. The measure corresponding to a statistic obtained from a population is referred to as a parameter. In most cases of interest, the values of parameters are unknown and must be estimated by the statistics derived from a sample.[3,4] Thus, the sample value is an estimate of the population value.

By definition, the mean, μ, and the variance, σ^2, of a random variable are constants characterizing the random variable's average value and its dispersion about is mean, respectively. The mean and variance can also be derived from the probability distribution function (PDF) of the random variable. If the PDF is unknown, however, the mean and variance can be estimated on the basis of a random sample of some, but not all, measurements on the random variable.[3,4]

There are several key terms that arise in a statistical analysis. The *median* is defined as the middle value (or arithmetic mean, of the two middle values) of a set of numbers. The *mode* is the value that occurs with the greatest frequency in a set of numbers. Thus, it is the typical or most common value in a set. One also clearly needs a measure of scatter, which can be used in samples of any size and in some sense, to make use of all the measurements in the sample. There are several measures of scatter that can be used for this purpose, and the most common of this is the *standard deviation*. The standard deviation may be thought of as the "natural" measure of scatter.

26.6 Maintenance

Maintenance activities fall in the major categories of preventive, emergency, and overhaul. Preventive maintenance (PM) includes such items as opacity monitor "window" cleaning or monitor calibration. These activities are included in a formal PM program. Emergency maintenance usually occurs based upon monitor calibration-check data or feedback from the various data review groups, and consists of any activity that corrects a malfunction in the monitoring system. Overhaul maintenance includes items that involve long monitor outages and are best done during annual plant overhaul outages when monitoring is not necessary.

26.7 Conclusions

As noted, the goal of sampling is to obtain a sample for analysis that is truly representative of the larger volume of media being sampled. The most critical factors necessary to achieve this goal are points of sampling, time of sampling, frequency of sampling, and maintenance of the integrity of the sample after sampling and prior to analysis. Homogeneity in the media being sampled is frequently lacking, necessitating multiple-point composite sampling. If it is impractical to use a single representative sampling point, it may be practical to determine and understand interrelationships among various sampling points so that results obtained with a minimum number of sampling locations may be extrapolated to the larger media volume and properly interpreted. With adequate interpretation, a nonrepresentative sample can yield valuable information about trends and can indicate locations where more representative data could be collected. Most samples collected from a single point in a system must be recognized as being nonrepresentative to some degree. Therefore, it becomes important to recognize the degree of representation in the sample and to make it a part of the permanent record. Otherwise, an artificial degree of precision is assigned to the data when it is recorded.

Finally, the following three general rules are applicable to all sampling procedures.

1) The samples must represent the conditions existing at the collection point.
2) The samples must be of sufficient volume and must be taken frequently enough to permit reproducibility for the desired objective, as dictated by the method of analysis to be employed.
3) The samples must be collected, preserved, labeled, packaged, shipped, stored, and handled before analysis in a manner that safeguards against change in the particular constituents or properties to be quantified.

References

1 **IARC Working Group on the Evaluation of Carcinogenic Risks to Humans**, *Outdoor Air Pollution*, International Agency for Research on Cancer, Lyon, France, 2016. (IARC Monographs on the Evaluation of Carcinogenic Risks to Humans, No. 109.) 1.3, Outdoor air pollution measurement methods https://www.ncbi.nlm.nih.gov/books/NBK368020/.

2 **US Environmental Protection Agency**. Exposure Assessment Tools by Approaches – Direct Measurement (Point-of-Contact Measurement). Available from: https://www.epa.gov/expobox/exposure-assessment-tools-approaches-direct-measurement-point-contact-measurement.

3 **S. Shaefer, and L. Theodore**, *Probability and Statistics Applications for Environmental Science*, CRC Press/Taylor & Francis Group, Boca Raton, FL, 2007.

4 **L. Theodore**, *Chemical Engineering: The Essential Reference*, McGraw-Hill, New York, NY, 2017.

5 **L. Theodore**, *Air Pollution Control Equipment Calculations*, John Wiley & Sons, Hoboken, NJ, 2008.

6 **J. Mycock, J. McKenna, and L. Theodore**, *Handbook of Air Pollution Control Engineering and Technology*, CRC Press/Taylor Francis Group, Boca Raton, FL, 1995.

7 **US Environmental Protection Agency**. Air Emission Measurement Center (EMC) Information and Guidelines. Available from: https://www.epa.gov/emc/emc-continuous-emission-monitoring-systems.

8 **G. Burke, R. Singh, and L. Theodore**, *Handbook of Environmental Management and Technology*, 2nd Ed., Wiley-Interscience, Hoboken, NJ, 2000.

9 **L. Theodore**, *Personal Notes*, Astoria, NY, 1974.

27

Optimization Considerations

Optimization? Here are some terms for optimal that can be found in a dictionary: most favorable or desirable; best, best possible option; most cheerful view of matters or expect the best outcome; bright side; to make the most of; develop or realize to the utmost extent; obtain the most efficient or optimum use of; the best or most favorable condition, or amount; etc.

As one might suspect, the term optimization has come to mean different things for different applications, i.e., it could involve a simple two-step calculation or one that requires the use of detailed numerical methods. To take this a step further, the authors were undecided as to how to present optimization in this chapter. After much deliberation and meditation, it was decided to present both qualitative and quantitative material. This decision was primarily influenced by the desire that this material not only be an introduction to optimization but also that which addresses the topic quantitatively.

The optimization problem has been described succinctly by Aris as "getting the best you can out of a given situation.[1]" Problems amenable to solution by mathematical optimization techniques generally have one or more independent variables whose values must be chosen to yield a viable solution and measure of "goodness" available to distinguish between the many viable solutions generated by different choices of these variables. Mathematical optimization techniques are also used for guiding the problem solver to the choice of variables that maximizes the aforementioned "goodness" measure (e.g., profit) or that minimizes some "badness" measure (e.g., cost).

It is also pertinent to understand who is the subject of the optimization: operations will favor a system that is less complicated to operate, environment will favor a system with the fewest emissions, project management will favor a system with the lowest capital cost, and maintenance will favor a system that is simple to maintain.

27.1 The History of Optimization

As to the origin of optimization, it depends on who provides the response, because there are so many aspects of optimization of interest to the practitioner. For example, some claim it was Thomas Edison when he developed a long-lasting, high-quality light bulb in the 1870s. His success was primarily the result of an extensive trial-and-error search for the optimum filament material. A few now refer to it as the Edisonian approach. One of the authors refers to it as the perturbation approach; on occasion, he has modestly termed it the Theodore approach.

Advances in computer calculation have not only reduced the cost and time required to perform each iteration but also provided a better understanding of how this method works. However, the success of this approach still relies heavily on the limitations of the user's intuition and experience; since one often cannot predict or even comprehend the effects of changing numerous variables in a complex system. Despite this barrier, the computer has expanded and improved the employment of the classical methods of optimization.

Today's computers allow optimization (or mathematics, computer science, operations research, mathematical optimization, or mathematical programming) to more easily select the best element (with regard to some criteria) from some set of available alternatives. In the simplest case, an optimization problem consists of maximizing (or minimizing) a real function by systematically choosing input values from within an allowed set of variables in order to compute the value of the function. Mathematically speaking, optimization allows one to find the best available value(s) of some objective function

Optimizing Air Pollution Control Equipment Performance: Operation & Maintenance, First Edition. Jay Richardson and Louis Theodore.

given a defined domain (or a set of constraints), including a variety of different types of objective functions and different types of domains. These various options are discussed in the literature.[2,3]

In addition to the optimization definition presented above by Aris, one might offer the following generic definition for many environmental engineers: "Optimization is concerned with determining the 'best' solution to a given problem.[4]" Alternatively, a dictionary would offer something to the effect: "to make the most of . . . develop or realize to the utmost extent . . . often the most efficient or optimum use of." The process of optimization in environmental practice is required in the solution of many problems and often involves the maximization or minimization of a mathematical function. As one might suppose many of the applications involve economic consideration.

27.2 Optimization Overview

Optimization is viewed by many as a tool in decision-making. It often aids in the selection of values that allow the practicing environmental engineer to better solve a problem. This brief answer provides a qualitative look at optimization.

As noted above, in its most elementary and basic form, one may say that optimization is concerned with the determination of the "best" solution to a given problem. This process is required in the solution of many general problems in environmental engineering and applied science in the maximization (or minimization) of a given function(s), in the selection of a control variable to facilitate the realization of a desired condition, in the scheduling of a series of operations or events to control completion dates of a given project, in the development of optimal layouts of organizational units within a given design space, etc.

One of the most important areas for the application of mathematical optimization techniques is engineering design. In engineering design, once a particular subject or process scheme has been selected for study, it is common practice to optimize the process from a capital cost and operation and maintenance (O&M) standpoint. This practice also applies to air pollution control equipment.

There are many optimization procedures available, most of them too detailed for meaningful application in a text of this nature. These sophisticated optimization techniques, some of which are routinely used in the design of conventional chemical and petrochemical plants, invariably involve computer calculations. Although the use of these techniques in the majority of environmental air pollution control applications was not warranted in the past, more real-world problems require the use of optimization techniques.

27.3 The Scope of Optimization

One can conclude from the above that the theory and application of optimization is mathematical in nature, and it typically involves the maximization or minimization of a function (usually known), which represents the "performance" of some "system": This is carried out by the finding of values for those variables, which cause the function to yield an optimal value.

Perhaps the most important tool employed in optimization by environmentalists is linear programming. Linear programming consists of a linear, multivariable function, which is to be optimized (maximized or minimized), subject to a particular number of constraints. The constraints are normally expressed in linear form. Integer linear programming refers to optimization problems in which at least some of the variables must assume integer values. The reader should note that the terms linear programming and nonlinear programming are essentially similar from an application perspective. There were problems around the middle of the last century because of some difficulties that arose in attempting to solve nonlinear programming problems. However, the arrival of the modern-day computer and sophisticated software (e.g., EXCEL) have removed these problems. Thus the reader need not be concerned with whether a system's pertinent describing equations and constraints are linear or nonlinear. They are simply not necessary and beyond the scope of this text but are available in the literature.[2,3]

One of the major responsibilities in optimization is to construct the correct objective function to be maximized. In a simple environmental problem, the objective function might be the profit associated with the operation of a landfill that treats two categories of air pollution control equipment waste: inert and non-inert. One optimization problem could be to maximize profit (from handling the waste) by employing an operating treatment schedule subject to the landfill's capacity and environmental regulatory requirements, or the problem might be rephrased in the following manner: what combination of the quantities of the two wastes will produce the maximum profit subject to the constraints connected with the landfill's processing capabilities and the different regulation requirements imposed on each waste.

The objective function may be quite simple and easy to calculate in some environmental applications, or it may be complicated and difficult to not only calculate but also specify and/or describe. The objective function may also be very illusive due to the presence of conflicting or dimensionally incompatible objectives; for example, one might be asked to optimize the profit for the aforementioned landfill that not only minimizes both air and water contamination but also is aesthetically appealing. Thus, it may not be always possible to quantify an objective function and hence be able to use any of the mathematical optimization procedures available.

There are a large number of mathematical optimization methods available in practice. Some of the simple ones are listed below:

1) A function of one variable with no constraints
2) A function of two variables with no constraints
3) A function of more than two variables with no constraints
4) Simple perturbation schemes
5) A function of one variable with constraints
6) A function of two variables with constraints
7) A function of more than two variables with constraints

27.4 General Analytical Formulation of Optimization Problems

When more than one independent variable is involved in determining the objective function, e.g., profitability, the objective function y, that depends on a system of $x_2, x_3, \dots, x_n$, independent variables, is written as follows:

$$y(\text{profit}) = f(x_1, x_2, x_3, \dots, x_n) \tag{27.1}$$

The problem is to specify $x_1, x_2, x_3, \dots, x_n$, independent variables so that y will be a maximum. In general, this will involve some type of trial-and-error calculation.

It is possible in some cases to set up an explicit function that will relate the objective function y to the independent variables. Differentiation will then result in a series of partial derivatives,

$$\left(\frac{\partial y}{\partial x_1}\right), \left(\frac{\partial y}{\partial x_2}\right), \textit{etc.} \tag{27.2}$$

Setting these partial derivatives equal to zero results in the same number of simultaneous equations as the number of variables involved. These equations, often subject to "limitations" and "constraints," can be solved for the corresponding optimum values of these variables. The maximum occurs at a point where the first derivatives equal zero, accounting for restrictions on the second derivative, continuity of the function throughout the range of independent variables, and absence of optimum conditions at limiting values of one or more of the variables. These conditions can, on occasion, restrict the applicability of this approach.[5] For example, for the case of a two-variable function where $y = f(x_1, x_2)$, points $x_1 = a$ and $x_2 = b$ exist for a maximum value of $y = f(x_1, x_2)$ if the following conditions hold:

$$\left(\frac{\partial y}{\partial x_1}\right) = \left(\frac{\partial y}{\partial x_2}\right) = 0 \tag{27.3}$$

with,

$$\left(\frac{\partial^2 y}{\partial^2 x_1^2}\right) < 0; \left(\frac{\partial^2 y}{\partial^2 x_2^2}\right) < 0 \tag{27.4}$$

and,

$$\left(\frac{\partial^2 y}{\partial x_1^2}\right)\left(\frac{\partial^2 y}{\partial x_2^2}\right) > \left(\frac{\partial^2 y}{\partial x_1 \partial x_2}\right)^2 \tag{27.5}$$

In contrast, if $\frac{\partial^2 y}{\partial x_1^2}$ and $\frac{\partial^2 y}{\partial x_2^2}$ are positive, and Equation 27.5 holds, there will be a minimum. However, if,

$$\left(\frac{\partial^2 y}{\partial x_1^2}\right)\left(\frac{\partial^2 y}{\partial x_2^2}\right) < \left(\frac{\partial^2 y}{\partial x_1 \partial x_2}\right)^2 \tag{27.6}$$

the point concerned will be a "saddle" point. The case where

$$\left(\frac{\partial^2 y}{\partial x_1^2}\right)\left(\frac{\partial^2 y}{\partial x_2^2}\right) = \left(\frac{\partial^2 y}{\partial x_1 \partial x_2}\right)^2 \tag{27.7}$$

is open and the value may be a maximum, a minimum, or neither.Now consider an objective function specified as:

$$f(x_1, x_2) = \left(x_1^2 - 2x_1 + x_2^2 + x_1 x_2 + 1\right) \tag{27.8}$$

To determine if the function has a maximum or a minimum, proceed as follows

$$\left(\frac{\partial f}{\partial x_1}\right) = 2x_1 - 2 + x_2 = 0 \tag{27.9}$$

$$\left(\frac{\partial f}{\partial x_2}\right) = 2x_1 + x_1 = 0 \tag{27.10}$$

When Equations 27.9 and 27.10 are solved, the following equations result:

$$x_1 = \frac{4}{3} \tag{27.11}$$

$$x_2 = \frac{1}{3} \tag{27.12}$$

27.5 Optimizing Performance

Gaseous and particulate pollutants discharged into the atmosphere can be controlled. As noted earlier, five generic devices available for particulate control include gravity settlers, cyclones (centrifugal separators), electrostatic precipitators, wet scrubbers, and baghouses (fabric filtration). The four generic devices for gases include condensers, absorbers, adsorbers, and incinerators.

There are a number of factors to be considered prior to selecting and operating a particular piece of air pollution control hardware. In general, they can be grouped into three categories: environmental, engineering, and economic. The latter category was discussed earlier but key economic factors include:

1) Capital cost (equipment, installation, engineering, etc.)
2) Operating cost (utilities, maintenance, etc.)
3) Expected equipment lifetime and salvage value

Gas absorption[6,7], as applied to the control of air pollution, is concerned with the removal of one or more pollutants from a contaminated gas stream by treatment with a liquid. The necessary condition is the solubility of these pollutants in the absorbing liquid. The design and performance characteristics of any particular control system are important considerations. Optimizing the performance of an air pollution control device is often a standard operating practice.

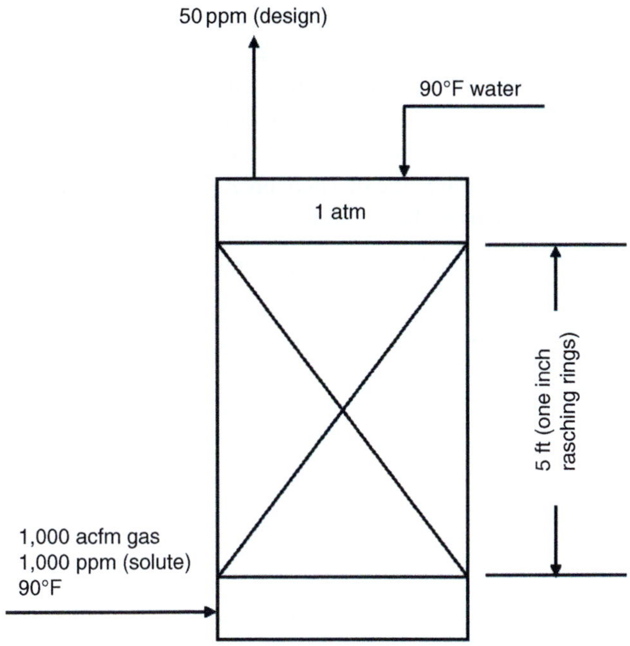

Figure 27.1 Faulty operation?

Consider the absorber system shown in Figure 27.1. It was designed to operate with a maximum pollutant/solute discharge concentration of 50 ppm. However, once the unit was installed and running, the unit operated with a discharge of 60 ppm. Rather than purchase a new unit, there are options available to bring the unit into compliance with the specified design concentration.

Theodore[8–10] proposed a host of possible solutions. Many of them are listed here:

1) Use a smaller packing size. A different size may produce a better packing factor.
2) Use a different type of packing. A different packing type may also produce a better packing factor.
3) Make sure there is no channeling inside the column. The packing should be randomly distributed and there should be no open spots where water will accumulate.
4) Increase the water flow entering the column. In effect, the slope of the operating line will increase and provide a greater driving force for mass transfer. Thus, there will be a greater mass transfer between the liquid and gas phases, and the efficiency of the column will increase.
5) Use a liquid other than water to scrub the gas. Another liquid may cause better mass transfer between the gas and liquid phase.
6) Check to see if flooding is occurring. If the column is near the flooding point, there will not be much mass transfer. The water will be pushed up the column by the gas rather than the water coming down the column.
7) Lower the initial concentration in the carrier gas (if possible).
8) Increase the height of the bed. This will provide more surface area for mixing and more time in the absorber.
9) Increase the pressure. This will reduce the slope of the operating line and lead to a greater driving force.
10) Decrease the temperature. This will reduce the slope of the operating line and lead to a greater driving force.
11) Increase the pressure and decrease the temperature.
12) Additional packing height could be added at the top of the column. This would increase the amount of the solute removed from the gaseous phase.
13) Add sprays at the bottom of the column.
14) Modify the process producing the problem. Since the details of this process are not included in the problem statement, no specific recommendations can be mentioned in this discussion.
15) Finally, before considering changes to the system, one should undertake a thorough inspection of the unit and surrounding components. The emission monitoring system should be recalibrated. All valves, fittings, and pipes should be checked for plugging or leaks. The liquid distributor should be checked to make sure it is functioning properly. The distribution of the packing should be inspected to make sure it is as uniform as possible. Any problems encountered during this inspection should be corrected immediately. Following the maintenance check, the performance of the unit should be reevaluated.

The reader should note that some of the previously mentioned recommendations could lead to higher pressure drops and potential problems with the flow. An example of this problem would be the implementation of suggestion.[1] The reader is left the exercise of determining what other steps could lead to flow/pressure drop problems. Hint: There are at least six suggestions that fall into the proposed solution, including firing the air pollution engineer who designed the absorber.

A similar situation could exist with a gaseous adsorption unit for pollution control. Once again, there are options available to bring the unit into compliance with the specified design concentration. Some of Theodore's[8–10] recommendations are listed below, assume the adsorber packing is activated carbon and there is no liquid flow.

1) Check/increase the depth of the adsorption bed. Increasing the depth of the bed will increase the removal capacity of the column.
2) Change the adsorbent type of size. Use an adsorbent with a higher saturation capacity.
3) Reducing the flow will increase the residence time and should allow for more adsorption.
4) Decrease the inlet temperature. This will favor the adsorption equilibrium process.
5) Increase the system pressure. This will also favor the adsorption equilibrium process.
6) Regenerate the adsorbent for a longer period.
7) Replace the adsorbent.
8) Make up any adsorbent lost to carryover (as necessary).
9) Consider changing the flow direction.

Again, pressure drop considerations should be included in the analysis. Finally, numerous optimization examples are available in the literature.[2,3]

27.6 Recent Developments

The one major development in optimization is the growth of artificial intelligence (AI) in automated control systems.[11] As products of combustion, fuel burners produce a variety of compounds including one or more pollutants such as sulfur dioxide, carbon monoxide, carbon dioxide, particulates, volatile organic compounds, and NO_x. To reduce emissions, various air pollution control equipment technologies are implemented. However, these control devices are affected by many factors such as hardware positions, load swings, and non-optimized or nonideal operating conditions (e.g., low pressures and high oxygen conditions). Variations in these factors can result in inefficient operation and increased pollutant emissions. The frequency and extent of these problems typically worsen with increasingly stringent emissions requirements, due to the narrowing operating zones.

In many systems, air pollution control devices provide efficient operation with few or no emissions excursions. Systems may include one or more sensors that monitor and produce data related to the environment, fuel, and/or operation. For example, sensors may monitor, feed, and/or log data related to flue composition, ambient environmental conditions, and selected processes (e.g., boiler, furnace, combustor, oxidizer, and kiln parameters) and/or burner parameters such as burner pressure and flame scanner signals.[11]

The AI includes instructions that when executed by a processing element are operative to learn, to solve complex problems, make predictions, or undertake human-like tasks like sensing (i.e., vision, speech, and touch), perception, cognition, planning, learning, communication, or physical action. The AI may receive and process data from one or more internal and/or external sources or sensors. In some embodiments, the AI may include an artificial neuron network (ANN). In some embodiments, the ANN may be trained to recognize air pollution control device performance based on sensor data and may optimize performance. Optimization of performance may include varying or adjusting one or more actuators such as a flow controller and damper. The AI may model, tune, and/or optimize performance, by making or recommending bias adjustments (i.e., changes to an output bounded by one or more limits) to one or more control outputs, such as by adjusting the trim control curve. The system may provide data analytics, system diagnostics, and/or burner optimization such as via the integrated-AI module.[11]

27.7 Conclusions

Most engineering optimization applications involve economics: maximizing profit, minimizing cost, or both. Environmental applications can involve minimizing toxic emissions, maximizing energy usage and conservation, minimizing health and safety risks, standard economic concerns, etc. The reader should not lose sight that most real-world industry applications involving optimization usually require simple solutions. Here is a comment from a retired engineer[12]: "generally, the consumer good industry where I worked does not rely on sophisticated models. Basically, optimization was conducted to determine overall consumer preference, to determine how to use minimum resources and lastly to determine how to minimize costs subject to constraints. In the determination of consumer preference, we typically used design of experiments to examine the broad range of attributes that led to maximum product liking or decision to purchase. Occasionally, we modeled processes and used regression models. In that case, we could search or compute the desired maximum. The last set of techniques used is what the business world calls Operations Research. Typically, linear or dynamic programming was used to evaluate and objective function subject to constraints. For example, we used liner programming in hot dog manufacturing. The goal was to come up with a mixture of meat cuts that minimized costs while meeting government requirements for protein, fat, and water content."

References

1 **R. Aris**, *Discrete Dynamic Programming*, Blaisdell, Boston, MA, 1964.

2 **L. Theodore, and K. Behan**, *Introduction to Optimization for Chemical and Environmental Engineers*, CRC Press/Taylor & Francis Group, Boca Raton, FL, 2018.

3 **C. Prochaska, and L. Theodore**, *Introduction to Mathematical Methods for Environmental Engineers and Scientists*, Scrivener-Wiley, Beverly, MA, 2018.

4 **L. Theodore**, *Personal Notes*, East Williston, NY, 2011.

5 **J. Hoppel**, *Chemical Process Economics*, John Wiley & Sons, Hoboken, NJ, 1958.

6 **L. Theodore, and F. Ricci**, *Mass Transfer Operations for the Practicing Engineer*, John Wiley & Sons, Hoboken, NJ, 2010.

7 **L. Thoedore**, *Chemical Engineering: The Essential Reference*, McGraw-Hill, New York, NY, 2014.

8 **L. Theodore**, *Personal Notes*, East Williston, NY, 1996.

9 **L. Theodore**, *Air Pollution Control Equipment Calculations*, John Wiley & Sons, Hoboken, NJ, 2008.

10 **L. Theodore, and R. Allen**, *Air Pollution Control Equipment, A Theodore Tutorial*, Theodore Tutorials (Originally Published by the USEPA/APTI, RTP, NC), East Williston, NY, 1993.

11 **P. Santangeli and J. Richardson**, *Burner System*, US Patent 63193982 20210527, 2022.

12 **R. Altomare**, Communications to L. Theodore, Bronx, NY, 2016.

28

Factors in Pollution Control Equipment Selection

There are a number of factors to be considered prior to selecting a particular air pollution control device[1–3] in general, they can be grouped into three "E" categories:

1) Environmental
2) Engineering
3) Economic

Each category is reviewed in the next section.

Proper selection of a particular system for a specific application can be extremely difficult and complicated. In view of the multitude of complex and often ambiguous pollution regulations, it is in the best interest of the prospective user to work closely with regulatory officials as early in the process as possible. Finally, previous experience with a similar application cannot be overemphasized.

28.1 Environmental, Engineering, and Economic Factors

Environmental, engineering, and economic factors that can impact the selection process for air pollution control equipment are detailed below.

1) Environmental
   a) Equipment location
   b) Available space
   c) Ambient conditions
   d) Availability of adequate utilities (i.e., power and water) and ancillary systems facilities (i.e., waste treatment and disposal)
   e) Maximum allowable emission (air pollution codes)
   f) Aesthetic considerations (i.e., visible steam or water vapor plume)
   g) Contribution of air pollution control system to wastewater and land pollution
   h) Contribution of air pollution control system to plant noise level
2) Engineering
   a) Contaminant characteristics (i.e., physical and chemical properties, concentration, particulate shape and size distribution – in the case of particulates, chemical reactivity, corrosivity, abrasiveness, and toxicity)
   b) Gas stream characteristics (i.e., volumetric flow rate, temperature, pressure, humidity, composition, viscosity, density, reactivity, combustibility, corrosivity, and toxicity)
   c) Design and performance characteristics of the particular control system (i.e., size and weight, fractional efficiency curves – in the case of particulates, mass transfer and/or contaminant destruction capability – in the case of gases or

Optimizing Air Pollution Control Equipment Performance: Operation & Maintenance, First Edition. Jay Richardson and Louis Theodore.

vapors, pressure drop, reliability and dependability, turndown capability, power requirements, utility requirements, temperature limitations, maintenance requirements, and flexibility toward complying with more stringent air pollution codes)

3) Economic
   a) Capital cost (equipment, installation, engineering, etc.)
   b) Operating cost (utilities, maintenance, etc.)
   c) Expected equipment lifetime and salvage value

The final choice in equipment selection is usually dictated by that equipment's capability of achieving compliance with the appropriate regulatory codes at the lowest uniform annual cost, i.e., amortized capital investment plus operation and maintenance (O&M) costs. The reader is referred to Chapter 23 for economic details on the general subjects of economics.

28.2 Comparing Control Equipment Alternatives

As noted in the previous section, the final choice in equipment selection is usually dictated by the equipment capable of achieving compliance with regulatory codes at the lowest uniform annual cost. In order to compare specific control equipment alternatives, knowledge of the particular application and site is essential. A preliminary screening, however, may be performed by reviewing the advantages and disadvantages of each type of air pollution control equipment. For example, if water or a waste treatment system is not available at the site, this may preclude the use of a wet scrubber system and instead focus particulate removal on dry systems, such as cyclones, baghouses, and/or electrostatic precipitators. If auxiliary fuel is unavailable on a continuous basis, it may not be possible to combust organic pollutant vapors in an incineration system. If the particulate size distribution in the gas stream is relatively fine, cyclone collectors most probably would not be considered. If the pollutant vapors can be reused in the process, control efforts may be directed to adsorption systems. There are many more situations where knowledge of the capabilities of the various control options, combined with common sense, will simplify the selection procedure.

General advantages and disadvantages of the most popular types of air pollution control equipment for gases and particulates are presented in Tables 28.1–28.9.

Table 28.1 Advantages and disadvantages of cyclone collectors.

Advantages

1) Low cost of construction.
2) Relatively simple equipment with few maintenance problems.
3) Relatively low operating pressure drops (for degree of particulate removal obtained) in the range of approximately 2–6 in. water column.
4) Temperature and pressure limitations imposed only by the materials of construction used.
5) Dry collection and disposal.
6) Relatively small space requirements.

Disadvantages

1) Relatively low overall particulate collection efficiencies, especially on particulates below 10 μm in size.
2) Inability to handle tacky materials.

Table 28.2 Advantages and disadvantages of wet scrubbers.

Advantages

1) No secondary dust sources.
2) Relatively small space requirements.
3) Ability to collect gases as well as particulates (especially "sticky" ones).
4) Ability to handle high-temperature, high-humidity gas streams.
5) Capital cost is low (if wastewater treatment system is not required).
6) For some processes, the gas stream is already at high pressures (so pressure drop considerations may not be significant).
7) Ability to achieve high collection efficiencies on fine particulates (however, at the expense of pressure drop).

Disadvantages

1) May create water disposal problem.
2) Product is collected wet.
3) Corrosion problems are more severe than with dry systems.
4) Steam plume opacity and/or droplet entrainment may be objectionable.
5) Pressure drop and horsepower requirements may be high.
6) Solids buildup at the wet–dry interface may be a problem.
7) Relatively high maintenance costs.

Table 28.3 Advantages and disadvantages of electrostatic precipitators.

Advantages

1) Extremely high particulate (coarse and fine) collection efficiencies can be attained (at a relatively low expenditure of energy).
2) Dry collection and disposal.
3) Low-pressure drop (typically less than 0.5 in. water column).
4) Designed for continuous operation with minimum maintenance requirements.
5) Relatively low operating costs.
6) Capable of operation under high pressure (to 150 psi) or vacuum conditions.
7) Capable of operation at high temperatures (to 1,300°F).
8) Relatively large gas flow rates can be effectively handled.

Disadvantages

1) High capital cost.
2) Very sensitive to fluctuations in gas stream conditions (in particular, flows, temperatures, particulate and gas composition, and particulate loadings).
3) Certain particulates are difficult to collect due to extremely high or low resistivity characteristics.
4) Relatively large space required for installation.
5) Explosion hazard when treating combustible gases and/or collecting combustible particulates.
6) Special precautions are required to safeguard personnel from the high voltage.
7) Ozone is produced by the negatively charged discharge electrode during gas ionization.
8) Relatively sophisticated maintenance personnel required.

Table 28.4 Advantages and disadvantages of fabric filter systems.

Advantages

1) Extremely high collection efficiency on both coarse and fine (submicron) particulates.
2) Relatively insensitive to gas stream fluctuation. Efficiency and pressure drop are relatively unaffected by large changes in inlet dust loadings for continuously cleaned filters.
3) Filter outlet air may be recirculated within the plant in many cases (for energy conservation).
4) Collected material is recovered dry for subsequent processing or disposal.
5) No problems with liquid waste disposal, water pollution, or liquid freezing.
6) Corrosion and rusting of components are usually not problems.
7) There is no hazard of high voltage, simplifying maintenance and repair and permitting
8) Collection of flammable dusts.
9) Use of selected fibrous or granular filter aids (precoating) permits the high-efficiency collection of submicron smokes and gaseous contaminants.
10) Filter collectors are available in a large number of configurations, resulting in a range of dimensions and inlet and outlet flange locations to suit installation requirements.
11) Relatively simple operation.

Table 28.4 (Continued)

Disadvantages

1) Temperatures much in excess of 550°F require special refractory mineral or metallic fabrics that are still in the developmental stage and can be very expensive.
2) Certain dusts may require fabric treatments to reduce dust seeping or, in other cases, assist in the removal of the collected dust.
3) Concentrations of some dusts in the collector (~50 g/m^3) may represent a fire or explosion hazard if a spark or flame is admitted by accident. Fabrics can burn if readily oxidizable dust is being collected.
4) Relatively high maintenance requirements (bag replacement, etc.).
5) Fabric life may be shortened at elevated temperatures and in the presence of acid or alkaline particulate or gas constituents.
6) Hygroscopic materials, condensation of moisture, or tarry adhesive components may cause crusty caking or plugging of the fabric or require special additives.
7) Replacement of fabric may require respiratory protection for maintenance personnel.
8) Medium pressure-drop requirements, typically in the range of 4–10 in. water column.

Table 28.5 Advantages and disadvantages of absorption systems (packed and plate columns).

Advantages

1) Relatively low-pressure drop.
2) Standardization in fiberglass-reinforced plastic (FRP) construction permits operation in highly corrosive atmospheres.
3) Capable of achieving relatively high mass-transfer efficiencies.
4) Increasing the height and/or type of packing or number of plates can improve mass transfer without purchasing a new piece of equipment.
5) Relatively low capital cost.
6) Relatively small space requirements.
7) Ability to collect particulates as well as gases.

Disadvantages

1) May create water (or liquid) disposal problem.
2) Product collected wet.
3) Particulates deposition may cause plugging of the bed or plates.
4) When FRP construction is used, it is sensitive to temperature.
5) Relatively high maintenance costs.

Table 28.6 Comparison of plate and packed columns.

Packed Column

1) Lower-pressure drop.
2) Simpler and cheaper to construct.
3) Preferable for liquids, with high foaming tendencies.

Plate Column

1) Less susceptible to plugging.
2) Less weight.
3) Less of a problem with channeling.
4) Temperature surge will result in less damage.

Table 28.7 Advantages and disadvantages of adsorption systems.

Advantages
1) Product recovery may be possible.
2) Excellent control and response to process changes.
3) No chemical disposal problem when pollutant (product) recovered and returned to process.
4) Capability of systems for fully automatic, unattended operation.
5) Capability to remove gaseous or vapor contaminants from process streams to extremely low levels.

Disadvantages
1) Product recovery may require an exotic, expensive distillation (or extraction) scheme.
2) Adsorbent progressively deteriorates in capacity as the number of cycles increases.
3) Adsorbent regeneration requires a steam or vacuum source.
4) Relatively high capital cost.
5) Prefiltering of gas stream may be required to remove any particulate capable of plugging the adsorbent bed.
6) Cooling of the gas stream may be required to get to the usual range of operation (less than 120°F).
7) Relatively high steam requirements to desorb high-molecular-weight hydrocarbons.

Table 28.8 Advantages and disadvantages of combustion systems.

Advantages
1) Simplicity of operation.
2) Capability of steam generation or heat recovery in other forms.
3) Capability for virtually complete destruction of organic contaminants.

Disadvantages
1) Relatively high operating costs (particularly associated with fuel requirements).
2) Potential for flashback and subsequent explosion hazard.
3) Catalyst poisoning (in the case of catalytic incineration).
4) Incomplete combustion can create potentially worse pollution problems.

Table 28.9 Advantages and disadvantages of condensers.

Advantages
1) Pure product recovery (in the case of indirect-contact condensers).
2) Water used as the coolant in an indirect contact condenser (i.e., shell-and-tube heat exchanger) does contact the contaminated gas stream and can be reused after cooling.

Disadvantages
1) Relatively low removal efficiency for gaseous contaminants (at concentrations typical of pollution control applications).
2) Coolant requirements may be extremely expensive.

28.3 Equipment and Material Specifications

It is safe to say that standard equipment should be selected whenever possible. If the equipment is standard, the manufacturer may have the desired unit and size in stock, in addition, the manufacturer can usually quote a lower price and provide better guarantees for standard equipment than for specific equipment. However, this is rarely feasible and most industrial and utility grade equipment require engineering. The engineer should evaluate the needs associated with the equipment and prepare a preliminary specification sheet for the equipment. These specifications can be used by the engineer as a basis for the preparation of the final specifications for the vendor. They can also be used by the vendor to provide suggestions to the purchasers.

Preliminary specifications for equipment should include some or all of the following:

1) Identification
2) Function
3) Operating constraints
4) Materials handled
5) Basic design data
6) Essential controls and instrumentation
7) Insulation requirements, if applicable
8) Allowable tolerances, if applicable
9) Special information
10) Special design details
11) Materials of construction
12) Delivery detail(s)

The engineer or applied scientist should also allow the vendor to make suggestions before final specifications are submitted. The final decision on the unit could therefore include any changes that reduce the cost with little to no decrease in the performance of the equipment.

The effects of corrosion and erosion must be considered by the chemical engineer in the design and selection of equipment. Chemical resistance and the physical properties of construction materials are also important factors to consider in the selection and analysis process. The materials of construction must be resistant to the corrosive action of any chemicals that may contact the exposed surfaces. Structural strength, resistance to physical or thermal shock, cost, ease of fabrication, necessary maintenance, and general type of service required, including operating temperatures and pressures, are additional factors that influence the final choice of construction materials.[4]

Materials in most engineering applications can be classified in the following five categories:

1) Ferrous metals and alloys, including steel, stainless steel, low-alloy steel, cast irons, alloy cast irons, medium alloys, and high alloys
2) Nonferrous metals and alloys, including nickel and nickel alloys, aluminum and its alloys, copper and its alloys, lead and its alloys, titanium, and zirconium
3) Inorganic nonmetallics, including glass and glassed steel, porcelain and stoneware, brick construction, cement and concrete, and soil
4) Organic nonmetallics, including plastic materials
5) Thermoplastics, including thermosetting plastics, rubber and elastomer, asphalt, carbon and graphite, and wood

28.4 Instrumentation and Controls

The applicable controls require careful consideration of all operational and maintenance requirements. The control system should also be designed to vary one or more of the process variables to maintain the appropriate conditions with the equipment in question. These variations are usually programmed into the system on the basis of past experience with the unit. The operational parameters that may vary include the system temperature and system pressure. The control system should be subjected to extensive analysis of operational problems and items that might become problematic. The system must identify the problem and integrate necessary actions and alarms or trips into the system. Any environmental regulatory limitations that are imposed on the unit must also be identified and monitored to ensure that they are not exceeded. In the end, the purpose of the control system is to ensure that the equipment will be operating in a safe reliable manner within the guidelines of the equipment design.

Process controls, together with their associated instrumentation, can be considered as the mechanical brain of the equipment of concern. Variables in the process must be measured and then controlled and integrated for optimum processing conditions. These components and systems have been designed by instrumentation and control engineers to reduce labor costs and improve performance capabilities of the equipment. The economic advantage of process control is well documented, thus accounting for its widespread use in the chemical process industry.[5]

Engineers and applied scientists not specializing in process control should have a working knowledge of this field so that they can appreciate its relative merits.

The final selection of equipment for a process requires some experience, particularly if the process is partially or completely new. If the process is a traditional one or in operation elsewhere, then the task consists chiefly of comparative calculations, while incorporating pertinent innovations and improvements that past experience suggests. Any new equipment requires a study of the process involved before a selection of the type and size of equipment required for guaranteed performance. Literature reviews and background reading plus applicable equipment books will assist in equipment selection.

A specification form for the equipment should be prepared after the chemical engineer has made the necessary equipment calculations. If outside bids are acquired, detailed specification sheets may be required. Standard forms are available from both vendors and equipment manufacturers' associations.

The value of using standard equipment was discussed earlier. The experience of others should be used whenever possible; significant information is usually available from vendors who usually see possibilities of additional orders for other equipment. Thus, in many instances, they are not anxious to enter into a bid where their equipment will not provide satisfactory service.

Engineers and applied scientists are often confronted with situations for which the equipment requires a special design and the use of special materials. In such cases, they must draw upon their training and expertise to design the requisite equipment.

Notwithstanding the above, the reader should note that traditional standard equipment has withstood the test of time and the rigorous test of use and service. These units have undergone long periods of experimentation and many modifications of their original design. Standardization can not only mean a minimum cost in manufacturing but also that standard units in standard sizes have usually been given the best of thought in their ultimate state of design.

28.5 Equipment Fabrication

The fabrication process varies with the equipment, equipment size, and the vendor. Thus, the process of equipment fabrication is site-specific, and no set procedure can be provided. One can say that fabrication expenses often account for a large fraction of the purchase cost for equipment. A chemical engineer should therefore be acquainted with the methods for fabricating equipment, and the problems involved in the fabrication should be considered when selected equipment specifications are prepared.

Many fabrication details are governed by various codes, such as the American Society of Mechanical Engineers (ASME) codes. These codes can vary depending on the complexity and type of equipment being reviewed. The following steps are often involved in the fabrication process.[4]

1) Material selection
2) Cutting to correct dimensions
3) Forming into desired shapes
4) Fastening, if applicable
5) Testing
6) Heat treating, if applicable
7) Final preparatory steps

28.6 Installation Procedures

Although nearly all the chapters in Part II contain a section on installation procedures, this topic is reiterated in this chapter because of the important role it plays in the selection procedure. Its role in future O&M should not be minimized. See individual chapters for details on specifics for each unit.

28.7 Equipment Purchasing Guidelines

Purchasing guidelines obviously with the equipment in question. This section provides information as it applies to an air pollution control device. However, the recommendations can be adapted or modified for other equipment.

Prior to the purchase experience showed that the following points should be emphasized:

1) Refrain from purchasing any device without reviewing certified independent test data on its performance under a similar application. Request that the manufacturer provide performance information and design specifications.
2) In the event that sufficient performance data are unavailable, request that the equipment supplier provide a small pilot model for evaluation under existing conditions.
3) Prepare a good set of specifications. Include a strong performance guarantee from the manufacturer to ensure that the device will meet all design criteria and specific process conditions.
4) Closely review the overall process, other equipment, and economic fundamentals.
5) Make a careful energy balance study.
6) Refrain from purchasing any device until firm installation cost has been added to the total cost. Escalating installation costs are the rule rather than the exception.
7) Give O&M costs high priority on the list of selection factors.
8) Refrain from purchasing any device until a solid commitment from the vendor(s) is obtained. Make every effort to ensure that the device is compatible with the (plant) process.
9) The specification should include written assurance of prompt technical assistance from the supplier. This, together with a completely understandable operating manual with parts list, full schematics, consistent units and notations, etc. This is essential and is too often forgotten in the rush to get the device operating.
10) Schedules can be critical. In such cases, delivery guarantees should be obtained from the manufacturers and penalties identified.
11) The device should be of fail-safe design with built-in indicators to show when performance is deteriorating.
12) Perhaps most importantly, withhold 10–15% of the purchase price until satisfactory operation is clearly demonstrated.

These 12 points have, in a real sense, become the "bible" to those involved with and are responsible for the purchase of equipment.

The usual design procurement, installation, and/or start-up problems can be further compounded by any one or a combination of the following:

1) Unfamiliarity with the device in general
2) New suppliers, frequently with unproven equipment
3) Lack of industry standards
4) Compliance schedules that are too tight
5) Vague specifications
6) Weak guarantees
7) Unreliable delivery schedules
8) Process reliability problems

These eight problems represent the standard that the purchaser should also be aware of.

Elan[6] offers the following four purchasing guidelines from a cash flow perspective. Engineers representing buying organizations can contribute to responsible and effective cash flow management by observing the following practices:

1) Clearly define the project scope and delivery schedule.
2) Ensure that the suppliers are qualified to support the project.
3) Help the suppliers understand the payment terms and process.
4) Approve or deny invoices promptly.

Elan[6] also suggests the following for project managers representing a selling organization:

1) Fully understand the project scope and deliverable schedule.
2) Order a credit report before writing the proposal.
3) Understand the payment terms.
4) Agree on the invoice schedule, format, and delivery method at the project outset.
5) Fully understand the project scope and deliverable schedule.
6) Order a credit report before writing the proposal.
7) Close payments when they are not received by the due date and stop work until payments are brought up to current status.

Thus, although project managers of procuring and selling organizations may sit across the table from one another, they all face similar pressures and are best served by similar approaches to cash flow management.

Proper selection of a particular unit for a specific application can be extremely difficult and complicated. The final choice in equipment selection is usually dictated by that unit capable of achieving the aforementioned design performance criteria and required process conditions at the lowest uniform annual cost.

As noted earlier, in order to compare specific equipment alternatives, knowledge of the particular application and site is also essential. A preliminary screening, however, may be performed by reviewing the advantages and disadvantages of each type or class of equipment. However, in many other situations, knowledge of the capabilities of the various options combined with common sense, will simplify the selection process.

28.8 Future Trends

While technology is always advancing, the selection process should stick to the thorough process outlined above.

References

1 **M.K. Theodore, and L. Theodore**, *Introduction to Environmental Management*, 3rd Ed., CRC Press/Taylor & Francis Group, Boca Raton, FL, 2021.
2 **L. Theodore**, *Air Pollution Control for Hospitals and Other Medical Facilities*, Garland Press, New York, NY, 1981.
3 **L. Theodore**, *Air Pollution Control Equipment Calculations*, John Wiley & Sons, Hoboken, NJ, 2008.
4 **M. Peters**, *Plant Design and Economics for Chemical Engineers*, McGraw-Hill, New-York, NY, 1958.
5 **P. Vasudaven**, *Process Dynamics and Control, A Theodore Tutorial*, Theodore Tutorials, East Williston NY (originally published by USEPA/APT I, RTP, NC), 1996.
6 **D. Elan, Jr.**, *Managing Cash Flow*, EM (Environmental Management), Air & Waste Management Association (AWMA), Pittsburg, PA, 2012.

29

Control Equipment for Specific Industries

Emma Parente

Childhood Education, Adelphi University, Garden City, NY, USA

The basic types of emission control devices are mechanical collectors, wet scrubbers, baghouses, electrostatic precipitators, combustion systems, condensers, absorbers, and adsorbers. All of these have been used to some extent to control emissions from a nearly infinite variety of processes, with the selection procedure almost always dictated by experience. The concluding chapter of this book provides information on suggested control measures for a number of industries. Details are provided in a table titled "control techniques applicable to unit process at important emission sources."

29.1 Continue Techniques Applicable to Specific Sources

Information on control techniques applicable to unit process is at specific emission sources in Table 29.1. Information includes:

1) Industry
2) Process or operation
3) Contaminants emitted
4) Control techniques

Optimizing Air Pollution Control Equipment Performance: Operation & Maintenance, First Edition. Jay Richardson and Louis Theodore.

Table 29.1 Control techniques applicable to unit processes at important emission sources.

Industry	Process or Operation	Air Contaminants Emitted	Control Techniques
Aluminum Reduction Plants	Materials handling: buckets and belt conveyor or pneumatic conveyor	Particulates (dust)	Exhaust system and baghouse
	Anode and cathode electrode preparation	Hydrocarbon emissions from binder	Exhaust systems and mechanical collectors
	1) Cathode: baking	Particulates (dust)	
	2) Anode: grinding and blending		
	Baking	Particulates (dust), CO, SO_2, hydrocarbons, and fluorides	High-efficiency cyclone, electrostatic precipitators, scrubbers, catalytic combustion or incinerators, flares, and baghouse
	Pot charging	Particulates (dust), CO, HF, SO_2, CF_4, and hydrocarbons	High-efficiency cyclones, baghouses, spray towers, floating-bed scrubbers, electrostatic precipitators, chemisorption, and wet electrostatic precipitators
	Metal casting	Cl_2, HCl, CO, and particulates (dust)	Exhaust systems and scrubbers
Asphalt Batch Plants	Materials handling, storage and classifiers: elevators, chutes, vibrating screens	Particulates (dust)	Local exhaust systems with a cyclone pre-cleaner and a scrubber or baghouse
	Drying: rotary oil- or gas-fired	Particulates and smoke	Proper combustion controls, fuel-oil preheating where required local exhaust system, cyclone, and a scrubber or baghouse
Cement Plants	Truck traffic	Dust	Wetting down truck routes
	Quarrying: primary crusher, secondary crusher, conveying, storage	Particulates (dust)	Wetting and exhaust systems with mechanical collectors
	Dry processes: materials handling, air separator (hot-air furnace)	Particulates (dust)	Local exhaust system and mechanical collectors and baghouse
	Grinding	Particulates (dust)	Local exhaust system with cyclones and baghouse
	Pneumatic conveying and storage	Particulates (dust)	
	Wet process: materials, handling, grinding, and storage	Wet materials, no dust	
	Kiln operations rotary kiln	Particulates (dust), CO, SO_2, NO_2, hydrocarbons, aldehydes, ketones	Electrostatic precipitators and baghouses, scrubber, flare
	Clinker cooling: materials handling	Particulates (dust)	Local exhaust system and mechanical collectors
	Grinding and packaging: air separator, grinding, pneumatic conveying, materials handling, packaging	Particulates (dust)	Local exhaust systems and mechanical collectors
Coral Preparation Plants	Materials handling: conveyors, elevators, chutes	Particulates (dust)	Local exhaust systems and cyclones
	Sizing: crushing, screening, and classifying	Particulates (dust)	Local exhaust systems and cyclones

Table 29.1 (Continued)

Industry	Process or Operation	Air Contaminants Emitted	Control Techniques
	De-dusting	Particulates (dust)	Local exhaust system, cyclone pre-cleaners, and baghouse
	Storing coal in piles	Blowing particulates (dust)	Wetting, plastic spray covering
	Refuse piles	H_2S, particulates, and smoke from burning storage piles	Digging out fire, pumping water onto fire area, blanket with incombustible material
	Coal drying: rotary, screen, suspension, fluid bed, and cascade	Dust, smoke, particulates, sulfur oxides, and H_2S	Exhaust systems with cyclones and venturi scrubbers
Coke Plants	By-product ovens charging	Smoke, particulates (dust)	Pipeline charging, careful charging techniques; portable hooding and scrubber or baghouses
	Pushing	Smoke, particulates (dust), SO_2	Minimize green coke pushing, scrubbers, and baghouses
	Quenching	Smoke, particulates (dust and mists), phenols, and ammonia	Baffles and spray tower
	By-product processing	CO, H_2S, methane, ammonia, H_2, phenols, hydrogen cyanide, N_2, benzene, xylene, etc.	Electrostatic precipitator, scrubber, flaring
	Material storage (coal and coke)	Particulates (dust)	Wetting, plastic spray, fire-prevention techniques
Fertilizer Industry (Chemical)	Phosphate fertilizers: crushing, grinding, and calcining	Particulates (dust)	Exhaust system, scrubber, cyclone, and baghouse
	Hydrolysis of P_2O_5	PH_3, P_2O_5, H_3PO_4, mist	Scrubbers, flare
	Acidulation and curing	HF, SiF_4	Scrubbers
	Granulation	Particulates (dust) (product recovery)	Exhaust system, scrubber, or baghouse
	Ammoniation	NH_3, NH_4Cl, SiF_4, HF	Cyclone, electrostatic precipitator, baghouse, and high-energy scrubber
	Nitric acid acidulation	NO_x, gaseous fluoride compounds	Scrubber, addition of urea
	Superphosphate storage and shipping	Particulates (dust)	Exhaust system, cyclone, or baghouse
	Ammonium nitrate reactor	NH_3, NO_x	Scrubber
	Prilling tower	NH_4, NO_3	Proper operation control, scrubbers
Foundries	Melting (cupola):		
Iron	1) Charging 2) Melting 3) Pouring 4) Bottom drop	Smoke and particulates Smoke particulates, and fume Oil, mist, and CO Smoke and particulates	Closed top with exhaust system, CO afterburner, gas-cooling device and scrubbers, baghouse or electrostatic precipitator, and wetting to extinguish fire
Brass Bronze	Melting:		
	1) Charging 2) Melting 3) Pouring	Smoke particulates and oil Zinc oxide fume, particulates, and smoke zinc oxide Fume and lead oxide fume	Low-zinc-content red Brass: use good combustion controls and slag cover; high-zinc-content brass: use good combustion controls, local exhaust system, and baghouse or scrubber

(Continued)

Table 29.1 (Continued)

Industry	Process or Operation	Air Contaminants Emitted	Control Techniques
Aluminum	Melting: charging, melting, and pouring	Smoke and particulates	Charge clean material (no paint or grease) and proper operation should be required, no air pollution control equipment if no fluxes are used and degassing is not required, and dirty charge requires exhaust system with scrubbers and baghouses
Zinc	1) Melting 2) Charging	Smoke and particulates	Exhaust system with cyclone and baghouse; charge clean material (no paint or grease)
	Melting	Zinc oxide fume	Careful skimming of dross
	Pouring	Oil mist and hydrocarbons from die-casting machines	Use low-smoking die-casting lubricants
	Sand handling shake-out	Particulates (dust), smoke, organic vapors	
	Magnetic pulley, conveyors and elevators, rotary cooler, screening, and crusher mixer	Particulates (dust)	Exhaust system, cyclone, and baghouse
	Coke-making ovens	Organic acids, aldehydes, smoke, hydrocarbons	Use of binders that will allow ovens to operate at less than 400°F, or exhaust systems and afterburners
Galvanizing Operations	Hot-dip galvanizing tank kettle, dipping material into the molten zinc, and dusting flux onto the surface of the molten zinc	Fumes, particulates (liquid), vapors: NH_4Cl, ZnO, $ZnCl_2$, Zn, NH_3, oil, and carbon	Close-fitting hoods with high indraft velocities (in some cases the hood may not be able to be close to the kettle, so that the indraft velocity must be very high), baghouses, electrostatic precipitators
Kraft Pulp Mills	Digesters: batch and continuous	Mercaptans, methanol (odors)	Condensers and use of limekiln, hog fuel boiler, or furnaces as afterburners
	Multiple-effect evaporators	H_2S, other odors	Caustic-scrubbing and thermal oxidation of non-condensibles
	Recovery furnace	H_2S, mercaptans, organic sulfides, and disulfides	Paper combustion controls for fluctuating load and unrestricted primary and secondary airflow to furnace and scrubber or electrostatic precipitator
	Weak and strong black liquor oxidation	H_2S	Packed tower and cyclone
	Smelt tanks	Particulates (mist or dust)	Demisters, venturi, packed tower, or impingement-type scrubbers
	Limekiln	Particulates (dust), H_2S	Venturi scrubbers
Municipal and Industrial Incinerators	Single-chamber incinerators	Particulates, smoke, volatiles, CO, SO_x, ammonia, organic acids, aldehydes, NO_x, hydrocarbons, odors, and HCl	
	Flue-fed		Settling chambers, scrubbers, afterburner, bypass flue, and ash cleanout
	Multiple-chamber incinerators: retort, inline	Particulates, smoke, and combustion contaminants	Operating at rated capacity, using auxiliary fuel as specified, and good maintenance, including timely cleanout of ash
	Flue-fed	Particulates, smoke, and combustion contaminants	Use of charging gates and automatic controls for draft
	Wood waste	Particulates, smoke, and combustion contaminants	Continuous-feed systems, operate at design load and excess air, and limit charging of oily material

Table 29.1 (Continued)

Industry	Process or Operation	Air Contaminants Emitted	Control Techniques
	Municipal incinerators: 50–100 tons/dy	Particulates, smoke, volatiles, CO, ammonia, organic acids, aldehydes, NO_x, hydrocarbons, SO_x, hydrogen chloride, and odors	Preparation of materials, including weighing, grinding, shredding, control of tipping area, furnace design with proper automatic controls, proper startup techniques, maintenance of design operating temperatures, use of electrostatic precipitators, scrubbers and baghouses, and proper ash cleanout
	Pathological incinerators	Odors and hydrocarbons	Proper charging
	Wood waste and industrial waste	Particulates, smoke, and combustion contaminants	Modified fuel feed, auxiliary fuel and dryer systems
	Box type	Particulates, smoke, and combustion contaminants	Allow proper startup, and charge material slowly, without overload
Nonferrous Smelters, Primary Copper	Roasting	SO_2, particulates, and fume	Exhaust system, settling chambers, cyclones or scrubbers and electrostatic precipitators for dust and fumes, and sulfuric acid plant for SO_2
	Reverberatory furnace	Smoke, particulates, fume, SO_2	Exhaust system, settling chambers, cyclones or scrubbers, electrostatic precipitators for dust and fumes, and sulfuric acid plant for SO_2
	Converters: charging, slag skim, pouring, air or oxygen blow	Smoke, fume, and SO_2	Exhaust system, settling chambers, cyclones or scrubbers, electrostatic precipitators for dust and fumes, and sulfuric acid plant for SO_2
Lead	Sintering	SO_2, particulates, and smoke	Exhaust system, cyclones and baghouse or precipitators for dust and fumes, sulfuric acid plant for SO_2
	Blast furnace	SO_2, CO, particulates, lead oxide, and zinc oxide	Exhaust system, settling chambers, afterburner and cooling device, cyclone, and baghouse
	Dross reverberator furnace	SO_2, particulates, and fume	Exhaust system, settling chambers, cyclone and cooling device, and baghouse
	Refining kettles	SO_2, particulates	Local exhaust system, cooling device, baghouse, or precipitator
Cadmium	Roasters, slag, fuming furnaces, and de-leading kilns	Particulates	Local exhaust system, baghouse, or precipitator
Zinc	Roasting	Particulates (dust) and SO_2	Exhaust system, humidifier, cyclone scrubber, electrostatic precipitator, and acid plant
	Sintering	Particulates (dust) and SO_2	Exhaust system, humidifier, electrostatic precipitator, and acid plant
	Calcining	Zinc oxide fume, particulates, SO_2, and CO	Exhaust system and baghouse
	Retorts: electric arc		
Nonferrous Smelters, Secondary	Blast furnaces and cupolas recover metal from scrap and slag	Dust, fumes, particulates, oil vapor, smoke, and CO	Exhaust systems, cooling devices, CO burners, and baghouses or precipitators
	Reverberatory furnaces	Dust, fumes, particulates, smoke, and gaseous fluxing materials	Exhaust systems and baghouses, precipitators, or venturi scrubbers
	Crucible furnaces	*See* Nonferrous smelters	

(*Continued*)

Table 29.1 (Continued)

Industry	Process or Operation	Air Contaminants Emitted	Control Techniques
	Sweat furnaces	Smoke, particulates, and fumes	Precleaning metal and exhaust systems with afterburner and baghouse
	Wire reclamation and autobody burning	Smoke and particulates	Scrubbers and afterburners
Paint and Varnish Manufacturing	Resin manufacturing: 1) closed 2) reaction 3) vessel	Acrolein, other aldehydes and fatty acids (odors), and phthalic anhydride (subl.)	Exhaust systems with scrubbers and fume burners
	Varnish cooking-open or closed vessels	Ketones, fatty acids, formic acids, acetic acid, glycerin, acrolein, other aldehydes, phenols and terpenes, from tall oils, hydrogen sulfide, alkyl sulfide, butyl mercaptan, and thiofene (odors)	Exhaust system with scrubbers and fume burners–close-fitting hoods are required for open kettles
	Solvent thinning	Olefins, branched chain aromatics and ketones (odors), solvents	Exhaust system with fume burners
Rendering Plants	Feedstock storage and housekeeping	Odors	Quick processing, washdown of all concrete surfaces, pave dirt roads, and proper sewer maintenance, packed towers
	Cookers and percolators	SO_2, mercaptans, ammonia, odors	Exhaust system, condenser, scrubber, or incinerator
	Grinding	Particulates (dust)	Exhaust system and scrubber
Roofing Plants (Asphalt Saturators)	Felt or paper saturators: spray section, asphalt tank, and wet looper	Asphalt vapors and particulates (liquid)	Exhaust system with high inlet velocity at hoods (>200 ft/min) with either spray scrubbers, baghouses, or two-stage low-voltage electrostatic precipitators
	Crushed rock or other minerals handling	Particulates (dust)	Local exhaust system, cyclone, or multiple cyclones
Steel Mills	Blast furnaces: charging, and pouring	CO, fumes, smoke, and particulates (dust)	Good maintenance, seal leaks, use of higher ratio of pelletized or sintered ore, CO burned in waste-heat boilers, stoves or coke ovens, cyclone, scrubber, electrostatic precipitator, or venturi scrubber
	Electric steel furnaces: charging, pouring, and oxygen blow	Fumes, smoke, particulates (dust), and CO	Segregate dirty scrap, and proper hooding, baghouses, venturi scrubbers, or electrostatic precipitator
	Open-hearth furnaces: oxygen blow, and pouring	Fumes, smoke, SO_x, particulates, (dust), CO, and NO_x	Proper hooding, settling chambers, waste-heat boiler, baghouse, electrostatic precipitator, or venturi scrubber
	Basic oxygen furnaces: oxygen blowing	Fumes, smoke, CO, particulates, (dust)	Proper hooding (capture emissions and dilute CO), scrubbers, or electrostatic precipitator
	Raw material storage	Particulates (dust)	Wetting or application of plastic spray
	Pelletizing	Particulates (dust)	Proper hooding, cyclone, and baghouse
	Sintering	Smoke, particulates (dust), SO_2, and NO_x	Proper hooding, cyclones, venturi scrubbers, baghouse, or precipitator

29.2 Control Techniques Applicable to Other Sources

Information on control techniques for petroleum refineries, chemical plants, mining, etc. is available in the literature.[1]

Reference

1 *Field Operations and Enforcement Manual for Air Pollution Control, Vol. 2: Control Technology and General Source Inspection*. EPA APTD-1101, Research Triangle Park, NC, 1972.

Index

Optimizing Air Pollution Control Equipment Performance: Operation & Maintenance, First Edition. Jay Richardson and Louis Theodore.